BEVERAGES

BEVERAGES

technology, chemistry and microbiology

Alan H. Varnam
Consultant Microbiologist
Southern Biological
Reading
UK

and

Jane P. Sutherland
Head of Food and Beverage Microbiology Section
Institute of Food Research
Reading Laboratory
Reading
UK

CHAPMAN & HALL
London · Glasgow · Weinheim · New York · Tokyo · Melbourne · Madras

Published by Chapman & Hall, 2-6 Boundary Row, London SE1

Chapman & Hall, 2-6 Boundary Row, London SE1 8HN, UK

Blackie Academic & Professional, Wester Cleddens Road, Bishopbriggs, Glasgow G64 2NZ, UK

Chapman & Hall GmbH, Pappelallee 3, 69469 Weinheim, Germany

Chapman & Hall Inc., One Penn Plaza, 41st Floor, New York NY 10119, USA

Chapman & Hall Japan, Thomson Publishing Japan, Hirakawacho Nemoto Building, 6F, 1-7-11 Hirakawa-cho, Chiyoda-ku, Tokyo 102, Japan

Chapman & Hall Australia, Thomas Nelson Australia, 102 Dodds Street, South Melbourne, Victoria 3205, Australia

Chapman & Hall India, R. Seshadri, 32 Second Main Road, CIT East, Madras 600 035, India

First edition 1994

Typeset in 10½/12½pt Garamond by Acorn Bookwork, Salisbury, Wilts
Printed in Great Britain by St. Edmundsbury Press, Bury St. Edmunds, Suffolk

ISBN 0 412 45720 2

A catalogue record for this book is available from the British Library

Library of Congress Catalog Card Number: 93-074892

∞ Printed on permanent acid-free text paper, manufactured in accordance with ANSI/NISO Z39.48-1992 (Permanence of Paper).

Contents

Preface

Beverages are a diverse group of commodities, which range from that most innocuous and essential drink, water, to the most ardent of spirits, such as Navy rum. Beverages may be consumed hot, cold or very cold indeed, and may or may not be carbonated. Despite these wide differences, there are many common factors, not least, as the more chemically minded will note, the occurrence of the same flavour-active compounds in beverages of an apparently very different nature. It is also noteworthy that the consumption of beverages can be totally unrelated to the basic biological function of slaking thirst. Coffee and tea - the cup that cheers, but does not inebriate - are often drunk for their mild stimulatory properties in quantities which far exceed those required to maintain bodily hydration. Sports drinks are drunk to assist athletic performance, while the motive behind consumption of alcoholic beverages is, of course, well known. Sadly a minority, albeit a significant minority, drink alcohol purely for the temporary escape from reality or, more alarmingly, to find the courage for violence and aggression.

The beverage industry, in its widest context, faces changing times, with the contraction of old markets offset by new developments and opportunities. The total value of the beverage market, especially that of soft drinks and some 'international' beers is very large indeed but, for a number of reasons, this tends not to be reflected in the content of food science and technology courses. The intention of this book is to provide, for persons with a basic knowledge of chemistry and microbiology, a technical view of beverages, which is both comprehensive in approach and yet of sufficient detail to be truly useful both for undergraduate and equivalent students and for persons entering the unfamiliar, and often alarming, world of industry. In common with the companion books in this Series, *Beverages*, is structured to meet the requirements of both undergraduates and the graduate in industry, and, to attain this end, provides a full discussion of manufacturing processes in the context of technology and its related chemistry and microbiology, as well as a more fundamental appraisal of the underlying science.

Information boxes and * points are used to place the text in a wider scientific and commercial context, and exercises are included in all chapters to encourage the reader to apply the knowledge gained from the book to unfamiliar situations. Where appropriate, an outline of quality assurance and control procedures is included.

A.H.V.
J.P.S.

A note on using the book

EXERCISES

Exercises are not intended to be treated like an examination question. Indeed in many cases there is no single correct, or incorrect, answer. The main intention is to encourage the reader in making the transition from an acquirer of knowledge to a user. In many cases the exercises are based on 'real' situations and many alternative solutions are possible. In some cases provision of a full solution will require reference to more specialist texts and 'starting points' are recommended.

Acknowledgements

The authors wish to thank all who gave assistance in the writing of this book. Special appreciation is due to:

Debbie and Phil Andrews for providing hand drawn and computer-generated illustrations respectively.

Those manufacturers of food processing equipment, food ingredients and laboratory equipment, who willingly provided information concerning the 'state of the art': Carlsberg-Tetley Brewing Ltd, Burton on Trent, UK; Courage Brewing Ltd, Reading, UK; Dal Cin SpA, Milan, Italy; Unipektin AG, Zurich, Switzerland; G.E.A. Wiegand, GmbH, Ettlingen, Republic of Germany, Schmidt-Bretten UK Ltd, London, UK.

The libraries of the AFRC Institute of Food Research, Reading Laboratory and the University of Reading for their assistance in obtaining information.

Our colleagues in Reading and elsewhere for their help and interest during the preparation of the book.

1

MINERAL WATER AND OTHER BOTTLED WATERS

OBJECTIVES

After reading this chapter you should understand

- The nature of the different types of bottled water
- The key importance of the source
- The means of water abstraction
- Post-abstraction processes
- Quality assurance and control
- The chemical constitution of bottled waters
- Potential public health risks associated with bottled waters
- The general microbiological status of bottled waters
- The chemical and microbiological testing of bottled waters

1.1 INTRODUCTION

Bottled mineral waters have been consumed for many years either as a 'safe' form of water in areas where mains supplies are of dubious quality or because of the perceived therapeutic effects of minerals present. In recent years, however, there has been a very large increase in sales of natural mineral waters in countries such as the UK where consumption was previously limited. This may be attributed to two main causes: firstly, adverse publicity, in many cases misplaced, concerning the safety of mains water and, secondly, the perception of spring, or other natural waters, not only as a 'healthful' drink but as representing a 'healthful' and sophisticated life-style. Despite this many non-carbonated bottled waters have been found to contain large numbers of viable micro-organisms (see page 12).

The perceived therapeutic properties of some natural mineral waters have been noted above. This perception stems historically

BOX 1.1 **Drink no longer water**

In a real sense, organoleptic differences between different brands of water are usually slight. Despite this various newspapers and magazines have published articles in which the relative hedonistic merits of different waters were discussed in a manner normally reserved for fine wines. At a time when a considerable proportion of the world population is desperately short of water of any kind, such articles must be seen as some of the absurdities of life.

from the great spas of Europe and the association of certain water sources with the cure, or alleviation, of specific diseases such as liver complaints. Therapeutic properties were usually associated with the mineral content of the waters although more esoteric properties such as naturally occurring radioactivity were also considered beneficial. It should be appreciated, however, that while the value of mineral waters in treatment of specific disease must be considered dubious, the presence of a high level of inorganic salts in some waters make these unsuitable for persons with kidney disease or for infant feeding.

Various descriptions are attached to bottled waters and these frequently vary on a national or regional basis. Natural mineral water is extracted from underground water-bearing strata via springs, wells or boreholes. It is characterized by its content of mineral salts and trace elements, and is subjected only to minimal treatment in order that its essential properties should be preserved. A full legislative definition of natural mineral water has been established within the EEC (Table 1.1). In many non-EEC countries no definition has been established and additional treatments, particularly disinfection, may be permitted. EEC regulations also permit certain indications of mineral content to be stated on labels providing that specified conditions are met (Table 1.2) but no indication relating to the prevention, treatment or cure of human diseases may be used.

Various types of bottled drinking water other than natural mineral waters are available. Spring water, in the context of EEC regulations, is potable water from a source which either does not meet the requirements for mineral water or for which no application for recognition has been made. Spring water is not currently legally

Table 1.1 EEC requirements for natural mineral waters

Natural mineral water is a water clearly distinguishable from ordinary drinking because:

(a) It is characterized by its content of certain mineral salts and their relative proportions and of the trace elements or of other constituents.
(b) It is obtained directly from natural or drilled sources from underground water-bearing strata.
(c) It is constant in composition and stable in discharge and temperature (due account being taken of the cycles of natural fluctuations).
(d) It is collected under conditions which guarantee the original bacteriological purity.
(e) It is bottled close to the emergence of the source with particular hygienic precautions.
(f) It is not subjected to any treatment other than those permitted by this standard.
(g) It is in conformity with all the provisions laid down in this standard.

Note: Modified from *European Regional Standard for Natural Mineral Waters*.

Table 1.2 Requirements for label indications of mineral content

Label indication	*Mineral constituent*	*Requirement*
Low mineral content	inorganic constituents	dry residue not above 500 mg/l
Very low mineral content	inorganic constituents	dry residue not above 50 mg/l
Rich in mineral salts	inorganic constituents	dry residue above 1500 mg/l
Contains:		
bicarbonate	bicarbonate above	600 mg/l
calcium	calcium	above 200 mg/l
chloride	chloride	above 200 mg/l
fluoride	fluoride	above 1 mg/l
iron	bivalent iron	above 1 mg/l
magnesium	magnesium	above 50 mg/l
sodium	sodium	above 200 mg/l
sulphate	sulphate	above 50 mg/l
Suitable for a low sodium diet	sodium	not above 20 mg/l
Acidic	free carbon dioxide	above 250 mg/l

Note: Based on *The Natural Minerals Waters Regulations* (1985)

defined but labelling must be such as to avoid any possibility of confusion with natural mineral water. Such water is subject to legislation relating to potable water. In the US, no distinction is currently made between bottled mineral water and other bottled waters. It is likely, however, that legal definitions of mineral and spring water will be introduced.

The recognition that water, particularly that of low mineral content, is an uninteresting drink, together with the desire of bottlers to add value to basic material has led to the development of a range of what may best be described as 'flavoured waters'. These range from simple products containing water and a small quantity of fruit juice to more complex beverages containing infusions of herbs and other ingredients. In some cases, especially where the ingredients include ginseng, therapeutic properties may be implied. Such products are intermediate between natural waters and soft drinks.

BOX 1.2 **Children's teeth are set on edge**

Fluoridation of drinking water is a recognized means of reducing dental decay amongst children and, in June 1993, the UK Health Secretary, Virginia Bottomley, announced plans to extend fluoridation of piped water supplies. There is, however, considerable opposition to fluoridation. This is based both on the moral contention that fluoridation amounts to forced medication and on the public health contention that fluoridation can lead to bone cancer and an impaired immune system. Fluoridation of piped water is thus a factor favouring selection of bottled drinking water, sales of which benefited when the process was introduced in some parts of the UK. In contrast, fluoridated bottled water is available in the United States and commands a high price in areas where the natural fluoride content of piped water is low.

1.2 TECHNOLOGY

The technology of bottled waters is straightforward, few treatments are permitted and the underlying philosophy, supported by legislation in the case of mineral waters, is to preserve the properties ascribed to the water at source up to the point of consumption.

1.2.1 The source

Natural mineral waters must be abstracted only from officially recognized sources which, in the case of EEC member countries, are published in the *Official Journal of the European Communities*. Details of the source required for recognition include a hydrological description, physical and chemical characteristics of the water, microbiological analyses, levels of toxic substances, freedom from pollution and stability of the source. Such legislation does not apply to other waters, but judgement of the suitability of a source should be based on similar criteria.

In general terms, the most suitable aquifer is deep, with a long transit time and few cracks or fissures. In the case of shallow aquifers the main concern is the possibility of surface water passing more, or less, directly into the source. No aquifer, however, is totally immune from risk of pollution. In many cases this results from activities which affect the geology and hydrology of the catchment area. Examples include the drilling of deep shafts or the extension of existing mining activities, diversion of watercourses, and the dumping of waste material into disused mineshafts and deep quarries. There are no known recent examples of pollution necessitating the closure of a source used for bottled water, but bores providing potable water for food processing have been affected. Instances include separate cases of microbiological and chemical pollution due to the dumping of cattle slurry and cheese whey into disused mineshafts, and lead pollution from ancient spoil heaps following diversion of a stream.

1.2.2 Abstraction

The means of abstraction depends on the nature of the source, spring water typically rises from the spring through a bed of gravel, while water from artesian wells and bores normally requires no pumping. Pumping is required from non-artesian wells and bores and submersible pumps are commonly used. Precautions must be taken against contamination of the source at the point of abstraction. Pumps, for example, can become colonized by micro-organisms or, if not properly maintained, become a source of chemical pollution. Precautions should also be taken to ensure that the source is protected from pollution arising from ancillary operations. Hard standing for lorries and other vehicles, for example, should be designed and constructed to prevent oil or fuel leaks being washed into the water-bearing strata.

1.2.3 Post-abstraction treatment

(a) Filtration and disinfection

Filtration or decanting of natural spring water, preceded where necessary by oxygenation, is permitted in EEC countries. The technological objective is removal of unstable elements and filtration must not be intended to improve the microbiological status of the water. Other types of bottled water may, depending on national regulations, be microfiltered with membranes of 0.1–10 μm pore size, to remove micro-organisms. Microfiltration is usually coupled with ultraviolet disinfection or is complemented by ozone treatment directly before bottling. Alternatively ozone treatment may be applied without prior microfiltration.

(b) Addition, or removal, of CO_2

The situation with respect to CO_2 content, and thus to effervescence or non-effervescence, is complicated in that CO_2 may be removed from naturally carbonated waters, or added to naturally uncarbonated. Further a distinction must be made between CO_2 derived from the source and that from another origin. This is reflected in the labelling requirements (Table 1.3). Equipment for carbonation is the same as that used for soft drinks (Chapter 3, pages 92–4). Carbonation is usually effective in reducing the population of micro-organisms and preventing subsequent growth, but *must not* be relied upon as a means of disinfecting water from an unsafe source.

(c) Bottling

According to EEC regulations natural mineral water must be bottled at source (with the exception in the UK of water which, prior to 17 July 1980, was being transported in tanks to a remote bottling plant) and sold in bottles which were those used at the time of bottling. There are no restrictions on other types of water.

---

* Additional treatment can be applied in the US and some other countries. ***Specially prepared drinking water*** is water in which the mineral content has been adjusted and controlled to improve the taste. The designation may be applied either to bottled water, or water supplied by a public utility. In addition ***purified water*** is widely available at retail level. This conforms to the US Pharmacopoeia standard, with a mineral content of less then 10 mg/l. Processing may be by distillation, ion exchange or reverse osmosis.

Table 1.3 Requirements for label indications of carbonation

Label indication	*Carbonation*
Natural mineral water	none (non-effervescent)
Naturally carbonated natural mineral water	CO_2 content after bottling the same as at source and any replaced CO_2 derived from the source
Natural mineral water fortified by gas from the spring	CO_2 content greater than at source but the CO_2 derived from source
Carbonated natural mineral water	CO_2 partially or totally derived from an origin other than the source
Fully decarbonated or partially decarbonated	CO_2 removed by physical treatment

Note: Based on *The Natural Minerals Waters Regulations* (1985)

BOX 1.3 **Sparkles near the brim**

In some cases the CO_2 is removed from the water at abstraction and then used to carbonate the water at bottling. In the UK, some supermarket retailers have considered the description of such water as 'naturally carbonated', although legal, to be misleading and have refused to stock water labelled in this way. In New York State the bottlers of Perrier(R) water were fined $40 000 for claiming, amongst other contentious items, that the water is 'naturally sparkling' and 'bubbles to the surface'. (Anon 1991. *Food Chemical News*, **August 26**, 34).

Conventional bottling plant is used and there are no technological difficulties although careful sanitization is required to prevent the colonization of the bottling plant by micro-organisms.

Bottles may be glass, polyethyleneterephthalate (PET) or high-density polythene. Problems of chemical taint from PET bottles have been attributed to the leaching of plasticizers, or their thermal degradation products, into the water. Problems have also been reported in the US with the development of 'plastic' flavours in high-density polythene bottles containing water disinfected by ozone treatment. This has been attributed to low molecular weight oxygenated species formed by oxidation of the high-density polythene by residual ozone.

Although plastic bottles are popular due to their light weight and cheapness, they are considered environmentally unfriendly and there is a trend back to glass bottles including, in continental Europe, the thick-walled multi-use type. Such bottles inevitably carry a higher risk of introducing contaminants than the single-use type, and a means of washing, sanitizing and monitoring returning bottles must be established.

1.2.4 'Flavoured' waters

The technology of 'flavoured' waters is simple and involves little more than mixing and blending. The presence of additional nutrient sources and the increased microbial load due to micro-organisms derived from ingredients often means that heat treatment is required for stability. This may be applied in-bottle or, more commonly, on a continuous basis using a plate heat exchanger.

BOX 1.4 **The fruits of originality**

Simple fruit flavours, such as raspberry and orange, tend to be associated with childrens' drinks. This inevitably restricts the market and limits the extent to which added-value can be reflected in higher price. A number of flavour combinations have been devised, which are aimed at sophisticated (or would be sophisticated) adults and which are seen, at least partially, as substitutes for alcoholic beverages. These drinks typically contain juices such as peach, which have a relatively subtle flavour, as well as herb extracts to provide a flavour contrast.

1.2.5 Quality assurance and control

The HACCP system is highly applicable to bottling of natural waters. Quality assurance and control may be considered to have three main components: suitability of the source and its protection, prevention of contamination at point of abstraction, and post-abstraction processes.

Suitability of the source and its continuing protection is essential to ensure safety. In many cases water receives no further purification and the source is therefore a CCP 1. It is necessary to ensure that the source is inherently capable of producing, on a consistent basis, potable water of acceptable microbiological and chemical quality. This requires a detailed knowledge of the source and its catchment area (see page 5). Particular attention should be paid to the stability of the source and, while allowances must be made for seasonal variations, sudden changes in the level of wells or the flow of springs, especially after heavy rain, may indicate that surface water is entering the source directly. Sudden and non-seasonal variation in temperature is also a cause for concern, while discolouration, smells or taints are almost certainly indicative of pollution.

It is necessary to maintain a knowledge of the catchment area and to evaluate the consequences of changes in agricultural or industrial practices. Particular attention should be paid to any activities which could affect the water-bearing strata, even when these are remote from the point of abstraction.

Control at point of abstraction involves ensuring that water is not contaminated by pumps, the fabric of the well, etc., and that the well or spring is itself protected. Detailed procedures will vary according to the physical nature of the source, but account must be taken of any particular difficulties, such as problems of access to submersible pumps.

Post-abstraction includes the entire operation from the wellhead, or spring, to bottling and distribution. A twofold approach is recommended. The first stage involves the preparation of an overall control plan to cover general aspects of technical management. These include personnel training and management policies, definition of tasks and responsibilities, and overall hygiene. Special attention should be given to prevention of biofilm formation in pipework. The second stage involves control of specific processes, such as filtration, carbonation and bottling. Any particular operator training should be specified, and individual procedures for cleaning and monitoring of performance should be designed and implemented. A formal system of record keeping is required for verification of the operation.

Verification also requires laboratory analysis. Water should be sampled as close as possible to the point of abstraction and after bottling. Microbiological analysis (see pages 19–22) is of prime importance except in carbonated water with a pH value of less than 3.5, but regular chemical analysis is required to ensure consistency of the source. Visual and organoleptic examination is seen as being a valuable complement to laboratory analysis.

Performance of individual pieces of equipment also requires verification. Examples include level of fill, headspace and carbonation, and performance of filters and ultraviolet lamps. Precautions against biofilm formation should be verified by periodic dismantling and inspection of water-contact surfaces and microbiological analysis of swabs.

1.3 CHEMICAL CONSTITUTION OF BOTTLED WATERS

1.3.1 Minerals

The mineral content of any bottled water is dependent on the nature of the strata from which the water is abstracted, the solubility of the minerals and the contact time with the water. Analyses for a range of bottled waters are listed in Table 1.4.

Table 1.4 Mineral salt and trace element composition of representative natural waters

	mg/l			
Type	*Buxton Mineral*	*Valvert Mineral*	*Glenburn Spring*	*Perrier Mineral*
Aluminium	0	NS	NS	NS
Bicarbonates	248	204	36.6	400
Calcium	55	67.6	11.2	145
Chloride	42	4	18.4	24
Iron	0	NS	NS	NS
Magnesium	19	2	4.6	4
Nitrates	<0.1	4	<0.1	17
Potassium	1	0.7	0.8	NS
Silica	NS	NS	11.2	NS
Sodium	24	1.9	7.8	8
Sulphates	23	18	4.0	33
Total dissolved solids[1]	280	201	NS	450
pH at source	7.4	7.7	NS	5.7

[1] At 180°C

Notes: (a) Figures obtained from bottlers' analyses as stated on labels.
(b) NS = not stated

1.3.2 Other constituents

Water sources do not exist in isolation from the environment and water will contain both dissolved and suspended substances acquired during precipitation and passage through the surface soil and vegetation. Suspended material will be removed during percolation through the strata but dissolved substances persist. The most important of these are agricultural chemicals including pesticides and nitrates. In most cases the catchment areas for natural waters are in remote regions where intensive agricultural practices are not used. Despite this nitrate levels have occasionally been reported which approach the maximum levels permitted in piped mains supplies. This is of particular concern when bottled water is used preferentially for preparation of infant feeds.

1.3.3 Analysis

Generally used methods of analysis are summarized in Table 1.5.

1.4 MICROBIOLOGY

1.4.1 General microbiological status

A number of surveys have been made of the microbiological status of bottled waters. Non-carbonated waters typically have 'total'

Table 1.5 Chemical analysis of bottled water

Total solids	drying at 105°C
Dissolved solids	drying at 180°C
pH value	electrometrically
Aluminium	colorimetry
Bicarbonates	titration
Calcium	EDTA titration or atomic absorption spectrophotometry
Chloride	silver nitrate titration
Copper	colorimetry
Fluorine	colorimetry
Iron	colorimetry
Magnesium	colorimetry or atomic absorption spectrophotometry
Nitrates	colorimetry
Potassium	Flame photometry or atomic absorption spectrophotometry
Sodium	flame photometry or atomic absorption spectrophotometry

viable counts ranging from *ca.* 10^3 to greater than 10^5 cfu/ml. Numbers in carbonated water were considerably lower, due to the lower water pH value and the direct antimicrobial effect of CO_2.

Published work has shown that numbers of micro-organisms in uncarbonated waters increase after bottling, usually reaching a peak within 1 week. Numbers then remain more or less constant for 6 months or more. The extent to which numbers increase varies. In waters with low levels of 'easily assimilable organic carbon', for example, numbers of micro-organisms increase only to a very limited extent after bottling.

Temperature of storage is also important in determining the extent of microbial growth in bottled water. Findings do differ, however, and while some workers have found greater increases during storage at 6°C than at 20°C, others consider that maximum multiplication occurs at temperatures between 15 and 20°C. This probably reflects differences in the composition of the microflora. In some studies it has appeared that the material from which the bottle is made is of considerable importance. For a number of years it was generally considered that water stored in polyvinyl chloride (PVC) bottles supported greater growth of micro-organisms than that stored in glass bottles. A number of reasons were suggested (Table 1.6), but none have been fully substantiated. In contrast to general opinion, some workers have found no significant differences between water stored in PVC or glass bottles. Others have suggested that differences were artefacts caused either by residues of sanitizing agents in glass bottles or due to discrepancies in counting methods. In some cases counts have been found to be actually higher in glass bottles than in PVC bottles. This has been attributed to the protective effect of the green-tinted glass against ultraviolet rays.

---

* It has been suggested that 'easily assimilable organic carbon' should be used as a means of assessing the biological stability of non-carbonated mineral waters. An extension of this concept is the possible imposition of a maximum value for organic carbon, to prevent excessive growth of micro-organisms during storage. At present, there is insufficient evidence to assess the real benefit of imposing a regulatory standard. At the same time it should be appreciated that water with a low organic carbon content tends to be old water with a high sodium content. Such water is likely to be unsuitable for consumption by babies and persons suffering renal disease and may be unpopular in terms of taste. (van der Kooij, D. 1990. *Rivista Italiana d'Igiene*, **50**, 375–92; Hunter, P.R. 1993, *Journal of Applied Bacteriology*, **74**, 345–52).

Table 1.6 Possible reasons for greater growth of micro-organisms in water stored in PVC than in glass bottles

1. PVC bottles have a rougher internal surface promoting adhesion and colonization.
2. PVC permits a higher level of oxygen diffusion into the water favouring the growth of most contaminating micro-organisms.
3. Organic material is leached from the PVC during storage and provides further substrates for microbial growth.

The size of bottles may be important in determining the extent of microbial multiplication, numbers being larger in small bottles than in large. This probably results from the lower volume:surface area ratio of small bottles, which permits a relatively high proportion of available nutrients to be absorbed to the bottle surface.

1.4.2 Origin of micro-organisms in bottled water

In some cases, at least, it has been assumed that micro- organisms in bottled water are derived from the water at source (the autochthonous microflora) and that significant multiplication only occurs after bottling. This is a convenient assumption, since it avoids any suggestion of shortfalls in plant operation and hygiene. In a number of cases, however, a significant increase in microbial numbers in the water occurs during transit through the plant from source to bottle. This appears to be due to shedding of micro-organisms from biofilms which have developed in pipework and equipment. In some cases the major component(s) of the biofilm is can also be isolated from the water at source, which may therefore be considered the ultimate origin. In other cases the major component of the biofilm cannot be demonstrated in the water at source.

* The autochthonous microflora is defined as the microflora which is indigenous to a given environment. This microflora maintains more or less constant numbers and biomass, which reflects a more or less constant level of nutrients. In many environments the nutrient level, although constant, is low. The allochthonous microflora is the non-indigenous microflora. In some discussions of the microbiology of bottled water the term zymogenous microflora is used to describe the population which increases in number after bottling. The correct definition of the zymogenous microflora, is predominantly transient and alien micro-organisms which exhibit an upsurge of growth on those occasions when nutrient levels increase, or a particular substrate becomes available. These latter events do not occur in a closed system, such as bottled water, and in this context, the use of the term zymogenous is incorrect. (Definitions based on Singleton, P. and Sainsbury, D. 1991. *Dictionary of Microbiology and Molecular Biology*, 2nd edn. John Wiley, Chichester).

Under these circumstances, it has been concluded that contamination was derived from the plant environment, or from personnel. In some cases this is likely to be true, especially in the case of non-mineral waters which have been subjected to 'sterilizing' filtration and ultraviolet disinfection. Care is needed before drawing such a conclusion, however, since environmental pressures at the biofilm may select for a very insignificant component of the microflora of the water at source.

The role of plant personnel as a source of micro-organisms in bottled water is an obvious source of interest and concern. Human commensals, such as *Staphylococcus epidermidis* and *Staph. hominis*, have been isolated from bottled drinking water, and it has been argued that this indicates post-extraction contamination by man (and by implication contravenes the European directive). In a study of a particular source, it was shown that staphylococci were present only when water was bottled by hand for experimental purposes. This, together with the absence of faecal index organisms, suggests that, under conditions of good management, the level of human contamination is low.

1.4.3 Main types of micro-organism in bottled water

(a) 'Autochthonous' microflora

A wide range of micro-organisms have been isolated from natural waters and assigned to the autochthonous microflora (Table 1.7). Gram-negative bacteria, especially species of *Pseudomonas* tend to be dominant, although *Acinetobacter*, *Alcaligenes* and *Flavobacterium* are also common. Yellow pigmentation is common even amongst bacteria normally considered non-pigmented, but this is a finding common to all types of water. The proportion of Gram-positive bacteria in the autochthonous microflora is usually small.

---

* Considerable care is required in assigning bacteria to the autochthonous microflora. A study of the microflora of the Campsie spring in Scotland, showed that *Xanthomonas* comprised 21% of isolates from plates incubated at 20°C and is included with the autochthonous microflora. *Xanthomonas*, however, is a plant pathogen, which survives only for short periods in soil and water in the proximity of infected plants. If the identification is correct, the presence of *Xanthomonas* implies contamination of the source with surface water at a point close to abstraction. (Bradbury, J.F. 1984. In *Bergey's Manual of Systematic Bacteriology*, vol.1 (eds Holt, J.G. and Krieg, N.R.). Williams and Wilkins, Baltimore, pp. 199–210; Mavridou, A. 1992. *Journal of Applied Bacteriology*, **73**, 355–61).

Table 1.7 Bacteria isolated from water and assigned to the autochthonous microflora

Gram-negative bacteria	
Acinetobacter	*Alcaligenes*
Alteromonas[1]	*Cytophaga*
Enterobacter	*Hafnia*
Flavobacterium	*Moraxella*
Pseudomonas	*Xanthomonas*
Gram-positive bacteria	
Arthrobacter	*Bacillus*
'coryneforms'	*Micrococcus*
Streptomyces	

[1] *Alteromonas* is generally considered to be of marine origin and this identification is thought unlikely.
Note: The term 'autochthonous' is often used loosely and some of the bacteria listed here, although of environmental origin, may not be indigenous to the water.

Species of *Micrococcus*, *Bacillus* and coryneform bacteria are most common. In some cases the proportion of Gram-positive bacteria is relatively high. This may suggest colonization of the source at point of extraction or post-extraction contamination.

(b) Recognized enteric pathogens and their index organisms

With the exception of *Vibrio cholerae*, isolated during an epidemic, there are no known isolations of recognized enteric pathogens from bottled water. In the industrialized world, index organisms also appear to be absent, or present in only a very small proportion of samples. Members of the Enterobacteriaceae are occasionally isolated during routine monitoring, but these are environmental strains of no public health significance. Index organisms, including *Escherichia coli*, are present at a relatively high frequency in water bottled in industrializing nations. This indicates faecal contamination of the water and an absence of effective antimicrobial treatment. The value of *E. coli* as an index organism for *Salmonella* has, however, been doubted.

(c) Aeromonas

Some strains of *Aeromonas* are putative pathogens and the genus is frequently associated with water. The organism has been isolated from both chlorinated and unchlorinated piped water supplies and from bored wells. Problems with *Aeromonas* in bottled water have been anticipated and a number of attempts to isolate the organism have been made. Despite this, the only known isolations of *Aeromonas* from bottled water were from a single Lebanese brand, 41% of which contained *Aer. hydrophila*. It has been suggested that the 'normal' flora of bottled water is inhibitory to aeromonads, but this has not been confirmed.

(d) Pseudomonas

In a number of surveys, species of *Pseudomonas* have been found to be the major constituent of the microflora. A wide range of species have been isolated, including *Ps. aeruginosa* and *Ps. cepacia*, which are of public health significance.

(e) Protozoa

The detection of protozoa in water is difficult and bottled water is only rarely examined for these organisms. Various free-living amoebae have been isolated from bottled water in Mexico. These include *Naegleria*, *Acanthamoeba* and *Vahlkampfia*. Species of *Naegleria* and *Acanthamoeba* are recognized as potential pathogens, especially in immunocompromised persons, while *Vahlkampfia* was originally isolated from a person suffering diarrhoea. A causal relationship was not, however, established. There is no information available concerning *Cryptosporidium* or *Giardia* in bottled water, although both protozoa have been implicated as a cause of waterborne disease.

BOX 1.5 **A walk on the wild side**

'Trekkers' trots' are a well known hazard for walkers in remote places. The most common cause of infection is stream water, contaminated not by the proverbial dead sheep just upstream (and just out of sight), but by rodents and other creatures of the wild. The most common micro-organisms involved are *Giardia lamblia* and species of *Campylobacter*. Co-infections involving both organisms have been reported. Disinfecting tablets are available, although some types are of limited efficiency against cysts of *G. lamblia*. The use of disinfecting tablets is, in any case, eschewed by those who seek a true 'wilderness experience'.

1.4.4 Bottled water and public health

Concerns over the safety of bottled water are essentially two-fold. The first involves the possibility that sources may become contaminated with recognized enteric pathogens, such as *Salmonella*, *Vibrio cholerae*, *Cryptosporidium* or diarrhoeal viruses. The second involves the possible role of the autochthonous microflora in disease, especially amongst infants and the immunocompromised.

The only known outbreak of illness directly related to bottled water was cholera. During an outbreak in Portugal in 1974, 82 of 2467 confirmed cases were attributed to consumption of contaminated bottled water, while a further 36 cases had visited a spa, which shared the same source as the bottling plant. *Vibrio cholera* was isolated from the water and it was concluded that the limestone aquifer had been contaminated by broken sewers in a nearby village. It is obvious that, where sewerage systems are rudimentary and poorly maintained, a risk of contamination with many other enteric micro-organisms exists. In all circumstances, the catchment area for natural waters should be situated away from centres of population, but it must be appreciated that even a single, poorly maintained septic tank system may present a significant risk.

The risk of contamination with specific pathogens does not arise solely from human activities. Many pathogens, such as most serovars of non-typhoid *Salmonella*, are of zoonotic origin and may be derived from farm animals and related agricultural practices, including slurry spreading. At the same time, wild animals and birds are important reservoirs of pathogens, such as *Campylobacter*, while the environment itself may be a source of *Aeromonas* and *Cryptosporidium*. It is obvious that, while agricultural practice can be controlled, there is no means by which human pathogens can be eliminated from the wild animal population, or the environment. There is thus an inevitable risk of contamination, which must be minimized by selection of a suitable deep source and continuing application of appropriate monitoring systems (see page 8–9). In the case of waters not classed as natural mineral waters, filtration, ultraviolet disinfection and carbonation reduce the risk, but are not primary means of assuring safety.

In many cases the risk posed by micro-organisms classified as members of the autochthonous microflora is more difficult to define. *Pseudomonas aeruginosa*, however, is a recognized pathogen, which has been a cause of infection among infants in a nursery when a well supplying drinking water was contaminated by sewage and the infiltration of stream water. The very young, the elderly and invalids are at particular risk from *Ps. aeruginosa*; infection and colonization of children suffering cystic fibrosis is particularly serious. Species of *Pseudomonas*, including *Ps. fluorescens* and *Ps. cepacia* have been associated with disease in immunocompromised and other high-risk persons. *Pseudomonas cepacia* has also been identified as a cause of serious chest infections in persons with cystic fibrosis.

Other bacteria isolated from bottled water, such as *Acinetobacter*, have been associated with human disease, especially amongst hospitalized patients.

It may be concluded that the major risk, in industrialized countries, of infection from consumption of bottled water affects fairly well defined groups. These are generally considered to be infants, the elderly and invalids, but immunocompromised persons, including persons with acquired immunodeficiency syndrome (AIDS) are also at increased risk. Paradoxically, it is persons in high risk categories who are more likely to consume bottled water, due to its perceived higher purity. In the case of infant feeding, it has been recommended that bottled water should be boiled before being used to make up food (it is also necessary to ensure that the mineral content is suitable). In the current state of uncertainty with respect to the public health risks associated with bottled water, this recommendation appears sound. Although directed at infants, it would appear wise to follow the recommendation for other risk groups.

---

* Colonization of persons with cystic fibrosis by species of *Pseudomonas* is a serious and potentially life-threatening problem. *Pseudomonas aeruginosa* has been most commonly involved, but *Ps. cepacia* is increasingly implicated. Pulmonary infection with *Pseudomonas* in this situation leads to an exacerbation of the underlying disease and invasive infection of the pulmonary parenchyma. Infection is often associated with antibiotic therapy, despite *Pseudomonas* being highly sensitive to the antibiotics used. Infection is often derived from environmental sources, including spa pools used in physiotherapy. The presence of *Ps. aeruginosa* and *Ps. cepacia* in drinking water, or foods, is also obviously unacceptable for persons with cystic fibrosis. In the case of *Ps. cepacia*, however, person to person spread also appears to be significant. This poses serious problems of management of adults with stable cystic fibrosis, since social contacts may have to be curtailed. This, in turn, leads to the loss of benefits obtained through group contacts. (Walters, S. and Smith, E.G. (1993) *The Lancet*, **342** (8862), 3–4.

1.4.5 Microbiological examination of bottled water

In EEC and several other countries the microbiological quality of the source of natural mineral waters must be verified by microbiological analysis. In EEC countries specified criteria are:

(i) Total viable count to conform to normal values indicating that the source is protected from contamination.
(ii) Absence of parasites, pathogenic micro-organisms, *E. coli*, coliforms and faecal streptococci (*Enterococcus*) in 250 ml; sporulated sulphite-reducing anaerobes in 50 ml; *Ps. aeruginosa* in 250 ml.

Similar criteria are recommended for spring water. In all cases a statistically based sampling schedule must be employed and extra samples should be taken after heavy rain or other events which increase the risk of source contamination.

It is also necessary to sample water after bottling. Microbiological concerns are largely restricted to non-carbonated waters and testing is not necessary if the pH value is less than 3.5. Legislative standards may not, however, distinguish between carbonated and non-carbonated. The same criteria may be used to verify the microbiological quality of the bottled water as are applied to the water at source. The water should be sampled within 12 h of bottling, at which time the total viable count should not exceed 100/ml when incubated at 20–22°C for 72 h, or 20/ml when incubated at 37°C for 24 h. Higher numbers are permitted after storage but these should be 'no more than that which results from the normal increase'. Examination for *Aeromonas* has also been suggested, although the threat posed by this organism is potential rather than actual.

* Concern over the quality of piped water supplies has led to the development of 'purifiers', fitted at point of use. These vary in complexity from simple filters to relatively sophisticated reverse osmosis devices. Purifiers are prone to colonization by micro-organisms and water emerging from these devices has a microflora similar, quantitatively and qualitatively, to that present in many bottled waters. One study has shown an association between gastrointestinal disease and consumption of water with high colony counts, it being suggested that a direct relationship exists between illness and consumption of 'purified' water with colony counts between 10^3 and 10^5 cfu/ml. (Payment, P. *et al.* 1991. *Applied and Environmental Microbiology*, **57**. 945–8).

Bottled water should also be examined according to a statistically designed sampling scheme. It is essential that the scheme chosen reflects the associated special public health concerns and the level of protection should be at least that required for piped community supplies.

In general, standard media and methods are used for examination of bottled water (Table 1.8). There is, however, some disagreement concerning the most suitable medium and incubation temperature for determination of 'total viable count'. In some cases a standard medium, such as nutrient agar or plate count agar, has been used, but there is evidence that higher counts are obtained with a

Table 1.8 Methods for enumeration of bacteria from bottled water

'Total' viable count
1/10 strength plate count agar: 5 day, 20–22°C
Plate count agar: 2 day, 37°C
'Coliforms'
Membrane filtration using membrane lauryl sulphate broth: 4 h, 30°C; 14 h, 37±0.5°C
Escherichia coli
Membrane filtration using membrane lauryl sulphate broth: 4 h, 30°C; 14 h, 44±0.25°C
'Faecal streptococcus' (*Enterococcus*)
Membrane filtration using membrane *Enterococcus* medium (Slanetz & Bartley glucose–azide medium): 48 h, 37°C; or 4 h, 37°C, 44 h, 44–45°C
Sulphite reducing clostridia and *Clostridium perfringens*
Pasteurize sample at 75°C for 10 min. Most probable number technique using differential reinforced clostridial medium: or membrane filtration using membrane clostridial medium: *Clostridium perfringens* is enumerated by sub-culture of presumptive clostridia into litmus milk and examining incubated tubes for formation of 'stormy clots'
Pseudomonas spp.
Cephaloridine–fucidin–cetrimide medium: 48 h, 25–30°C

[1] A defined substrate method, Colilert, has been developed in the US for simultaneous determination of 'coliforms' and *E. coli*. There are reports, however, that a high level of false-negative results may occur with Colilert when numbers of 'coliforms' and *E. coli* are small.

[2] Biochemical confirmation is required for presumptive coliforms and *E. coli*. In the case of *Pseudomonas*, confirmation and identification to species level may be carried out using the API 20NE™.

nutrient medium diluted 10-fold. Such a medium is also more effective at recovering the true autochthonous microflora of water, although it must be appreciated that environmental contaminants, such as soil bacteria, are often also able to grow at low nutrient levels.

Recovery of micro-organisms from water is generally enhanced by incubation at low temperatures. Incubation temperatures as low as 10°C have been used in combination with incubation times as long as 28 days. In a quality assurance context, however, incubation at 20–22°C for 5 days would seem to be adequate, A second count on full strength medium incubated at 37°C for 2 days may be considered useful in assessing contamination from sources other than the water itself.

Changes in the composition of the microflora may also be important in the quality assurance of bottled water. For this reason the normal microflora of a source should be characterized as fully as possible and changes monitored. The choice of criteria to characterize the source may, however, be difficult. A simple flora analysis, based on colonial and cellular morphology, Gram-reaction, and catalase and oxidase tests is always valuable and should be applied on a routine basis. The value of full identification of the components of the microflora is doubtful under normal conditions, unless specific micro-organisms are being sought. In a number of cases, identification is likely to be difficult and, especially where conventional media are used, time consuming. Commercial identification kits, such as the API 20NE™ are easier and more economical to use, but may not identify the more poorly defined bacteria, especially the yellow-pigmented, Gram-negative rods. The profile generated by the API 20NE™ is itself, however, useful in characterizing the microflora. The Biolog™ system, based on utilization of carbon sources, has been successfully used to identify bacteria from a number of aquatic sources and would appear to have considerable value in characterizing the microflora of bottled water. The consumable cost, however, is relatively high and a dedicated computer program must be purchased. A nutrient-tolerance (NT) test has been proposed for the characterization of bacteria from bottled water. This arbitrarily classifies isolates into three growth types, facultative or obligate oligocarbotolerant and oligocarbophilic, and three nutrient types, eutrophic, mesotrophic and oligotrophic. The NT test has the advantage of being directly related to the ecology of the micro-organisms. The test has been found to be very useful in some published reports, but to lack discrimination when used on a routine basis. This may be due to differences in the microflora of different waters, but further work is required.

Microbiological examinations should not be made in isolation to other testing. Organoleptic assessment is of importance not only in ensuring that the characteristics of the water are maintained, but in providing an early warning of microbiological contamination. Taints or odours due to microbial colonization of plant or to pollution of the source may not be immediately reflected in increased microbial numbers.

EXERCISE 1.1.

In an attempt to diversify, the owner of a farm in southern England has identified bottled well water as a source of income and has asked you to conduct a preliminary feasibility study. The proposed source is a Victorian built, brick-lined well which previously served a small village. The depth of the well is unknown, but is thought to be shallow. The well is situated in a small downland valley, some 25 m above and 350 m to the west of the village. The wellhead is protected by a brick and tile building, which is preserved by a local group and in good condition. The well itself was capped when mains water reached the village, but has since been used to supply irrigation water. The well has occasionally flooded during heavy rainfall. The catchment area is a fold of chalk downland, which is used both for grazing sheep and for growth of cereals. A portion is 'set-aside'. The catchment area is crossed by a long distance footpath and an area is occasionally used for motorcycle scrambling. The farmer would like to extend leisure use and is considering a camping and caravan site.

What additional information would you require before being able to assess the suitability of the source? Specify supportive laboratory analyses.

Assuming that the well itself is suitable, outline a management plan for the catchment area to minimize the possibility of pollution. To what extent are leisure activities, including the long distance footpath, compatible with protection of the source?

EXERCISE 1.2.

You are employed in a regulatory authority concerned with the legality of claims made for the health-promoting properties of foods and beverages. Two bottled spring waters have been drawn to your attention. The first, a domestic water, is marketed through health clubs, sports centres, etc., and claims that 'The mineral balance is such that the water is rapidly adsorbed by those parts of the body most affected by dehydration . . . water alone is proven to be many times more effective than isotonic sports drinks'. The analysis of the major minerals is (mg/l): bicarbonates 220, calcium 74, chloride 50, magnesium 24, sodium 18, sulphate 22.

The second water is imported from a volcanic zone and claims that 'Unique minerals, found only in volcanic rock ensure the pH is perfectly balanced at 7.2. These minerals restore the body's balance and, when consumed daily, stimulate the system and build sexual stamina'. The analysis of minerals is (mg/l): aluminium 0.5, bicarbonates 220, calcium 63, chloride 49, iron 2, magnesium 31, sodium 27, sulphates 31.

Comment on the scientific validity of these two claims. Do you consider that claims of this nature should be permitted? Is more stringent government legislation and enforcement required to protect consumers from misleading claims concerning health- promoting properties (of foods and beverages in general), or should emphasis be placed on self-regulation through trade associations?

Further details concerning sports drinks may be obtained in Chapter 3, pages 98–101.

EXERCISE 1.3.

In the UK, piped water supply utilities continue to specify that employees must undergo the Widal test. This test is an agglutination test for detecting serum antibodies to those salmonellas which cause human enteric fevers, including typhoid fever. Anti-H, but not anti-O, antibodies persist for many years.

Discuss the value of the Widal test in protecting public water supplies in both the industrialized countries of western Europe and developing countries, such as Mexico. Should the test also be applied to employees in water bottling plants? Under what circumstances would you recommend stool testing for employees in water bottling plants?

EXERCISE 1.4.

'Absence of parasites and pathogenic micro-organisms' is a common stipulation in regulations concerning bottled water. How would you, as microbiologist in a medium-sized mineral water bottling company, ensure that your products comply with these regulations?

Parasites, such as *Giardia* and *Cryptosporidium*, are notoriously difficult to detect. In the specific case of *Cryptosporidium*, do you consider that laboratory examination for cysts of the organism, using current methodology, is of any value in reducing the risk from this organism. Do you consider that the situation would be significantly changed by application of gene-amplification techniques, such as the polymerase chain reaction?

Further information and references may be obtained from Casemore, D.P. 1989. *PHLS Microbiology Digest*, **6**, 54-66; Current, W.L. and Owen, R.L. 1989. In *Enteric Infection* (eds Farthing, M.J.G. and Keusch, G.T.). Chapman & Hall, London, pp. 223-49; Erlich, H.A. *et al.* 1989, *The Polymerase Chain Reaction*, Cold Spring Harbor Laboratory Press, Cold Spring Harbor, NY; Vesey, G. *et al.* 1993. *Journal of Applied Bacteriology*, **75**, 87-90).

2

FRUIT JUICES

OBJECTIVES

After reading this chapter you should understand

- The definitions of fruit juices and nectars
- The types of fruit juice
- The major species of fruit used in juice production
- The technology of juice extraction and processing
- Quality assurance and control
- The nutritional importance of fruit juice
- The main flavour constituents of the major fruit juices
- Chemical changes in juice during extraction, processing and storage
- Potential public health problems associated with fruit juice
- The microbial spoilage of fruit juice

2.1 INTRODUCTION

It is probable that fruit juices have been consumed in one form, or another, for many years. Until the 19th century, however, fermentation and consequent conversion to wine or cider was the only means of preservation. The commercial juice industry dates from 1869, when the Welch company of Vineland, New Jersey commenced bottling of unfermented grape juice. This introduced the principle of fruit juice preservation by pasteurization. The industry was slow to develop, although in the US a trend to expansion which commenced in the 1920s accelerated in the 1930s. The nutritional value of fruit juices, especially with respect to vitamin C, was recognized during the hungry years of the Great Depression and demand was further stimulated by technological developments, such as flash pasteurization, which greatly improved product quality. At this time, large scale consumption was largely a North American phenomenon and, in the case of orange juice, more than 75% of world production was concentrated in the US.

Consumption of fruit juice continued to expand in the US, production being stimulated by demand from the military, who fully recognized the nutritional value of juice. Further impetus resulted from the development of technology for the production of frozen concentrates.

BOX 2.1 **National health**

The nutritional importance of fruit as a source of vitamin C was fully recognized in Europe during World War II. Citrus fruits were largely unobtainable, but in the UK school children were mobilized to harvest wild rose hips. These were used to manufacture rose hip syrup to provide vitamin C for infants. In the post-war years the new National Health Service introduced national orange juice for infants. This rather oddly flavoured juice played an important role in ensuring the well being of a generation of infants raised during the shortages resulting from the war.

During the 1950s, fruit juice became established as an integral part of the North American diet. The industry was much slower to develop in Europe and, in the UK, fruit juice consumption was largely limited to small bottles or cans of juice, often sweetened. Fruit juices tended to be thought of as 'special occasion' drinks and increases in consumption were often associated with social trends, such as increased use of restaurants and changes in drinking habits. A dramatic increase in consumption occurred, however, in the years after the mid-1970s. This was driven by a demand for drinks which were compatible with perceptions of a healthy life-style, but this impetus was reinforced by the application of ultra-high-temperature (UHT) sterilization and aseptic packaging to fruit juices. This technology permitted the production, often from concentrate, of a high quality, long life consumer product, fully in accordance with perceptions of a 'healthful' product.

Citrus juices are most widely consumed and account for over 50% of juice in international commerce. Apple juice is also popular, and there are significant sales of grape and pineapple juices. A wide variety of other juices are produced, but volumes are relatively small and in many cases the juice is used in mixtures with other juices. In some cases, including passion fruit and kiwi fruit, juice production is seen as an important means of increasing utilization and enlarging the market.

In addition to juice packaged as a consumer product, recent years have seen large increases in usage as an ingredient of other products. Such uses include carbonated and uncarbonated soft drinks (see page 81), unsweetened canned fruits, ice cream products and 'flavoured' water (see page 4).

2.2 TECHNOLOGY

2.2.1 Legal definitions of fruit juice and nectars

In many countries fruit juice and nectars are fairly strictly defined. This is considered necessary to avoid confusion between fruit juices and fruit juice-containing beverages such as squashes and carbonates. Fruit juice is, further, a relatively expensive commodity, which requires legislative protection from adulteration.

In the EEC, a directive defines fruit juice as juice obtained from fruit by mechanical processes, fermentable but unfermented, having the characteristic colour, odour and flavour typical of the fruit from which it comes. The definition has been extended to include the product obtained from a concentrate, which must have sensory and analytical characteristics equivalent to those of juice obtained directly from the fruit. In the UK and some other countries this must be labelled 'Made from concentrated juice'. Fruit used in production of juice must be in good condition and free from deterioration and contain all the essential ingredients required to obtain a satisfactory product. Under EEC legislation, tomatoes are not classified as fruit.

Fruit nectars, under the EEC directive, are defined as unfermented, but fermentable, products obtained by the addition of water and sugars to fruit juice, concentrated fruit juice, fruit puree or concentrated fruit puree, or a mixture of these that conform with the specification given. Nectars may contain up to 20% added sugar (or honey). The minimum content of juice, or puree, and the minimum total acid content for each type is stipulated in legislation and the minimum juice or puree, content must be declared on packaging.

Although many consumers perceive fruit juices as 'additive-free', the EEC directive does permit various additional ingredients (Table 2.1). In some cases, quantities added and labelling requirements are at the discretion of individual governments.

Legislation in the US is broadly similar to that in EEC countries, although tomatoes are classified as fruit. The legislation is, however, more detailed in that standards are laid down for citrus and pineapple juices, and also for canned, frozen and pasteurized juices, and frozen, concentrated juices.

2.2.2 Types of juice

A number of types of fruit juice are produced on a commercial basis. These are primarily defined by the post-extraction processing.

Single-strength direct juice is technically the simplest and undergoes no concentration. Single-strength direct juice can be produced on a very small scale, for immediate consumption in cafes, etc., or

Table 2.1 Additional ingredients permitted in fruit juice in EEC countries

Ascorbic acid (as antioxidant)
Volatile components of the same fruit
  (need not be included in ingredients list)
Sulphur dioxide
    apples
    citrus fruits
    pineapples
  (need not be included in ingredients list if SO_2 content is no greater than 10 mg/l)
Sugar[1]
    All juice except grape and pear
  (must be labelled 'sweetened' if sugar content is greater than 15 g/l)
Acids[1]
    type and content varies according to fruit
Dimethypolysiloxane (antifoaming agent)
    pineapple only
Carbon dioxide
  (must be labelled 'carbonated' if CO_2 content is greater than 2 g/l)

[1] No juice can contain both added sugar and added acid.

packed without heat treatment for consumption within a few hours. Single-strength direct juice is also produced on a full commercial scale for much wider distribution. Preservation is necessary by pasteurization, or by in-container or UHT sterilization. Single-strength direct juice is primarily a consumer product, and the high costs of transport and storage has tended to restrict its role in international commerce. More recently, however, the introduction of aseptic packaging of UHT sterilized single- strength juice in bulk containers, which may be stored for long periods without refrigeration, has led to an increase in commercial usage.

Concentrated juice is the mainstay of international commerce. Juice is concentrated, usually by thermal evaporation, and may either be preserved by heat treatment or freezing. Frozen concentrated juice is also available as a consumer product. Juice from concentrates is a distinct category prepared by dilution of concentrated juice with water. Dried fruit juice is prepared from concentrated juice. Spray drying is most common, but some is freeze dried.

2.2.3 Citrus juices

Sweet orange (*Citrus sinensis*) juice is the most popular citrus juice followed by grapefruit (*C. paradisii*). Other species, the most important of which are bitter orange (*C. aurantium*), lemon (*C. limon*), limes (*C. aurantifolia*, *C. latifolia*) and mandarin (*C. reticulata*), are primarily used in blends or as ingredients of other products. Citrus hybrids also exist, 'tangelos' and 'tangers' (*C. sinensis* × *C. reticulata*) being permitted in small quantities in some countries to enhance the colour and flavour of orange juice.

Citrus cultivars are grown throughout the world between the latitudes 35° north and 35° south. This represents tropical and subtropical regions where soils are suitable, there is sufficient moisture to sustain the tree and a low incidence of frost. Citrus fruit have a good adaptation to different soil types, but the best are sandy loams which water readily penetrates. Fertilization and irrigation are important with well drained soils, where problems can arise with leaching of nitrogen and minerals.

The US is, for obvious reasons, a major grower of citrus fruit for juice production. Citrus fruit are produced on a large scale in three states: California, Florida and Texas. For a number of years juice production has been of greatest importance in Florida where, unlike California, climatic conditions mean that fruit, particularly oranges, tend to be of poor appearance and of lower value as dessert fruit. No less than 90% of Florida-grown oranges and 55% of grapefruit are used for juice production.

Brazil became a major producer of citrus fruit for juice production following frosts in Florida. Brazil is primarily a producing country, exporting *ca.* 95% of juice and accounting for 70–80% of world trade. In contrast, the US is a producer–consumer. Juice is imported from Brazil and only *ca.* 10% of the domestic production exported.

BOX 2.2 **All the trees they are so high**

Yield of oranges in Brazil has been lower than anticipated in recent years due to trees stopping fruiting short of the normal lifespan. This, inevitably, reduces the amount of juice available in commerce and leads to higher prices. The trend towards higher prices has been reinforced by increasing demand from Europe and, in some years, poor harvests in the US.

A number of other countries are important producers of citrus fruits, but in many cases the bulk of the fruit are exported. Israel is a significant exporter of frozen concentrated citrus juice, although volumes are low in comparison with Brazil. Most of the juice is exported to the European market. Citrus juice is also produced in southern Europe and North Africa, but overall Europe is heavily reliant on imports.

(a) Harvesting and pre-extraction processing

As the most important citrus crop, oranges tend to dictate agricultural and processing practice. In order to employ plant as fully as possible, each producing area grows several different varieties to ensure that harvesting is spread over many months. In Florida, four main varieties of orange are grown. Of these Hamlin and Parson Brown are earlies, harvested between September/October and December. Pineapple oranges are harvested between January and March, and late season oranges (Valentia) between March and June.

Harvesting commences when the fruit reaches defined maturity standards. These vary somewhat according to producing country, but are usually based on Brix:acid ratio, colour, oil content, etc.

In the early part of the season, oranges tend to be used for single-strength juice, production of concentrates commencing when the soluble solids (sugar) content is *ca*. 12% (12 °Brix). Concentrating juices of lower soluble solids content results in an uneconomic yield of concentrate.

Fruits may be picked by hand or, increasingly, mechanically and are trucked to the processing plant or, in some cases, the railhead. Unloading is invariably mechanized in large-scale operations. The increasing use of mechanical harvesting leads to greater problems with extraneous material, including leaves, stems, dirt and small branches. To avoid expensive manual removal of this material it is now usual practice to pass the fruit through a 'trash eliminator'. In this equipment, oranges pass in single file through a system of flexible belts and rollers, which both drops loose extraneous material and pulls off most of the stems remaining attached to the fruit. The fruit then pass to a grading station for inspection and culling of poor quality oranges. Any remaining stalks are removed at this stage, before the oranges are transported by conveyer to storage bins. In Florida and some other producing areas, the conveying systems are equipped with automated sampling devices to remove a representative sample of fruit, which is then transferred direct to an analytical laboratory.

Storage bins are constructed of steel mesh in a wooden or steel frame. Flies and other insects are inevitably attracted to the bins, which are sited some distance from the main processing operation, to minimize problems with flying insects. Fruit enter by inclined ramps, which impart a spiral motion and minimize damage, and which are also designed to spread the load within the bins and avoid crushing.

* Brix is the percentage (w/w) of soluble solids expressed as sucrose. It is often considered to be synonymous with total soluble solids, but strictly this is only true in the case of pure sucrose solutions. Brix may be measured by hydrometer or, more usually, a refractometer fitted with a Brix scale. Brix:acid ratio correlates well with flavour quality in both fresh fruit and processed juice.

Oranges are transported by elevator to a 'surge bin'. This is situated just outside the main processing plant and acts as a buffer to ensure a constant supply of fruit. The surge bin feeds a variable speed conveyer, which permits fruit flow rate to match demand by the extractors. Before feeding to the extractors, however, oranges are washed by rotating brushes, pass through a second grading station and are finally automatically sorted by size. The need for water economy in many juice producing areas means that condensate from evaporators is used for washing the fruit.

(b) Extraction

Two main types of extractor are in use for citrus fruit, the FMC and the Brown types. These operate on different principles, but each produces a high yield of good quality juice. In either case different extractors are required to deal with fruit of different sizes.

The FMC machine is equipped with a feeder head that loads fruit into the extractor cups. The cups are arranged in five rows in a five-head machine used for normal sized oranges. Eight-head machines are used for small oranges, lemons, limes and mandarins, and three-head machines for large oranges and grapefruit. Individual fruit are deposited in the bottom half of the cup, the top half then descending and applying pressure. As contact is made, a sharpened, perforated stainless steel tube, in the bottom half of the cup, cuts a plug in the base of the orange. The cups contain interlocking fingers which mesh, pressing the fruit inward and forcing juice into the perforated tube, which has a restriction in the bottom to prevent loss of juice. The resulting internal pressure forces the juice and smaller pieces of pulp through the perforations in the tube wall, straining out the remaining pulp and seeds. At this point the perforated tube, now containing plug, some pulp and seeds, rises to further compress its contents and eject the plug. Extracted juice is

* In contrast to the FMC extractor, the Brown-type does not significantly rupture the oil cells. It is therefore necessary to use additional equipment to recover the cold pressed oil. Two types are available: the 'oil extractor' and the 'peel skimmer'. The oil extractor, which is situated upstream of the juice extractors, punctures the oil cells and then washes the oil off the outside of the fruit for collection. The peel skimmer is situated at the peel discharge, where a knife quarters the emerging half oranges. The peel pieces then pass through a series of rollers and blades, which separate the albedo from the oil-containing flavedo (outer part of rind) and expels the oil by pressure. The oil is recovered from the final rollers by water sprays.

collected by a manifold and flows out of the extractor. Oil cells in the skin are ruptured by the pressure exerted by the extractor cups. Oil, together with small pieces of skin, is washed off the outside of the cups and passes to the cold pressed oil recovery line.

Most extractors of the Brown-type employ a reaming action. Fruit cup halves on travelling chains are fed by a rotating feeder wheel. Each passes sequentially over a knife where the fruit is cut in half. Each half is held in position by a slider plate while it travels, separately, to a reaming wheel. The reaming wheel and fruit cups are synchronized and as the two rotate, the juice is extracted, along with pulp, rag and seeds, and falls to the juice trough.

In an alternative Brown design, a feeder places fruit in three single lanes. As the fruit drops into the extractor it is caught by a series of slightly conical, rotating discs mounted on horizontal shafts. The discs carry the fruit across a knife, where it is bisected, to a stationery perforated screen. The cut faces of the fruit are forced against the screen; juice, pulp and some seeds passing through the perforations.

(c) Finishing

Juice from either the FMC or the Brown-type machine requires further processing to separate cloudy, but otherwise 'clean' juice from pulp, rag, peel and seeds. This processing is commonly referred to as finishing. Finishers separate juice by the action of an auger rotating within a cylindrical screen. The screen diameter is usually in the order of 35 cm and the mesh 0.02–0.03 cm. The juice passes through the perforations, while the rotation of the auger forces the pulp to the discharge end of the cylinder, where back pressure is applied by a mechanically or pneumatically controlled valve. A high back pressure forces more juice from the pulp through the cylinder walls and increases yield. At the same time, more pulp is forced through into the juice. This leads to off-flavours and general poor quality in single-strength juice, while in concentrated juices, the associated high pectin content leads to problems of excessive viscosity. For these reasons it is not usual practice to seek to obtain the maximum juice yield at the finishing stage and, in Florida, juice yields for frozen, concentrated orange juice are set by the State testing laboratory. This ensures that competitive pressures to obtain higher yield do not compromise the quality of the juice.

In many factories, two finishers are operated in parallel. The primary (upstream) finisher has a lower extraction rate and a higher throughput than the secondary (downstream) finisher. The disparity in throughput means that two secondary finishers operated in parallel may be required. Juice can also be taken from the primary finisher to supply markets that require a low pulp content.

(d) Pulp washing

Pulp may be utilized entirely as a by-product. In many plants, however, juice is recovered from pulp by back-wash leaching. This involves mixing the pulp with water and serial passage through finishers. These are operated at very low back- pressures to prevent further disintegration of the pulp and passage into the pulp wash.

BOX 2.3 **Information leading to the arrest . . .**

Adulteration of citrus juice became a major problem in the late 1970s, following a steep rise in prices in 1977. The Florida Orange Juice Commission, alarmed by the quantity of adulterated juice in the market place, established, during 1979, a fund of $500 000 to support action against adulterators. Adulteration had progressed from simple dilution and substitution of low cost materials to use of sophisticated 'chemical cocktails' designed to mask evidence of adulteration. One of the most common adulterants, however, was pulp wash solids. These can be detected by instrumental analysis, but the equipment is expensive and the methodology complex. To overcome this problem Florida introduced, in 1981, a state law which requires pulp wash to contain a low level of sodium benzoate as marker. This legislation has been highly successful in reducing adulteration juices imported into Florida from outside the state. (Nagy, S. *et al.* 1988. *Adulteration of Fruit Juice Beverages*. Marcel Dekker, New York).

Pulp wash usually contains 5–6% soluble solids and is of high pectin content. Pulp wash is generally considered to be of poor quality and is not normally used directly for beverage purposes. In some countries, including the US other than Florida, pulp wash may be combined with juice before concentration. Elsewhere this is prohibited and pulp wash must be segregated from juice during processing and used for other purposes.

(e) Debittering and reduction of acidity

Excessive bitterness can be a major problem in juices. Grapefruit causes particular problems, although some bitterness is both expected and desirable, but bitterness is also a problem with Navel oranges. Limonin, a triterpenoid dilactone formed from the precursor, limonoic acid α-ring lactone (Figure 2.1), is the major bittering agent and a number of methods for debittering juice have been developed. These include the use of immobilized bacterial enzymes, chemical masking and removal by adsorption. Enzymes can be obtained from *Arthrobacter* species or *Acinetobacter* species. The most satisfactory result is obtained using limonoate dehydrogenase, but the process has not been fully commercialized. A commercial debittering process is, however, available using adsorbents. A styrene divinylbenzene polymeric adsorbent has been in use in Australia since the mid-1980s and a similar process has subsequently become available in the US. In addition to limonin, grapefruit and bitter orange contain a second major bittering compound, the flavonone glycoside naringin. This may be debittered using the enzyme naringenase, derived from citrus pectin preparations, or *Aspergillus* species.

It is considered that some 20% of the population of the US reject citrus products due to their high acidity. A number of methods have been used, on an experimental basis, for acidity reduction, but most were unsatisfactory. A commercial process has, however, been developed using a weak base anion exchange resin, which preferentially absorbs citric acid. Reduced acid frozen concentrated juice is recognized as a distinct category of juice in the US.

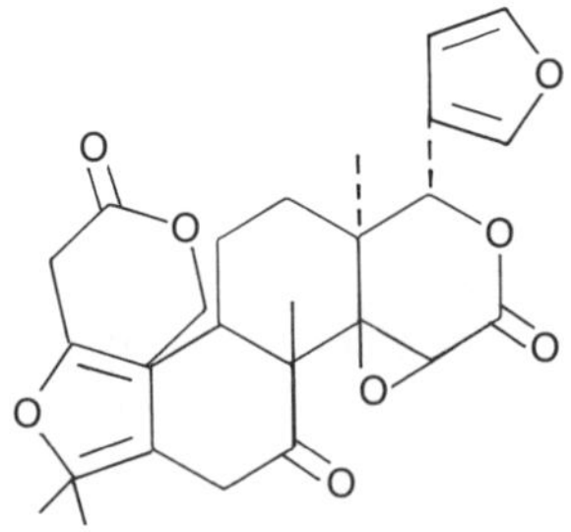

Figure 2.1 Structure of limonin.

(f) Processing of single-strength juice

In small-scale operations, such as catering or in-store juice bars, the juice receives no further processing and is either consumed immediately or filled into plastic bottles or cartons. According to the type of extractor, the juice may pass through a filter to remove seeds and large pieces of pulp. The chemical and microbiological stability is very limited and such a product is obviously unsuited to industrial-scale processing.

In industrial-scale operations the juice flows to holding tanks, equipped with an agitator to maintain the pulp in suspension. The main technological processes before packaging are de-oiling and pasteurization which, with a low pH value product, ensures shelf stability. For some markets the juice is depulped and ultrafiltration has been proposed as a means of clarification. Such processing, however, is more usually applied to juice for concentrating.

A certain quantity of citrus oil enters the juice during extraction. It is generally considered that levels of oil should not exceed 0.015–0.02%, but juice direct from the extractor may contain as much as 0.05%. In most processing plants, removal of the excess oil, by steam distillation (stripping), is necessary. In commercial practice, juice is heated in a heat exchanger and pumped into a vapour separator operated under reduced pressure. The juice is either broken up into small droplets by a spinning vane, or formed into a thin film by tangential entry into the separator. The quantity of oil removed is controlled by the temperature and degree of vacuum within the separator, the extent of evaporation increasing with increasing temperature and pressure differentials between the heater and separator. In older plant relatively high temperatures were used, in order that juice passing from the separator to the filling machine was at pasteurization temperature. This system could lead to marked changes in taste and it is now usual to use lower de-oiling temperatures and to re-heat the juice directly before filling. In typical plant, the incoming juice is heated in a heat exchanger to *ca.* 71°C and flashed in the vapour separator under a vacuum of *ca.* 28 mm Hg. The temperature of the juice is reduced to *ca.* 38°C and *ca.* 6% of the juice is vapourized. The juice is then re-heated before hot filling into cans or bottles. Direct heating by injected steam is efficient and compensates for water lost as vapour. It is sometimes considered, however, that off-taints are introduced by carry-over from steam generation equipment and indirect heating may be preferred.

In recent years the thermal load applied to single-strength citrus juice has been minimized by maintaining aseptic conditions after the initial heating. Modified ultra-heat treatment equipment can be used to 'sterilize' the juice. The vapour is flash evaporated under aseptic conditions, the cooled juice then being aseptically packed. Vapour is used to pre-heat incoming juice and then condensed. The water and oil phases are separated, the water being returned to the juice to compensate for vapour loss, while the oil is recovered. The juice may either be aseptically packed into Tetra Brik^{TM}, or similar retail packs, or aseptically packed in bulk containers of up to 1000 l for ingredient use.

Conventional hot filling may involve packaging in cans, or glass bottles. Cans are hot filled at *ca*. 80°C. The headspace is flushed with steam to remove air and the sealed can is inverted to pasteurize the inner surface of the lid with hot juice. Cans are then cooled to *ca*. 20°C and stored at ambient temperature. Glass bottles are filled at 70–75°C. Care must be taken to minimize stresses during cooling and bottles pass first through warm, and then cold water. Cooling is planned to lower the temperature to *ca*. 25°C after 1 h. Bottled citrus juice is less stable than canned and should be kept at 'cool' temperatures to reduce the rate of development of off-tastes and colour changes.

(g) Processing of concentrated juice

De-oiling is not required for juice for concentrating, since peel oils are removed during thermal evaporation. Varying degrees of pre-concentration processing are, however, required depending on the properties of the juice leaving the finishers and the end-market.

Where juice is of high pulp content, or where a clarified juice is required, the pulp content is reduced by centrifugation. This may

---

* In contrast to many other applications of ultrafiltration technology, permeate fluxes during citrus juice processing are largely independent of membrane type or molecular weight cut-off. It appears likely that the membrane acts partly or primarily as a support for the development of a fibrous deposit. It has been postulated that the deposit functions as a dynamic membrane and, in some circumstances at least, is responsible for ultrafiltration.

be combined with treatment with pectinolytic enzymes to prevent excessive concentrate viscosity. Ultrafiltration and microfiltration have also been suggested as a means of clarifying juices before concentration and have been stated to produce a juice of superior sensory qualities.

As noted above, pulp wash may be processed separately or, where permitted, combined with juice before evaporation. In either case, the pectin content must be reduced to prevent excessive viscosity in the concentrate. As with juice, treatment with pectinolytic enzymes (polygalacturonases) is used to reduce the pectin content. *Aspergillus niger* is the most common source of the enzymes, which are added to juice or pulp wash and incubated at *ca.* 60°C for 2 to 4 h. During this period the chains of galacturonic acid molecules are cleaved to yield products of lower molecular weight and less viscosity.

Juice also contains naturally occurring pectinolytic enzymes. These, however, are pectinesterases, which remove the methyl group from carboxylic acid side chains. Demethylated pectins form gels in concentrated sugar solutions and, at lower concentrations, react with calcium and magnesium ions to form insoluble floccular precipitates. In most circumstances pectinesterase activity is undesirable and the enzyme is destroyed by heating. This is achieved during concentration in modern evaporators, but where juice is stored for long periods before concentration, heat treatment should be applied as soon as possible after finishing. In products such as fully clarified lemon and lime juices, pectinesterase activity is promoted as a means of enhancing clarification. Pectinesterase activity is also desirable in some citrus products, such as gellified citrus salads.

Concentration of citrus juices normally involves evaporation under vacuum. Freeze-concentration has been investigated, but the increase in quality does not justify the higher expense. Membrane concentration techniques have also been investigated. Little use has been made of ultrafiltration at this stage, but interest is being taken in using reverse osmosis as a pre-concentration stage before evaporation.

Vacuum evaporation is an extremely widely used process in the food industry. Many types and configurations are available, but all have four basic components:

1. A vacuum evaporator acting as a heat exchanger in which the product being evaporated is heated;
2. A means of separating vapour and concentrate;
3. A vapour condenser;
4. A means of producing a vacuum and removing the condensate.

In addition, evaporators utilize ancillary equipment, such as pre-heaters, pumps, etc. Most modern equipment for fruit juice concentration incorporates a means of recovering volatile flavour and aroma compounds (see pages 42-3).

In a simple evaporator (single effect), the enthalpy of the evaporated vapour is approximately equal to the heat input. In an ideal situation of water evaporation, 1 kg/h of live steam will produce 1 kg/h vapour. The energy consumption of plants can, however, be markedly reduced by using the enthalpy (condensation heat) of the vapour to heat a second effect. Multiple effect evaporators may be built, with energy savings proportional to the number of effects. Thus in a three effect evaporator, 1 kg/h steam will produce 3 kg/h vapour; a specific steam consumption of 33%. The total temperature difference (the maximum heating temperature in the first effect minus the lowest boiling temperature in the last effect) is, however, distributed equally between individual effects. With a large number of effects the temperature difference for each effect is very small and large, and expensive, heating surfaces are required to achieve a given evaporation rate. This, together with increasing plant complexity and increasingly difficult operation, means that evaporators usually have no more than seven effects. Further improvements in efficiency may be obtained using vapour recompression. Vapour recompression is not normally used in evaporators for the concentration of citrus juices, which are of a distinctive type, but recompression is a feature of evaporators used for apple juice, etc.

* Some care is needed in the correct usage of evaporator technology. *Effect* is used to describe heat flow, the first effect always being at the highest temperature and the final effect at the lowest. Effect should not be confused with *stage*, which defines product flow. In *forward flow* evaporators, the product enters at the first effect (first stage) and flows sequentially through to the final effect (final stage). In *reverse flow* evaporators, product enters the final effect (first stage) and flows sequentially to the first effect (final stage). *Mixed flow* evaporators do not conform to either the forward, or reverse flow types and product flow between effects is not sequential.

Early citrus juice evaporators operated at relatively high temperatures. The concentrate was hot-filled into cans and stored at ambient temperature. Major usage was the European market and US forces overseas. To a large extent, subsequent developments reflected that in the dairy industry, in that the operating temperature of evaporators was reduced to minimize thermal damage to the product. Steam-driven, multiple effect evaporators fitted with thermal vapour recompression were common, although refrigeration cycle evaporators were also widely used. Evaporators of this type produced a high quality product, which was often enhanced by the addition of a small quantity of single-strength, unpasteurized juice ('cut-back') to replace volatiles lost during evaporation. The juice was stored frozen, but problems associated with poor temperature control highlighted the need for a high temperature processing stage to control microbial numbers and inactivate pectinesterases. The application of high temperature–short time pasteurization, without serious effects on quality, led to the realization that high temperatures are acceptable for short periods and, ultimately, to the development of the thermally accelerated short time (TASTE) evaporator.

TASTE evaporators are of the multi-effect type, usually employing a forward, or mixed flow pattern. Many TASTE evaporators are of the falling film type, which consists of a bundle of tall tubes, surrounded by the heating medium. Juice enters at the top of the evaporator through a distributor, flows downwards in an even thin film and is partially evaporated. Falling film evaporators are of low liquid content and high flow velocity, and consequently have a very short residence time.

Evaporators of the plate type are popular, especially in Europe. These consist of a series of alternately connected product and heating plates mounted in a frame. Plate evaporators are compact in size and do not usually require to be housed in specially con-

---

* Refrigeration cycle evaporators were usually two effect. The first effect received heat as the condenser side of a refrigeration system. Vapour produced by evaporation of juice in the tubes of the first effect passed to the heating side of the second effect. Here the vapours condensed, giving up heat to the evaporating juice in the tubes of the second effect. Vapour from the second effect was condensed in a vapour condenser, transferring heat to boiling refrigerant. The refrigerant vapours were compressed and passed to the first effect. Here the cycle was completed as the condensing heat of the refrigerant was used to heat juice.

structed buildings. Juice enters the lower part of the evaporator plate and rises upward as a high velocity thin film. Residence time is low, and plant is able to handle viscous juice and juice of high pulp content.

Early TASTE evaporators had three to four effects and five to six stages. It was necessary to place two to three stages in the final effect to obtain the necessary heat transfer rate. As long as energy costs were low, there was little impetus for increased efficiency. The sharp increase in fuel prices in the 1970s, however, led to a greater concern for energy cost and development of seven effect TASTE evaporators. Such equipment was direct heated without vapour recompression but, especially in European designs, there is now a tendency to simpler, three to five effect evaporators fitted with thermal vapour recompression.

The actual design and configuration of evaporators varies according to the requirements of the customer and the different approaches taken by manufacturers to solving engineering problems. There are, however, many features common to all plant. Juice normally flows to the evaporator at a concentration of 10–11% total solids. In some plant a degassing unit is placed upstream of the evaporator to minimize undesirable oxidative changes. Degassing should be carried out at the lowest possible temperature to avoid off-flavour development, but can be followed directly by pasteurization. Alternative procedures include degassing in the first stage of a mixed-flow evaporator. Pasteurization may also take place in the evaporator, usually in the steam-driven heat exchanger, which functions both as pasteurizer and pre- heater before the juice enters the highest temperature (first) effect. This usually operates at *ca.* 90°C. Supplementary pre-heaters are required with some, but not all, designs. The partly concentrated juice then flows to lower temperature effects, leaving the final effect at a concentration of *ca.* 65% total solids. Rapid cooling is essential and this is achieved by flash cooling from *ca.* 50°C to 12–13°C.

Volatile aroma compounds (essence) are removed from the juice during evaporation. Their loss leads to an unbalanced and unnatural flavour and it is now usual practice to recover the aroma compounds for subsequently adding back to the concentrate.

In some older plants aroma recovery equipment is situated upstream of the evaporator, but this arrangement is obsolescent and, in modern plant, aroma recovery is an integral part of the evaporator. Aroma recovery equipment has three main components; a rectifying column (see pages 403–4), an aroma scrubber and an aroma cooler. Vapours containing aroma are sometimes taken from the first evaporator stage (third or later effect) operating at *ca.* 45°C. In some cases the juice is also deaerated at this stage. Aroma-containing vapour is passed through a partial condenser, aroma-enriched vapour passing to the rectifying column. In forward flow evaporators aroma-containing vapours from the first effect are used to heat the second effect. Water condenses first, enriching the aroma content of remaining vapours which pass to a rectifying column. Inert gas may be used to extract aroma compounds from the evaporator. The aroma compounds are cooled to near freezing point in the aroma cooler. On condensing, the aroma compounds separate into an aqueous and an oil-based phase. The two phases are separated by decanting centrifuges, graded and stored.

Not all of the aroma evaporates in the first stage and recovery is possible from the condensed vapour using an aroma scrubber. Dissolved, non-condensable gases, removed during deaeration, also contain highly volatile aroma compounds. These are recovered by washing out with cold aroma concentrate.

The quantity and quality of essence varies. Early fruit, for example, produces low levels of essence, which is often of poor quality. Addition of such essence to concentrate has a serious adverse effect on quality. Essence grading usually involves a combination of expert tasters with a taste panel, rather than chemical analysis. More essence is produced than is required in concentrate manufacture and much is sold for other purposes, including flavouring.

After leaving the cooler, the concentrate is pumped to batch blenders, standardized and pumped through chillers to storage tanks. Blending is also possible at this stage and cut-back juice or, more commonly, aroma concentrates are added if required. A storage temperature of less than −7°C is required to prevent non-enzymic browning and concentrate is normally stored in bulk at −8 to −10°C. For retail use, concentrate is canned and blast frozen.

(h) Juice from concentrate

Juice from concentrate is an extremely popular retail commodity and has done much to enhance fruit juice consumption in countries remote from producing areas. The basic commodity is 65% total solids concentrate. Frozen concentrate is thawed, taking care to control conditions to prevent localized areas rising in temperature sufficiently to support microbial growth. The concentrate is then blended with water at 2–3°C. Aroma concentrate can be added at this stage and pulp may also be added to impart a 'fresh' feel. Considerable care is required to ensure that the water is of satisfactory quality. Water must be potable and be free from contaminants, especially chlorine, sulphides and iron. Soft water is preferred as calcium bicarbonate in hard water can react with flavour components. These requirements often mean that water treatment is required at the re-processing plant. Water should be disinfected, if required, using ultraviolet or ozone sterilization, possibly in combination with filtration. It is usual to soften hard water and to remove any residual chlorine by charcoal filtration. Sulphides can also be removed by sprays and sand filtration.

In the past, reconstituted juices were pasteurized and packed into glass bottles or cartons. Pasteurization is at 80–90°C for 12 s. Tubular heat exchangers capable of handling pulp-containing juice are used. Some measure of post-process contamination is inevitable and the juice requires storage at 2°C for a limited period. More recently, aseptic packaging in a variety of pack sizes has become very widely used. Storage life is limited by non-enzymic browning. Reaction rate is slower than in concentrates, however, and storage lives are in excess of 6 months.

2.2.4 Apple juice

Apples are amongst the most widely grown and consumed temperate fruit, being second only to grape in overall production. World production is of the order of 40 million tons, of which 5 million tons is processed into juice. Apple juice is perceived as a healthy drink and is popular amongst children, who may dislike the sharpness of fresh orange juice. In addition large quantities of apple juice are used in other beverages and in low-sugar canned fruit.

Fresh apple juice, a colourless and virtually odourless liquid, is a most unstable product. Truly fresh juice does not exist in commerce, since within seconds of extraction it undergoes a series of enzymic changes. Distinct types of apple juice are, however, available commercially and reflect the processing applied. Raw apple juice is the basic product and is available as a farm-gate product especially in the US.

Raw juice has a storage life of only a few days under refrigeration. The product is brown in colour, tends to sediment and has the distinction of being the only type of fruit juice identified as a vehicle of bacterial food poisoning (see page 69). The life of raw juice may be extended by pasteurization or, rarely, addition of preservatives. Clarification is required to produce a juice of acceptable appearance and the resulting product is the conventional bottled apple juice. Natural-style apple juice is a premium product, which has an aroma relatively close to that of fresh apples. This type of juice has enjoyed considerable popularity in recent years and is highly suited to small-scale production. Post-extraction changes are minimized by addition of ascorbic acid or flash heating as soon as possible after extraction. The juice is light in colour and of high turbidity. It is usually, however, relatively stable to sedimentation and the opalescence associated with turbidity is considered to be a quality attribute. Although considerable quantities of single-strength apple juice are produced, the greatest volume is processed into a concentrate of 70% total solids. This is a commodity product, which is used for manufacture of juice from concentrate and as an ingredient in other products.

(a) Harvesting and pre-extraction handling

Apples are not generally grown specifically for juice making, exceptions being some areas of central Europe and, in some other countries, fruit from organically cultivated orchards used for making premium juice. In most cases, second-grade fruit is used. This is fruit which is in good condition, but mis-shapen. Whatever the source of the fruit, it should be sound and free from gross damage and contamination, especially mould and bacterial infection. Fruit should be picked when fully ripe, with seeds dark brown and no residual starch. Slightly under-ripe fruit should be allowed to mature before use.

The best quality apples are obtained by hand-picking. For economic reasons, large growers increasingly use automated picking systems, many of which shake apples to the ground, from where they are recovered by machine. The risk of bruising is significantly greater than with hand-picking and extraneous matter tends to be picked up with the apples. Apples harvested in this way are also exposed to a greater risk of contamination with environmental micro-organisms including pathogens. Risk from pathogens is significantly increased where animal manure is used for fertilization. Windfalls are also subject to contamination. Such apples are often of poor quality and are generally considered to be unsuitable for juice production, but may be used by small pro-ducers. Apples of any type which have been subject to contamination with soil or other debris should be washed with clean water before leaving the farm.

On receipt at the factory, apples are tipped into concrete storage pits or onto concrete aprons. Bucket elevators are sometimes used for transport to the mill and these usually incorporate a spray washing. Where practical canals or flumes are an effective means of transport and provide an additional means of washing the apples. The water is, however, recycled with only minimal purification and a final rinse spray using clean water should be installed.

Concentrated apple juice is a commodity and juices from different sources can be blended to the desired characteristics. This is not possible with juices packed immediately after pressing and this can lead to variation in juice characteristics. This can be overcome by mixing different types of apple before pressing. In the UK it is common to mix dessert varieties, such as Worcester or Cox's with Bramley cooking apples, to achieve a balanced sugar:acid ratio. Cold storage of apples is common to extend the availability and the processing period.

Apples require milling to a pulp before juice extraction. Traditionally, a horse- or water-powered edge runner mill was used, but these have long been superseded, except for cosmetic purposes at some small producers. Hammer mills are widely used in North America and give a very high juice yield. Flavour problems may

* Cold stored apples may have different characteristics and it is necessary to adjust the processing accordingly. The total soluble protein content of Granny Smith apples falls during long-term (9 month) storage at $-1°C$ and juice from such apples is more prone to haze formation than juice from apples cold stored for short periods (3 months). Haze in juice from apples cold stored for short periods is due to proteins of molecular weight 21 to 31 kDa, but in juice from apples stored for long periods, phenols and traces of neutral polysaccharides are also present. More rigorous clarification procedures are required where apples have been cold-stored for long periods and heat treatment is usually also necessary. (Hsu, J.C. *et al.* 1989. *Journal of Food Science*, **54**, 660–2).

occur, however, due to seeds being broken. In Europe, fixed knife mills are very common. These consist of a circular, gravity-fed chamber containing a three-armed rotor. The rotor spins at high speed, forcing the apples against fixed shredding knives. The pulp falls into a hopper and is pumped to the press, passing through screens where stalks and other debris are removed.

(b) Extraction (pressing)

A number of types of press have been developed, operating as both batch and continuous processes. Of these, the rack and frame press is the oldest, but is still used by smaller processors. In this press, apple pulp is placed in a heavy cotton or nylon cloth positioned in a frame mounted on a wood or metal rack. The pulp is levelled and the cloth folded over the top to make a 'cheese' and the frame removed. The process is then repeated with a second rack placed on top of the cheese. When the required number of cheeses have been prepared, the stack is placed under a hydraulic ram, which applies pressure to express the juice. The rack and frame press can produce a high quality juice, but is labour-intensive. Capital costs are relatively low, but great care is required in cleaning and sterilizing the cloths.

Highly automated hydraulic presses have been developed, which have superseded the rack and frame press for large-scale production. The Stoll press, which is widely used in the US, operates on a batch basis and consists of a vertical chamber with bar screens at the sides. Pressure is applied in two stages. Labour costs are low, but press aids must be added.

The Bucher–Guyer press is also operated on a batch basis and is popular in Europe. The press consists of a rotatable cylinder with a hydraulic ram. Inside the cylinder are a large number of flexible rods. Each rod is covered with a knitted synthetic fabric and has a serrated surface, which allows juice to flow to the discharge end after passing through the fabric. In operation, the juice is filled with pulp under pressure, a considerable quantity of juice being

---

* Press aids are added where required to provide firmness to the mash and channels for the juice to exit. They may also reduce cloudiness in the juice. Common types of press aid, which may be used singly or in combination, include ground wood pulp, sterilized rice hulls and defribrated bleached Kraft fibre. In all cases relatively long fibres are required for efficient action.

expressed at this stage. Continuous hydraulic pressure is applied for a predetermined time, the ram is withdrawn and the cylinder rotated to break up the press cake. The pressing cycle is then repeated several times before the press cake is discharged and the press rinsed with clean water. The Bucher–Guyer press produces juice of a high quality with only small quantities of suspended particulate matter. Press aid is not required. The enclosed construction of the press means that operations can be conducted under a blanket of nitrogen during production of natural-style juice.

Air pressure can be used as an alternative to hydraulic pressing. A modern design of this type, the Atlas-Pacific is an automated batch press. Pulp is placed on a mesh belt and transported into position under a pressing head. This is locked onto the belt using a hydraulic drive, before air pressure is applied to a flexible membrane forming part of the pressing head. Air pressure is usually applied in two stages. The press has low labour requirements, but press aid is necessary.

Screw presses were developed in Europe, but are now popular in the US. Several designs are available, but all consist of a heavy screw, of graduated pitch, fitting closely within a cylindrical screen. Press aids are essential and equipment for grinding Kraft paper or wood pulp and blending with the apple pulp is usually incorporated. Most screw presses operate in two stages, the first stage involving draining off the easily extracted juice. The pulp then passes to the screw press itself, where it is compressed, the action of the screw being aided by interaction with compressor bars incorporated into the press. Back-pressure is applied by a conical exit valve, which is maintained under hydraulic, or pneumatic, pressure and which rotates at a different rate to the screw. Screw presses are efficient, but produce a juice high in suspended solids content. The solids content can be reduced by using higher levels of press aid, but this increases materials cost and leads to significantly higher power consumption.

---

* In addition to the major methods of extraction, other approaches have been taken, including the use of various types of centrifuge and vacuum filters. Vacuum filters form the basis of the Murch process, in which apple pulp is enzymatically depectinised, heated to *ca.* 85°C and screened to remove seeds and stalks. Juice is then extracted by filtration through a rotary vacuum filter. The process can be modified to produce a concentrate by removing 30% of the water from the pulp in a preconcentrator, before filtration. The juice is then concentrated to 70% total solids in a conventional evaporator.

Although most continuous presses are of the screw type, use is also made of belt presses. Several designs are available, but all operate in a similar way. Apple pulp is blended with press aid and deposited on a continuous mesh belt. In most designs a second, upper belt seals the pulp, which is then subject to increasing pressure during transit through a series of rollers. The juice passes through the lower belt into a collecting channel. Variant designs exist, in which a single belt is used, folded to seal the pulp during passage through the rollers. Belt presses are highly effective with firm fruit, but tend to produce a juice of high solids content with softer apples.

Although pectinolytic enzymes are widely used for clarification of apple juice after extraction, application as a means of improving extraction efficiency is less consistent. In general the addition of pectinolytic enzyme to the pulp increases yield (usually 78-85%) by *ca.* 5%. Enzyme pretreatment is rarely used in the US, where yield increases have been obtained by developments to press aids, but has found wider application in central Europe. Enzyme pretreatment involves dosing the pulp and then holding for 24 h at 5°C or 1 h at 15-30°C before pressing. In central Europe, enzyme pretreatment usually takes place within the widely used Bucher-Guyer HP press. Pretreatment has a number of disadvantages which limit application (Table 2.2), but the process is useful in specific situations, such as pressing Golden Delicious and other soft cultivars after a long period of cold storage.

(c) Processing of conventional single-strength and concentrated juice

Apart from juice for immediate consumption, clarification is required for conventionally processed juices. Clarification normally

Table 2.2 Disadvantages resulting from enzyme pretreatment of apple pulp

1. The process disrupts production and is expensive.
2. Partial pectin breakdown can lead to formation of arabinans and storage haze.
3. Heating to *ca.* 85°C may be necessary to inactivate polyphenol oxidase activity and prevent inhibition of pectinases. This can result in poor flavour.
4. The methanol content of the juice can be as high as 400 mg/l (usual level *ca.* 50 mg/l), resulting in problems during essence production.

involves a combination of enzyme treatment, fining, centrifugation and filtration. Treatment with pectinolytic enzymes is useful, but not essential, for single-strength juice providing that the fruit used is in good condition and not over-ripe. In the case of juice concentrated to 70% solids, however, enzyme treatment is essential to prevent formation of a pectin gel.

Apple pectin is highly methoxylated and forms a stable colloidal suspension. Pectin methyl esterase (PME) partly demethoxylates the pectin resulting in release of some free, negatively-charged galacturonic acid groups. These may either combine with strong complexing cations, such as Ca^{2+}, to form a readily sedimenting floc or alternatively combine with weak complexing cations to form a relatively stable hydrated floc. This forms a pectin cloud with native protein, which is subsequently set by heat treatment. For this reason it is necessary to add polygalacturonase (PG), which breaks down long chains and markedly reduces viscosity. The chain breakdown alters the charge on the pectin–protein complexes causing aggregation into large particles which readily sediment, improving the filtration properties of the juice.

Soluble arabinose oligomers may be split from the pectin backbone by the action of pectinolytic enzymes. During concentration, these condense, forming small particles, which superficially resemble yeast. Arabinase is now incorporated into commercial pectinolytic enzyme preparations to minimize this problem.

Enzyme preparations are available commercially either as a liquid concentrate or adsorbed onto a powder carrier. Pectin lyase is sometimes included to rupture the polygalacturonate chains of methoxylated pectin, while amylase is added for treatment of juice from early season fruit, which can contain significant quantities of starch. Use of an amylase, however, requires heating of the juice to 60°C, since the enzyme is not active against native starch. This can

* The ultimate use of enzymes is to incorporate cellulases and hemi-cellulases to totally degrade the fruit structure. Under these conditions the press has no role other than that of a drainage basket. On an experimental basis, yields of up to 96% have been obtained together with a three-fold increase in output. The use of cellulolytic enzymes, however, introduces further problems. Degradation of fruit leads to an increase in levels of non-sugar solids and of free galacturonic acid and its degradation products with consequent adverse effects on taste and colour. The degradation products are not normal constituents of apple juice and for this reason, the use of the descriptor 'pure' for enzyme-extracted juice is of dubious legality.

affect quality and removal of starch granules by centrifugation and filtration is often preferred if possible.

Enzyme treatment takes place in a holding tank, typical processes being 8 h at 15–20°C or 1 h at 45°C. Temperatures between 20 and 40°C should be avoided to minimize risk of yeast growth. Treatment with enzymes after aroma stripping at 70°C is advantageous, since at this stage only very small numbers of yeast are present and oxidizing enzyme activity is minimal.

Centrifugation is often used as a preliminary treatment to remove material of high, or low, density and is often also applied after fining. Fining is a major stage in apple juice processing and may be the only means of clarification available to the small producer. Gelatin is the traditional fining agent. The protein is positively charged at the pH value of apple juice and forms an insoluble floc with negatively charged juice debris, which can be removed by decanting, centrifugation or filtration. Fining efficiency is greatest where the juice contains a high concentration of procyanidins, which form hydrogen-bonded complexes with gelatin to increase the density of the floc. Where the juice is of low procyanidin content, tannic acid may be added to the juice before fining.

In use gelatin is dissolved in demineralized water at 40°C, aged for 4–5 h and dosed into the juice with gentle stirring before standing at 15°C. It is necessary to determine the optimal quantity of gelatin to use, since excessive quantities cause haze after heat treatment. An even greater problem is caused by the phenomenon of 'charge-reversal', when the gelatin–tannin complex is stabilized as a positively-charged colloid.

Although gelatin is a highly effective fining agent, its action is slow and the floc may take many hours to form. The long holding period can have adverse effects on flavour and microbiological quality.

* Poor fining efficiency can result from failure to specify the correct type of gelatin, which must be specially prepared for clarification. The gelatin must have a positive residual surface charge at pH 3.5 and must lack low molecular weight peptides to avoid haze formation. Type A gelatin made by acid hydrolysis is preferred. It is necessary to be aware that gelatin is a potential source of *Salmonella* and care is required to obtain a product of satisfactory microbiological quality and during handling at the juice processing plant. If juice is not heat treated after fining, gelatin is a critical control point 1.

Bentonite is frequently used as an adjunct and has a negative charge, which neutralizes the charge on the protein, forming a floc which enmeshes debris. Bentonite has a very high surface area and also removes some suspended material by adsorption and entrapment. The synthetic silica sol kieselsol, which is sold under various trade names, is an effective alternative to bentonite and is widely used in central Europe. When added either before, or after, gelatin, kieselsol forms compact flocs within minutes and introduces the possibility of clarification on a continuous basis.

The water-soluble compound chitosan (deacylated chitin; poly-β(1-4)*N*-acetyl-glucosamine) is an alternative to gelatin and can be more effective, the glucosamine structure resulting in both positively- and negatively-charged groups being present. At present, however, chitosan is very expensive and use tends to be restricted to solving particular problems.

Many types of filter have been used in clarification of apple juice. The objectives of filtration can vary from removal of finings to final polishing of the juice. Plate filters, using kieselguhr or perlite as filter aids, and filter candles are popular for general purpose use. Sheet filters, utilizing a combination of cellulose and kieselguhr or perlite, have become widely used, especially for polishing, and 'paper' cartridge filters of varying porosity are also suitable for final stage filtration. In recent years considerable use has been made of rotary vacuum filters for clarifying juice direct from the press or recovery of lees from tank bottoms. Filters of this type are efficient, but require skilled operation and must be operated on a continuous basis. The juice is also prone to oxidation and microbial growth can be a significant problem.

In recent years there has been considerable interest in ultrafiltration. This technique offers the possibility of effectively combining fining and filtration. In contrast to fining, ultrafiltration cannot, however, remove soluble haze constituents and aeration and temperature cycling must be minimized to prevent post-bottling, or post-concentration haze. If possible polyphenol levels should be reduced by oxidation onto pulp before pressing.

Early attempts at use of ultrafiltration met major problems due to low flux, rapid membrane fouling and microbial growth. To a considerable extent, these problems can be overcome by use of adequately depectinized and starch-free juice, but operation at 55°C is still required to reduce viscosity. Turbulent or aerated flow must be avoided to reduce polyphenol oxidation, which contributes to fouling. For this reason, crossflow membranes are preferred. These may be either constructed of polysulphones or be a ceramic material, such as zirconia.

Single-strength juice may be sold without heat treatment, but shelf life is very short. Addition of permitted preservatives, such as sodium, or potassium, benzoate, is an alternative to heat treatment and long-term storage is possible. Addition of preservatives is viewed by many manufacturers (and consumers) as being incompatible with a 'natural' product, such as apple juice. Where used, potassium benzoate is favoured, since the sodium salt reportedly generates off-flavours. Sodium benzoate is, however, effective in inactivation of *Escherichia coli* O157:H7 and its use very significantly increases the safety of non-heat treated apple juice (see page 69).

Most conventionally processed single-strength apple juice is heat treated. In small-scale operations the juice is cold filled into bottles and heated in a water bath. This method is inefficient and can have adverse effects on taste and colour. A more common procedure is heat treatment in either a plate, or tubular, heat exchanger followed by hot filling into bottles, or occasionally cans, and sealing under 'clean' conditions. Some processors use a secondary in-bottle heating, but this is not usually considered necessary. Single-strength juice can also be aseptically packed after heat treatment, but the quantities involved are currently small.

Thermal evaporation (see page 40) is used for concentration of apple juice. In Europe, considerable use is made of falling film evaporators or plate evaporators of the rising film type. Aroma recovery may constitute a separate stage, but it is more common now to incorporate an aroma recovery stage into the evaporator. Four stage plants, incorporating vapour recompression, are common. An example of a high performance plant widely used for apple juice is the Sigmastar^{R}, which permits aroma recovery at only

---

* Microbiologically stable apple juice has been produced on an experimental basis by ultrafiltration through a sterilizable sintered stainless steel membrane. Mould spores and yeast cells were removed, but retention of bacteria was not absolute. Despite this the likelihood of microbial spoilage is considered low. (Barefoot, S.F. *et al.* 1989. *Journal of Food Science*, **54**, 408–11).

70°C. The second stage operates at 85°C, which completely gelatinizes starch and fining takes place between the second and third stages. After fining, the juice is heated to 90°C, which is an effective heat treatment. In all cases, the concentrate must be rapidly cooled to minimize deterioration through the Maillard reaction.

Apple concentrate at 70% solids may either be frozen or aseptically packed and is a product in international commerce. Considerable quantities are used for dilution to single-strength drinks and the juice is also used as an alternative to syrup in canning low sugar fruit and as an ingredient in a range of products. Juice is usually blended for beverage use and aroma essence added. The quantity of aroma essence tends to determine the quality of the juice and none may be added to the cheaper types. Juices must be blended before final filtration to prevent haze formation, but the aroma essence must be added post-filtration, to avoid loss of aroma compounds by adsorption to filter material.

In common with other reconstituted fruit juices, most apple juice is heat treated and aseptically packed in 1-2 l bulk packs or individual packs.

(d) Processing of natural-style juices

Natural-style juices are the closest to the fresh apple. Processing depends on the addition of up to 500 mg/l ascorbic acid to the fruit being milled or immediately after pressing. The juice is then heat treated as soon as possible. These procedures prevent oxidation of polyphenols, a major cause of browning and a contributory cause of flocculation. Ascorbic acid also inhibits pectinolytic enzymes and maintains the fresh apple aroma, possibly by preventing oxidation of volatile aldehydes. Careful control of addition is required and a residual level of *ca.* 250 mg/l is required in the packaged juice.

Additional measures used to minimize polyphenol oxidation are pressing of fruit after pre-chilling to 4°C and maintaining an atmosphere of nitrogen in presses of the Bucher-Guyer design. Juice must be protected during bulk transport by low temperatures and a blanket of an inert gas.

In some countries, ascorbic acid must be declared as an additive.

Vitamin C may be used as a descriptor and this reinforces the healthy perceived image of natural-style apple juice.

2.2.5 Pineapple and other tropical juices

Pineapple is the most important of the tropical fruit juices, although until recently juice production was secondary to processing fruit for canning. Pineapples are cultivated in a large area of the tropics with a relatively mild climate, the fruit being unusual in that very little grows wild.

> BOX 2.4 **The stately homes of England**
>
> Pineapples enjoyed considerable popularity amongst the English upper classes during the 18th and 19th centuries. The pineapple was favoured for its appearance rather than its organoleptic appeal and fruit were commonly used as decorative centre-pieces for lavish cold buffets. Very large pineapples were preferred, which were often grown in the hothouses of large estates. Something of a 'pineapple cult' resulted, artefacts of which may be seen as the stylized stone pineapples found on the gateways of some country houses.

For many years Hawaii supplied a very high proportion of the world market for pineapples and pineapple products. More recently, however, output in the Phillipines, Thailand and Brazil has become greater than that in Hawaii, very little of the Hawaiian crop being used for processing. Other significant producers include Malaysia, Taiwan, Mexico and some countries in sub-Saharan Africa.

The most widely used pineapple for juice production, Smooth Cayenne, is also favoured for canning. The juice is dark yellow in colour and tends to have a higher acidity than other varieties. A second variety, Abacaxi (Permambuco), is increasing in popularity and produces a whitish-yellow juice of low acidity, which is preferred by many consumers. The fruit has a high juice yield and is economical for processors.

Pineapple juice was originally made from pineapple waste resulting from preparation of fruit for canning. The waste was pressed and combined with juice expelled during fruit cutting. The juice was partially clarified by centrifugation, deaerated, pasteurized and concentrated. During the early 1980s, a market developed for premium quality juice made from whole, Brazilian pineapples. Juice from whole fruit is of significantly better quality, having improved body and flavour and a more natural taste. Purpose-built extractors have been developed, such as the Polypine(R), which cuts the juice in half against a screen. The juice is deaerated and pasteurized before concentration to 60–72% total solids and either frozen or aseptically packed. Conventional evaporators are used, but operating conditions must take account of the high soluble solids content of the juice and its high Brix:acid ratio. Considerable quantities of concentrated juice are used to prepare reconstituted, aseptically packed drinks. Foaming is a problem and requires the addition of silicon antifoaming agents.

Juice is produced commercially from a wide range of other tropical fruits (Table 2.3). The market, however, is relatively undeveloped and most juice is used in nectars or in combinations. Difficulties also arise from the relatively small scale of production, which makes investment in sophisticated plant uneconomical.

Table 2.3 Types of tropical fruit juice in commercial production

Acerola	Banana	**Guava**
Kiwifruit[1]	Lulo	**Mango**
Papaya	**Passion fruit**	Soursop
Umbu		

[1] Kiwifruit are not strictly tropical, but are classified as such commercially.
Note: Most important types are printed in **bold**.

Many tropical fruit juices are of very high pulp content and are effectively purees. This means that some juices are unsuitable for concentration and are shipped as frozen, single-strength juice, with associated high transport costs. Where concentration is possible, problems also occur due to loss of flavour volatiles, although these have been partially overcome in recent years.

* Pineapples are subject to pineapple pink disease caused by the acetic acid bacteria, *Acetobacter* and *Gluconobacter* and *Erwinia herbicola*. Affected pineapples are of normal appearance when raw, but develop a pink or brown discolouration when heated during processing. This is believed to be due to the presence of the bacterial metabolite, 2,5-diketogluconic acid.

2.2.6 Soft fruits

Although a great many soft fruits (in the broadest sense) are used for juice production, the quantity retailed directly for beverage use is generally small. The main exception is grape juice, which is consumed in large quantities in central Europe and the US. The processing of soft fruits for juice is generally similar to that of apples, although modifications are required according to the nature of the individual fruit.

Grapes and berry fruit require only light crushing before pressing, although ancillary processes, such as removal of stems from grapes, may also be required. Machinery in which intermeshed, rotating arms expose the fleshy interior of berry fruits and grapes is in common use. Stone fruits, however, must be crushed without damage to the stone to avoid adversely affecting the taste and stability of the juice. In this case, lobed wheels of hardened rubber are used to strip the flesh from the stone.

Presses used for juice extraction from most soft fruit are similar to those used for apples. Grape is the exception and, in Europe, a pneumatic press is most commonly used. The most common design consists of a horizontal, stainless steel cylinder, with perforations covering 70% of the outer surface. The remaining 30% is occupied by a diaphragm, which is deflated during filling to allow the cylinder to be fully charged with grape mash. Compressed air is then used to pressurize the mash and expel the juice. Enzyme pretreatment is not normally necessary in extraction of grape juice, but is essential for raspberries, blackcurrants, strawberries, cherries and plums.

Grape juice is produced as a single-strength consumer product, but large quantities of this and other soft fruit juices are concentrated. The rather delicate flavour of some juices can be affected by thermal evaporation and some use has been made of centrifugal evaporators, which have a very short heating period. The most commonly used is the expanding flow type, which consists of a stack of rotating, steam-filled, hollow cones. Juice is sprayed onto the inner side of the cone close to the apex and spreads across the cone to the outer edges, from where the concentrate is thrown onto the stationary outer walls for collection. Centrifugal evaporators produce a very high quality juice, but improvements to falling film and plate evaporators, which also have very short heating periods, have limited their application. Grape juices present particular problems in that salts of tartaric acid may precipitate onto heat exchange surfaces and reduce thermal efficiency. Prevention requires the use of high evaporation temperatures, followed by rapid cooling to minimize product damage. Grape juice may also contain sulphite, added as a preservative at time of pressing, and evaporators often incorporate a desulphurization stage. Desulphurization involves heating to *ca.* 150°C and then flash cooling to *ca.* 100°C. Free sulphur is then removed with the flash vapour.

Reconstituted concentrated grape juice may be aseptically packed after heat treatment or hot filled into glass containers. Grape juice packed in glass containers, including single-strength juice, may also be carbonated.

2.2.7 Quality assurance and control

Production of fruit juice can be maintained in the virtual absence of control or assurance procedures. In such cases, however, the operations are of small scale, usually producing untreated juice, where 'freshness' is perceived as the most important sales factor and where variability is acceptable. Production of processed juices, however, requires a high level of control from selection and handling of the fruit to final packaging. The adoption of a HACCP-type system is recommended. Particular attention must be paid, in all cases, to ensuring effective heat treatment, adequate plant hygiene, control of evaporator and aroma-stripping operations, and prevention of contamination during packaging. In the case of apple juice processing, production of natural-style juices requires special control, especially with respect to addition of ascorbic acid and other precautions against oxidation. The possibility that pathogenic micro-organisms, such as *E. coli* O157:H7 can survive in apple juice, means that special precautions must be taken in production of non-heat-treated juice. These include the washing of apples in good quality water. The use of apples, which have had contact with soil should be avoided, especially if animal manure has been used in the orchard, or animals have been allowed to graze.

The reconstitution and packing of concentrated juices is a relatively simple operation, although a formal quality assurance system should be instituted. The high cost of fruit juice means that adulteration remains a problem and considerable care is required in sourcing of concentrate.

End-product testing may involve no more than assessing taste and colour of the product, but total solids content (or °Brix), pH value and microbiological status are recommended. Level of fill and, if required, carbonation should also be monitored along with package integrity in the case of flexible aseptic packs. Determination of adulterants is beyond the scope of most small importers of concentrates, but analysis is available at central institutions such as, in the UK, the Leatherhead Food Research Association. In some cases, however, detection of adulterants is not absolute.

2.3 NUTRITIONAL VALUE OF THE MAJOR FRUIT JUICES

The major nutritional value of fruit juice lies in the ascorbic acid (vitamin C) content of citrus juices. Fresh orange juice typically contains 50–60 mg/100 g total ascorbic acid (ascorbate plus dehydroascorbate). The quantity of ascorbic acid varies, however, due to differences in soil, location of orchards, cultivar and fruit maturity. Variation can also be caused by differences in processing methods. Loss during processing is generally small, heat-treated juices retaining *ca.* 97% of the initial ascorbic acid content.

Loss of ascorbic acid occurs during storage of heat processed juices. The extent of loss is primarily a function of storage temperature and time. At an ambient temperature of *ca.* 22°C, a monthly loss of 1–2% may be expected from juice stored in metal cans. At temperatures of 8–9°C, however, losses are normally insignificant, but increase rapidly at temperatures above 25°C. Exposure to light in glass-packed products increases the rate of loss, but loss is greatest in oxygen-permeable plastic and cardboard packing, where up to 80% ascorbic acid may be lost during 3–4 weeks storage at 10°C. In addition to ascorbic acid, citrus juice also contains small, but significant quantities of vitamin A and B group vitamins.

Fresh apples usually contain *ca.* 10 mg/100 g ascorbic acid, although levels as high as 32 mg/100 g have been reported. In conventionally processed juice, however ascorbic acid is progressively destroyed by coupled oxidation with polyphenols and final levels are nutritionally insignificant. In natural style apple juice, however, the addition of large quantities of ascorbic acid during processing, means that levels in the final juice may approach those of citrus juices.

Fruit juices can also be valuable sources of minerals. Potassium and calcium are present in highest quantities in both citrus and apple juice. Magnesium and, in citrus juices, phosphorus are also present in significant quantities. Fruit juices can also be important sources of minor elements, including iron, copper, zinc and manganese, although in some cases metals are primarily derived as contaminants during packaging and processing.

Fruit juices contain significant quantities of sugars and can be of importance as sources of dietary energy, especially amongst invalids. Protein and amino acid levels are low and of only supplementary nutritional significance, except in restricted circumstances, such as extreme dietary regimens.

2.4 CHEMISTRY OF THE MAJOR FRUIT JUICES

2.4.1 Citrus juice

(a) Organic acids

Citric acid is by far the most important acid in citrus fruits, the content being particularly high in lemons and limes. Malic acid is also present in significant quantities and there are traces of other acids. Acidity is an important determinant of acceptability of citrus juice. Orange juice is generally considered acceptable at *ca.* 1% acid, pH 3.5, grapefruit at 1.5–2% acid, pH 3.0, and lemons and limes at 5–6% acid, pH 2.2.

(b) Carbohydrates

The three simple sugars, sucrose, glucose and fructose, represent *ca.* 80% of the total soluble solids of orange juice. Relative quantities vary somewhat according to cultivar, climatic conditions and location of orchard, but levels are usually in the range of: sucrose, 49–59%; glucose, 20–25%; and fructose 20–25%. The ratio of sucrose:glucose:fructose of 2:1:1 is usually fairly constant. Inversion of sucrose may occur during storage, with a corresponding increase in the contents of the reducing sugars, glucose and fructose.

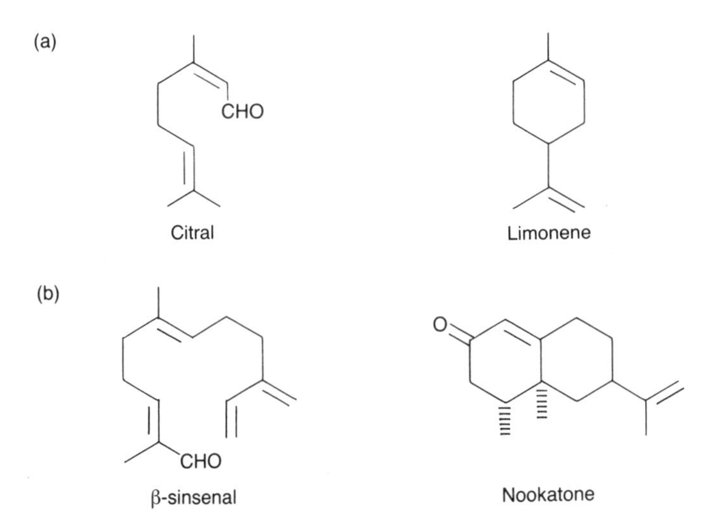

Figure 2.2 Major terpenes and sesquiterpenes of citrus juice.

(c) Flavour and aroma compounds

A great many compounds contribute, to a greater or lesser extent, to the flavour and aroma of citrus juice. Terpenes and sesquiterpenes are important character-impact compounds, which can often be related to the aroma of orange and other citrus juices (Figure 2.2). The monoterpene, limonene, which comprises up to 95% of orange essence oil (Table 2.4), imparts a fruity aroma important in characterizing oranges and limes, while citral plays a similar role in lemons. The sesquiterpene, β-sinsenal, also contributes to the character of oranges, while nookatone is of importance in grapefruit.

A complex of aldehydes, esters, ketones, alcohols, organic acids and hydrocarbons are also important contributors to the aroma and flavour of citrus juices. These are present in the aqueous fraction of

Table 2.4 The major constituents of orange essence oil

Acetal	Decanal	Ethanol
Ethyl acetate	Ethyl butyrate	Geranial
Hexanal	*trans*-2-hexanal	*d*-Limonene
Linalool	Myrcene	Neral
Octanal	α-Pinene	Sabinine
Valencene		

Table 2.5 The major constituents of the aqueous fraction of orange essence

Acetal	Acetaldehyde	Ethanol
Ethyl acetate	Ethyl butyrate	Hexanal
trans-2-hexanal	Limonene	Methanol
1-Propanol		

the essence produced by aroma recovery (Table 2.5). In most cases these are not character-impact compounds, but contribute to the overall flavour and aroma spectrum. Hydrolysis, especially of esters, occurs during storage and results in a loss of aroma and a 'flat' flavour.

Heat treatment inevitably imposes changes to the flavour of citrus juice, primarily due to Maillard reaction products, such as furfural and hydroxymethyl furfural. These are often immediately apparent after hot filling, but incipient reactions, initiated by less severe heat treatment, lead to flavour deterioration during storage. Changes occur most rapidly in concentrated juice, although the rate of reaction is minimized by storage at $-18°C$. Flavour deterioration also proceeds more rapidly in single-strength juice and chilled, or at least cool, storage is recommended for maximum acceptable life.

(d) Colour

The pigments of citrus fruit are concentrated in minute structures, the plastids. Pigments of orange and mandarin are concentrated in the juice sacs; while in blood oranges, limes and some grapefruit, the cell sap is also pigmented. Carotenoid pigments, including lycopene and β-carotene are of greatest importance in most citrus fruit, but the major pigment in blood oranges is an anthocyanin. The quantity of pigment present, and thus the depth of colour, varies according to a number of factors, including cultivar, maturity and climate.

2.4.2 Apple juice

(a) Organic acids

Malic acid is the dominant organic acid in apple juice and accounts for at least 80% of the total acids. Malic acid typically constitutes

0.5% of the total content of apple juice. Small quantities of quinic acid and traces of citric, citramalic and shikimic acids are also present. Galacturonic acid results from pectin breakdown, while lactic acid has also been reported, but is purely of microbial origin and is absent in high quality juice. Losses of up to 20% of malic acid can occur during storage of concentrated juice. This has been explained by formation of dimeric lactides, but formation of sugar–acid esters appears more likely. The effect on perceived acidity is not known.

(b) Carbohydrates

Sugars are the main soluble constituents of apple juice and comprise 7–14% of the fruit before pressing. As with citrus juice, fructose, glucose and sucrose account for virtually all of the sugar content, xylose being present in trace quantities. Fructose is the major sugar, being present at two to three times the concentration of glucose, while glucose and sucrose levels are similar. The state of maturity is important in determining levels, fructose and sucrose concentrations increasing markedly during ripening, while glucose concentration falls. The concentration of each sugar increases during storage of apples, in accordance with continuing starch breakdown.

Inversion of sucrose occurs during juice storage, with corresponding increases in fructose and glucose concentration. This occurs primarily in concentrates and up to 66% of sucrose can be lost during storage at 20°C for 3 years.

In apples and pears, sugars are transported from the leaves to the fruit as the sugar alcohol sorbitol (sucrose is the transport sugar in all other plants). A certain amount of free sorbitol remains in the juice, usually at levels of *ca.* 4 g/l.

(c) Flavour and aroma compounds

Flavour extracts of apples and apple juices have been found to contain several hundred components, although the number directly involved in flavour and aroma is probably relatively small. Three major chemical groups are present: esters, aldehydes and alcohols (Table 2.6), together with a miscellaneous group of compounds, which includes those produced during heat treatment, such as furfural and hydroxymethyl furfural ('process vola-

Table 2.6 The major flavour and aroma compounds of apple juice

Alcohols
ethanol
butanol
isoamyl
hexanol
trans-2-hexanol
Aldehydes
trans-2-hexanal
hexanal
benzaldehyde
Esters
ethyl 2-methyl butyrate
ethyl butyrate
isopentyl acetate
hexyl acetate
hexyl 2-methyl butyrate
Miscellaneous
furfural
hydroxymethyl furfural
6-methyl 5-hepten-2-ol
damascenone

Note: Within each group, compounds are listed in decreasing concentration.

tiles'). To some extent, the apple used affects the aroma composition of the juice and juices made from apples of high ester content, such as Red Delicious (but *not* Golden Delicious) also have a high level of esters. Milling of apples, however, initiates a number of rapid enzymic changes, which affect aroma and flavour. Unsaturated fatty acid precursors, especially linolenic and linoleic acid are converted to C_6 aldehydes and both saturated and unsaturated alcohols through a series of lipoxygenase-mediated reactions. At the same time esters, which are predominant in the aroma of fresh apples, are broken down to the parent acids and alcohols. Benzaldehyde is formed, probably from precursors in the apple pips and there is also some breakdown of terpenes. Changes are modified by addition of ascorbic acid during processing of natural style juice and, in this case, unsaturated alcohols and aldehydes are dominant, producing a fresh, 'green' character.

Character-impact compounds are not readily identified in apple juice and it is likely that much of the flavour and aroma is derived from a balance of many different compounds. Alternatively, it has been suggested that many of the important aroma compounds remain unidentified. Despite these difficulties, several compounds have been identified as being of major importance in apple juice flavour and aroma. These include, hexyl butyrate, ethyl butyrate, ethyl 2-methyl butyrate, hexanal, *trans*-2-hexenal and damascenone. In addition, 4-methoxy allyl benzene appears to be a flavour-impact compound in apple cultivars identified as having a particularly 'spicy' aroma.

Some attempts have been made to relate the presence of specific volatiles with desirable, or undesirable, organoleptic characteristics. As usual, direct correlations were difficult to elucidate, although a general relationship has been found between good quality, high levels of carbonyls and low levels of alcohols. This tends to reflect the composition of juice made from high quality apples. More specifically, the presence of butanol, *trans*-2-hexanal, hexanol and ethyl 2-methyl butyrate are considered to impart good flavour characteristics.

Maillard reaction products ('process intermediates') are undesirable in apple juice, adversely affecting the taste and colour of the juice. Maillard browning continues during storage, especially in concentrated juices stored at temperatures above 5°C. Detectable changes can occur in as little as 2 weeks during storage at 20°C and hydroxymethyl furfural levels of 100 mg/l, or higher, are common in concentrates stored under poor conditions. Breakdown of ascorbic acid in natural style juices enhances Maillard browning through release of carbonyls. The reaction can be inhibited by addition of SO_2, but the presence of this additive is considered undesirable. The best protection against Maillard browning is rapid cooling of juice after heating and storage at low temperature.

Although polyphenols are not generally considered as flavour compounds *per se*, they play an important role in imparting astringency. A certain degree of astringency is a desirable characteristic of apple juice, low levels of polyphenols leading to blandness. Excessive polyphenol levels must also be avoided. Acidity is also an important factor in determining juice acceptability, higher acidity juices being generally preferred in Europe, but disliked in North America.

(d) Colour

Apple juice, immediately after pressing, is of very pale colour. Red apples contain anthocyanin pigments, but this is retained in the peel and not extracted into the juice. The characteristic yellow-brown colouration of apple juices results, primarily, from the post-extraction oxidation of polyphenols.

Apples contain four major classes of polyphenols; procyanidins, phenolic acids, dihydrochalcones and catechins (Figure 2.3). The composition of specific polyphenols can vary according to cultivar. Chlorogenic acid, for example, is usually the principal phenolic acid, but in some cultivars 4-*p*-coumaroyl quinic acid is of equal, or greater, importance. Analogues of the principal dihydrochalcone, phloridzin, are also found in some cultivars, while (+)-catechin is present in one cider cultivar in place of (−)-epicatechin.

Enzyme-mediated oxidation of polyphenols commences immediately after pulping and, under some conditions, continues until the enzyme responsible, polyphenol oxidase, is inhibited by oxidation products. Oxidation products of phloridzin and epicatechin are yellow-orange in colour and account for *ca*. 25% in each case of the total colour. The remaining 50% is derived from oxidation products of procyanidins, which are brown in colour. Procyanidins themselves are not a substrate for polyphenol oxidase and their oxidation involves a coupled reaction with the quinone group of chlorogenic acid. Chlorogenic acid and other phenolic acids are not directly involved in colour formation.

The extent of colour formation can be modified by a number of factors and, in processing of natural style juices, minimized by addition of ascorbate. Polyphenol oxidase activity is at a maximum at *ca*. 40°C and is minimal at temperatures below 2°C. Enzyme activity is also significantly reduced in high acidity apple juice, which can have a pH value as low as 3.0. Colour formation can also be limited by unrelated reactions in the juice, which decrease the amount of substrate available. Procyanidin concentration can be very markedly reduced by tanning onto the pulp and, to a lesser extent, by polymerization. Coloured oxidation products are also

---

* Polyphenols may be referred to, generically as 'tannins'. This, however, is incorrect and only the procyanidins are 'true' tannins, able to combine with proteins.

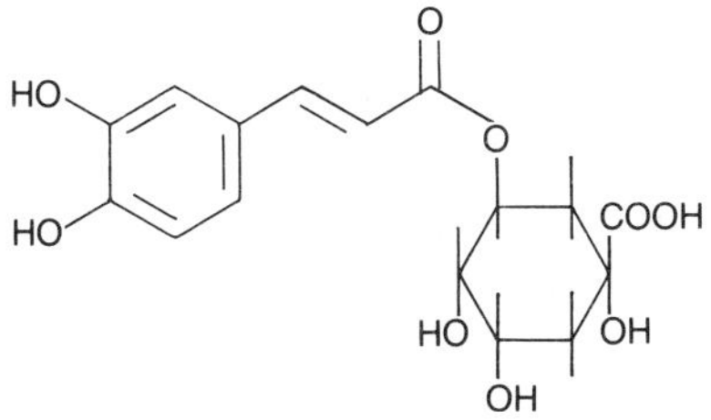

Phenolic acid (chlorogenic acid)

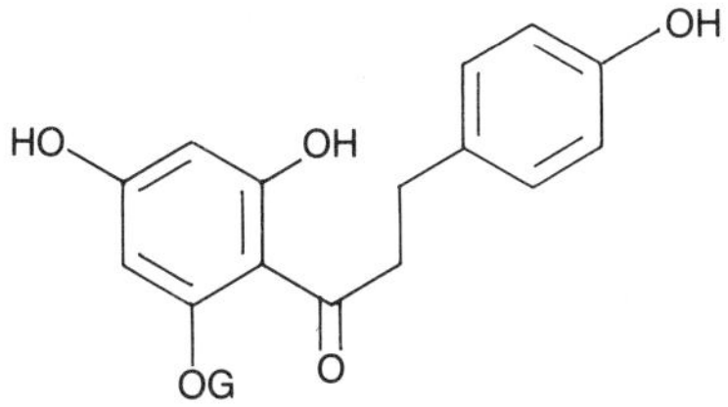

Dihydrochalcone (phloridizin)

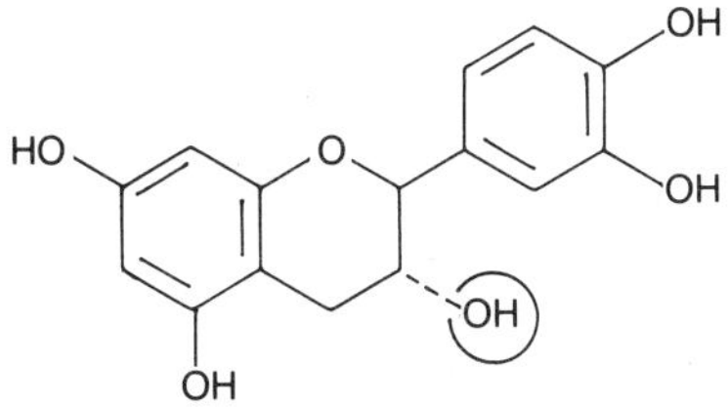

Catechin ((−) epicatechin)

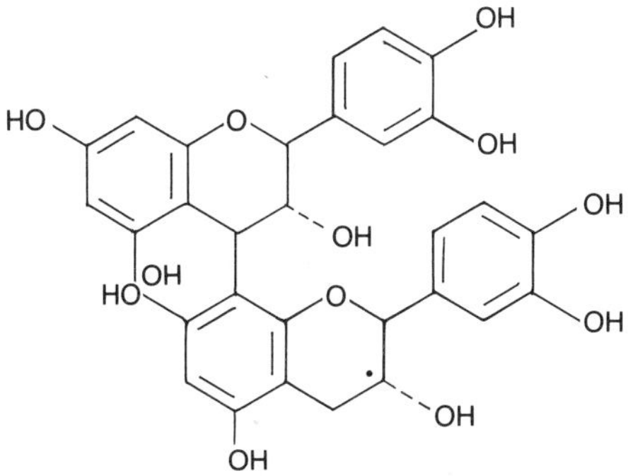

Procyanidin (procyanidin B2)

Figure 2.3 The major classes of polyphenols in apple juice.

removed by adsorption onto pulp. Chlorogenic acid levels may also fall, although this is usually limited in extent. Chlorogenic acid may, however, be split into caffeic and quinic acids by the secondary activity of depectinizing enzymes and, under extreme conditions, levels may fall to as little as 10 mg/l.

(e) Haze formation

A number of factors can contribute to haze formation in apple juice, but the most important is that arising from the polymerization of polyphenols. Procyanidins ('tannins'), of molecular weight 580 to *ca.* 2500, are the only polyphenols involved. Under oxidizing conditions, at low pH value, or in the presence of aldehydes, procyanidins polymerize to form insoluble compounds. Although these may sediment out of suspension, this is rare, and most commonly a colloidal suspension is formed, which appears as haze. The colloidal suspension is often stabilized by formation of hydrophilic complexes with proteins. Colloid formation may be temperature dependent, appearing on cooling to *ca.* 6°C. Rate of cooling is important, colloids being formed more readily when the temperature is reduced rapidly. Although fining may be used to remove complexes (see pages 51-2), sufficient polyphenols should remain to impart the desired degree of astringency.

Haze associated with pectin, or starch, is an ever-present problem for processors of clear juices and is routinely controlled by addition of pectinolytic and amylolytic enzymes (see pages 51-2). Difficulties can arise where treatment has not gone to completion and verification is required by testing for residual pectin or starch. A fairly reliable alcohol precipitation test is available for pectin, but while intact starch can be detected by an iodine-based reagent, partly degraded starch cannot. Partially degraded starch may reform and gelatinize during subsequent heat treatment, forming a heavy cloud that is resistant to further amylase treatment. For this reason, amylase treatment should be continued for a significant period after disappearance of starch is indicated by the iodine reaction.

2.5 MICROBIOLOGY

2.5.1 Fruit juice as an environment for microbial growth

With the exception of the small quantities of juice which contain benzoate or which are carbonated, the main (and in many cases the only) factor controlling growth is pH value. The pH value of juice varies, but in all cases is sufficiently low to select for yeasts, moulds, lactic acid bacteria and acetic acid bacteria. Many of these micro-organisms are able to grow rapidly in juice and in non-heat-treated juices, refrigeration is required as an extrinsic control. In some cases, fruit juice is nutritionally deficient for yeasts, moulds and lactic acid bacteria, although in practice this is of very limited importance.

2.5.2 Micro-organisms of public health significance

As may be expected with a product of low pH value, the safety record of fruit juice is very good. The only juice involved in food poisoning by a recognized pathogen is apple juice, which has been associated with *Salmonella* and *E. coli* O157:H7 infections. The most recent reported outbreak occurred in 1991 in New England, and involved diarrhoea and haemolytic uraemic syndrome as a result of infection with *E. coli* O157:H7. In this and earlier cases, the juice was not heat-treated and had been made from unwashed apples contaminated with animal manure.

2.5.3 Fruit juice and mycotoxins

Fruit may be spoiled by mycotoxigenic moulds and use for juice extraction inevitably means a risk of introducing mycotoxins into the diet. Apple juice is again of particular concern and significant levels of patulin have been detected in commercial juice samples on several occasions. Risk is associated with long-term consumption and is difficult to determine, but there is now a growing body of opinion which suggests that patulin is neither carcinogenic nor acutely toxic to man. Various methods have been proposed to remove patulin from apple juice, but control must lie in selection of sound fruit. In this context, the greatest value of assays for patulin may lie in indicating the use of poor quality fruit. Tolerance

* *Escherichia coli* O157 is the most common enterohaemorrhagic serotype and has most commonly been associated with undercooked burger-type products. Two distinct types of illness, haemorrhagic colitis and haemolytic uraemic syndrome, can result from infection. Haemolytic uraemic syndrome is characterized by a triad of microangiopathic haemolytic anaemia, thrombocytopenia and acute renal failure. Less commonly a related syndrome, thrombocytopenic purpura occurs. This can result in serious kidney damage and death several years after acute symptoms have abated.

levels for patulin in apple juice have been set in many countries and currently range from 0 to 75 μg/l.

Alternaria is a common spoilage mould of fruit, including those commonly used for juice production, such as citrus, apples and soft fruit. *Alternaria* produces over 30, potentially toxic, metabolites, but relatively little is known of their occurrence in foods or drinks. Further work is required to assess fully the risk posed by *Alternaria* toxins in juices, but process control procedures to minimize the risk should already be in place.

2.5.4 Spoilage of fruit juice

The micro-organisms involved in spoilage of fruit juice are largely those involved in spoilage of soft drinks (Chapter 3, pages 121–2), although the less selective nature of juice means that a wider range of species may be involved. In the case of heat-treated juices, growth of micro-organisms in the juice before heating can result in off-flavours in the end product. Examples include production of alcohol and other metabolites by yeasts and production of diacetyl in citrus juice by lactic acid bacteria.

2.5.5 Microbiological analysis of fruit juice

Methods used for soft drinks are also suitable for fruit juice (Chapter 3, pages 122–3). In addition, analysis for diacetyl has been used as a means of detecting prior growth of lactic acid bacteria in orange juice.

EXERCISE 2.1.

During the 1980s, in-store production of single strength orange juice became a popular feature of supermarkets. The juice is prepared from whole oranges using proprietary equipment and hand-filled into plastic bottles or cartons. It was originally intended that the juice would be produced more or less on demand, but undue delays at peak times necessitated filling bottles in advance and storing under refrigeration. In some cases juice production was poorly controlled and problems of spoilage have arisen due to poor cleaning of equipment and storage of juice for excessive periods.

You are employed as a food hygiene specialist for a large company operating a mixture of medium size suburban supermarkets and very large superstores on greenfield sites. The company is attempting to upgrade its downmarket image and is installing in-store features including a juice bar selling fresh produced orange juice. Develop a quality assurance programme, to be operated by store personnel with central guidance, for the production of fresh orange juice. Pay particular attention to selection of oranges (including detection of fruit with internal mould infections), cleaning schedules and the storage of juice. What training would you recommend for personnel involved in juice production, including supervisory staff (up to senior management)?

Flushed with the success of fresh orange juice, your directors wish to extend the range of fresh juice and have suggested that over-ripe and damaged fruit from the produce section, should be used where possible. Prepare a paper outlining your response to these proposals. Consider the legal and moral aspects as well as the technical.

EXERCISE 2.2.

In the case of tropical juices of high pulp concentration, producing a clear juice presents no particular technical difficulty. The juice, however, is insipid and usually lacks any of the characteristics of the fruit. Consider the implications with respect to the distribution of flavour volatiles and use your knowledge of food processing technology to devise a means by which a clear juice of good flavour characteristics could be produced. Estimate the capital investment and operating costs and determine if your process would be economically viable, given the current relatively small market for tropical juices.

EXERCISE 2.3.

You are employed as a microbiologist for a large supermarket retailer. A customer has complained that own label, aseptically packed orange juice (reconstituted from concentrate) contains a preservative. This complaint is based on the failure of the starter yeast to grow during home wine making. The customer has influential connections in broadcasting and, as a precaution, a fairly full investigation was undertaken, which showed that the starter yeast had a high nitrogen requirement, which could not be met by the orange juice. This explanation has not been accepted and your company has now been warned that the case is to be 'exposed' on a national whistleblowing television programme. You are asked to present your company's case on the programme. Prepare an outline of your proposed approach, bearing in mind the difficulties of presenting technical information to a, possibly sceptical, television audience. Be prepared to explain, in lay terms, the nature of the processing the juice has received, the reasons preservatives are not used and the purpose and meaning of your experimental work.

3

SOFT DRINKS

OBJECTIVES

After reading this chapter you should understand

- The various types of soft drink
- Principles of formulation
- The role of the different ingredients
- Processing of soft drinks
- Quality assurance and control
- The basic chemistry of ingredients
- The microbiology of soft drinks

3.1 INTRODUCTION

The term 'soft drinks' is open to various interpretations and careful definition is necessary. The widest interpretation is that the term encompasses all non-alcoholic drinks (including non-alcoholic beer and wine and water), but in common usage tea, coffee, etc., and milk-based drinks are usually excluded. In the UK, this interpretation is supported, to some extent, by the legal definition (Table 3.1). This, however, excludes fruit juices and fruit nectars (with the exception of lime juice), which are commonly described as soft drinks. In the current work, soft drinks are discussed in the context of the UK regulations, juices (and nectars) being described separately in Chapter 2.

Historically, soft drinks were derived from two main sources, fruit-flavoured sparkling waters associated with the popularity of the great European spas and non-alcoholic versions of domestically brewed herb beers. The temperance movement of the late 19th century provided a major stimulus for the soft drinks industry which, in the UK, grew to a very large size. Although relatively

Table 3.1 Legal definitions of the various types of soft drinks

Soft drink	Any liquid intended for sale for human consumption, either without or after dilution, but excluding water, fruit juice, milk or milk preparations, tea, coffee, cocoa, etc., egg products, meat, yeast or vegetable extracts, soups, vegetable juices, intoxicating liquor.
Squash	A soft drink containing fruit juice, not being a comminuted citrus drink, intended for consumption after dilution. The minimum fruit content varies from 10 to 25% according to type.
Crush	A soft drink containing fruit juice, not being a comminuted citrus drink, intended for consumption without dilution and including any cordial intended for consumption without dilution. The minimum fruit content varies from 3 to 5% according to type.
Comminuted citrus drink	A soft drink produced by a process involving the comminution of the entire citrus fruit. The minimum fruit content varies from 7 to 10% for juices consumed after dilution and from 1.5 to 2% for juices consumed without dilution.
Lemonade and other 'fruit-ades'	Soft drinks which do not conform to the minimum fruit content. Labels must not include illustrations of fruit.
Cordials	This term has no general statutory meaning, but may be applied to any clear citrus squash or crush.

Note: Based on Regulation 2 of the *Soft Drinks Regulations* (1964) and its subsequent Amendments.

small-scale manufacture continues today at a local level, much of the industry operates on a national and supra-national level, and the efforts of the Coca Cola™ and Pepsi Cola™ companies to establish a global hegemony is often discussed in the phraseology of international power politics. New products have been developed within established brand names, such as the variants of Coca Cola™, but over the years many traditional products, especially those derived from herbal brews, have disappeared. In some cases, however, products with strong regional associations have not only survived but expanded into wider markets. UK examples include Vimto™, popular for many years in the north of England and now exported to the Middle East, and Irn-Bru™, a Scottish product allegedly made from iron girders, which has achieved success on a national scale.

BOX 3.1 **The best brand in the world**

In recent years the Coca Cola company has disposed of its non-core businesses and may now be seen simply as a beverage company. The company, however, has no less than 45% of the world market in carbonated drinks and is the sixth most valuable company in the US. Despite this the product range remains relatively limited. Diet Coke, introduced in 1982, was the first extension of the trademark and was opposed by company lawyers as a risk to the copyright. Diet Coke is now the third most popular drink after Coke Classic and Pepsi Cola and was probably the most successful consumer product launch of the 1980s. In contrast, the 1985 debacle involving New Coke has been described as 'probably the most embarrassing consumer product launch ever'. A lesson is to be learnt in the acceptance of full responsibility by the Chief Executive Officer and Chief Operating Officer and the rapid move to limit damage by re-introducing Coke Classic rather than pursuing a lost cause. (Huey, J. 1993. *Fortune*, **127 (11)**, 24–32).

3.2 TECHNOLOGY

The technology of soft drinks manufacture is relatively straightforward, in contrast to development of the initial formulation, which can be very complex. Indeed, in many cases, the manufacture of the syrup, the most important stage with respect to product characteristics and quality, is carried out at a central plant and shipped to subsidiary plants whose major role is bottling and distribution.

3.2.1 Carbonated drinks

Carbonated soft drinks are invariably consumed without dilution, and include crushes, citrus comminutes and lemonade and other drinks of the latter category, including colas and mixer drinks. The ingredients are summarized in Table 3.2 and the manufacturing process in Figure 3.1.

Table 3.2 Ingredients of carbonated soft drinks

Water

Carbon dioxide

Syrup
- flavouring
  - fruit juice
  - essences
  - vegetable and nut extracts
  - herb extracts
  - product specific flavourings such as quinine (tonic water)
- sweetener
  - sugar
  - glucose syrup
  - high fructose corn syrup
  - high intensity sweeteners such as saccharine, aspartame and acesulfame[1]
  - bulk sweeteners such as sorbitol and mannitol[2]
- acidulants
  - ascorbic acid
  - citric acid
  - lactic acid
  - malic acid
  - tartaric acid
  - acetic acid[3]
  - phosphoric acid[3]
- colours[4]
  - many used including tartrazine, quinoline yellow and sunset yellow
- preservatives
  - benzoic acid
  - methyl-4-hydroxybenzoate }
  - ethyl-4-hydroxybenzoate } Parabens
  - propyl-4-hydroxybenzoate }
  - sorbic acid
  - sulphur dioxide
- antioxidants
  - ascorbic acid
  - butylated hydroxy anisole
  - butylated hydroxy toluene
  - ascorbyl palmitate and its salts
  - natural and synthetic tocopherols
- emulsifiers
  - proteins
  - sucrose esters
- stabilizers
  - extract of Quillaia
  - guar gum

clouding agents
guar gum
foaming agents
extract of quillalia
extract of yucca

---

[1] Low calorie drinks.
[2] Dietetic drinks.
[3] Not permitted in comminuted citrus drinks.
[4] Not permitted in countries of continental Europe.

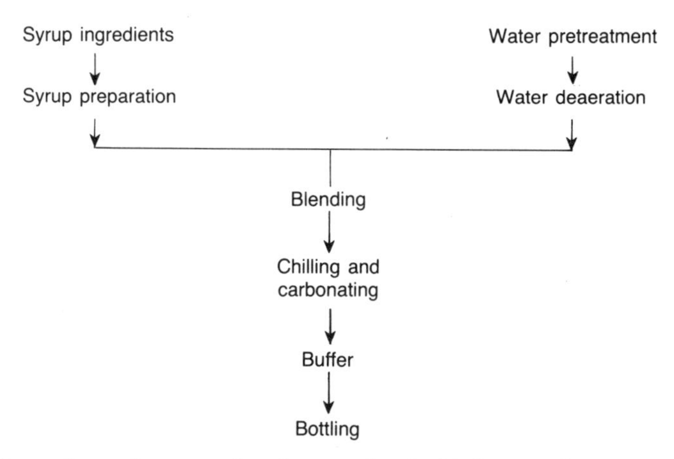

Figure 3.1 Flow diagram of carbonated soft drink manufacture.

(a) Water

Water is the major ingredient of carbonated soft drinks, comprising some 90% of the total. The quality of water used in manufacture has direct implications for the quality of the end-product (Table 3.3) and pretreatment is invariably required. The nature of the pretreatment varies according to the source of the water and its chemical composition. Removal of microscopic and colloidal particles by coagulation and filtration, softening and pH adjustment (alkalinity reduction) may all, however, be necessary where water supplies are of poor quality. In recent years improvements in the quality of municipal supplies has reduced the need for coagulation and permitted the introduction of ion exchange purification techniques. Considerable future use of reverse osmosis may also be anticipated, although present applications are restricted.

Disinfection is required, even where municipal water supplies are used, to remove bacteria which have entered the system during distribution and storage. Chlorination remains the preferred method and, in addition to the destruction of vegetative micro-organisms, is technologically advantageous in removing oxidizable materials, such as dissolved organic matter and soluble iron compounds, and in aiding coagulation processes. Chemical aspects of treatment require superchlorination with doses above 2 mg/l. This level is too high for consumption and necessitates removal of excess chlorine by passage through granular activated carbon.

---

* Children are major consumers of soft drinks and concern exists over the safety of a number of ingredients. These include:

Sweeteners

Saccharin: possible carcinogen, banned in many countries and must carry a 'health warning' in the US.

Cyclamates: possible carcinogens, banned in many countries, including the UK and the US, but may be re-instated.

Aspartame: allegations have been made concerning the safety of aspartame, although evidence appears to be anecdotal. The sweetener contains traces of phenylalanine and should be avoided by persons with phenylketonurea.

Colours

Synthetic colours: synthetic colours have been associated with adverse effects, including breathing difficulties, nettle rash and blurred vision, in sensitive people (including persons sensitive to aspirin and also, possibly, asthmatics and persons with eczema) and with hyperactivity in children. Tartrazine has received particular attention in the context of hyperactivity, but, with the exception of patent blue V and food green S, all artificial colours used in soft drinks in the UK are amongst the additives which the Hyperactive Children's Support Group recommend should be avoided.

A number of synthetic colours of the azo group have been banned due to proven carcinogenicity. Recent studies at Nottingham Trent University suggest that others currently considered safe become genotoxic following metabolism by gut bacteria.

Caramel: the safety of caramel has long been questioned, although the type currently used in soft drinks is considered 'safe'.

Preservatives

Benzoic acid: implicated as source of unacceptably high levels of benzene.

All preservatives: various allergic responses

Antioxidants

Butylated hydroxy toluene/anisole: reports of various adverse reactions including elevation of blood lipid and cholesterol levels. May protect against some carcinogens.

Table 3.3 Relationship between impurities in water and quality defects in carbonated drinks

All water used must be potable and must have no adverse effect on the character or quality of the finished product.

Impurity	*Standard*	*Defect*
Taste	tasteless	off-flavour
Odour	odourless	off-odour, off-flavour
Colour	5 hazen units	off colour, off-flavour
Turbidity	1 mg/l[1]	off-colour, off-flavour
Sediment	none	sediment, off-odour, off-flavour
Free chlorine	0.05 mg/l	off-flavour
Manganese	0.3 mg/l	off flavour, sediment
Lead	0.1 mg/l	toxic
Copper	0.5 mg/l	toxic
Fluoride	2.0 mg/l	mottles teeth
Nitrate	10 mg/l[2]	possible illness in young children, damage to cans
Nitrite	1 mg/l[2]	possible illness in young children, nitrosamine formation

[1] Measured against Fuller's earth.
[2] Expressed as nitrogen.

Although concern has been expressed over the possible risk to infants from high nitrate levels in drinking water, this is not seen to be a major problem in the context of soft drink manufacture since consumption by at-risk infants is very low. High levels of nitrate, however, intensify the corrosion of tin plate and cause perforation of the lacquer lining of cans. A large outbreak of foodborne illness in Japan was attributed to an orange drink containing 300–700 mg/l tin as a result of nitrate-induced corrosion and it has been

* In the US concern has been expressed over the formation of trihalomethanes and the unacceptably high levels of chloroform found in some soft drinks. This has led to concern over the efficiency of granulated activated carbon treatment and interest in the use of alternative disinfection methods such as ultraviolet irradiation and ozonation. The need to balance safety concerns has, however, been stressed and measures to reduce levels of trihalomethanes should not be introduced without ensuring that microbiological safety is not compromised.

suggested that nitrate levels must be as low as 4–5 mg/l to avoid problems in tin-plated, or lacquered cans. If necessary nitrate may be removed from water by ion exchange.

Deaeration of water is required to facilitate subsequent carbonation and filling operations and to improve the perceived quality of the dispensed drink. At low levels of air the partial pressure to be overcome to dissolve CO_2 is less and foaming problems due to the gas coming out of solution during filling are minimized. High levels of air also lead to excessively rapid release of CO_2 on pouring, resulting in a 'flat' and insipid drink, and more rapid deterioration during storage. The extent of deaeration is usually assessed by determining dissolved oxygen levels, current practice requiring dissolved oxygen to be reduced from 8–9 to 1 mg/l or less.

Deaeration equipment either uses a vacuum to remove all dissolved gases or CO_2 stripping, in which case some CO_2 remains in solution after deaeration. The most commonly used type is the vacuum spray deaerator in which water is sprayed into an evacuated vessel in which the partial pressure of air is very low. An alternative system, the linear tube deaerator, is based on the fundamentally different cavitating jet principle, in which air is removed from solution as water passes at high velocity up a stepped tube. The linear tube deaerator achieves a final oxygen content of less than 0.5 mg/l compared with *ca.* 1.0 mg/l in the vacuum spray type, although performance of the latter may potentially be improved. Deaeration can only be effective when strict precautions are taken to avoid uptake of air during subsequent processing operations.

(b) Preparation of sugar syrup

Once a suitable formulation has been developed, preparation of the syrup is a relatively simple procedure involving mixing of ingredients, measured either manually or automatically, in stainless steel tanks fitted with top-driven agitators. The process must be protected against microbial contamination and is usually carried out in a separate 'clean' room, ideally equipped with air filtration equipment and maintained at a slight positive air pressure. It is now usual practice to heat treat the sugar syrup using a plate heat exchanger.

Flavouring

The flavouring component of the sugar syrup is obviously the major influence on the flavour of the final product, although the actual concentration may only be 0.015%. It should, however, be appreciated that water, carbonation, acidity and sweeteners also impinge on flavour to an extent which depends on the nature of the product.

The nature of the flavouring obviously varies according to the nature of the product. Fruit is most commonly used, with the exception of the colas, which are flavoured by extract of cola root together with as much as 10% caffeine and a mixture of essences. Fruit flavour may be added as juice, as a comminute (in the case of citrus fruit) or as an essence. Juice is normally used as a concentrate, citrus fruits, especially orange being most popular. Citrus juice is debittered where necessary (see Chapter 2, page 36). Significant quantities of pineapple juice are also used, as well as smaller quantities of berry fruit and apple.

Citrus comminutes are made from the whole fruit, in contrast to juice which is expressed from the pericarp. The original process used for production of comminutes involved chopping the fruit in sugar, which dissolves and absorbs a high proportion of the peel oils, and milling and screening the resulting slurry. This process is now too expensive for the vast majority of applications, and current methodology involves preparing concentrated juice and adding back some of the milled peel and part of the endocarp, or flavedo, which contains the oil sacs. Such comminutes are often considered to be of poor quality compared with those prepared by earlier methods.

Essences may be prepared from artificial or natural sources. Artificial flavouring has connotations of poor quality and dubious safety. Use has markedly decreased, although to some extent this is due to recognition that some flavours classed as synthetic are actually nature-identical.

Natural citrus essences are composed largely of essential oils from the peel of the fruit. Hydrocarbons, mostly limonene, constitute more than 90% of the oil, but contribute little, or nothing, to flavour, acting primarily as a carrier. Water-soluble flavourings may

* Lemon juice has tended to be unpopular as a flavouring due to its acidity. A method for removal of acid by selective adsorption to a weak-base resin has been developed and offers potential for wider beverage use.

BOX 3.2 Green, green, its greener still . . .

The use of natural colours, flavours and other ingredients is attractive to a manufacturer eager to demonstrate 'green credentials' to ecology conscious consumers. It is rarely considered, however, that extended cultivation of the source-plants in undeveloped countries may not only disturb established ecosystems, but divert land and other resources from production of food staples. In this context the use of nature-identical ingredients rather than their natural equivalent is most desirable.

be prepared by 'washing', in which the oil is extracted in dilute alcohol solutions and separated from the hydrocarbon fraction, or by simple distillation. Distillation permits concentration by up to five-fold, characteristic flavour being retained by adding back the first fraction of distillate to the concentrated oil. A higher quality essence is obtained using counter-current extraction with a mixture of methanol and pentane or hexane. Oxygen containing compounds dissolve in the methanol and hydrocarbons in the pentane or hexane. A concentrated citrus oil is then obtained by distilling-off the methanol. Maceration is a traditional method, in which ground material is steeped in solvents for periods ranging from several hours to several months, before decanting and filtering the solvent. Interest in this method has revived as a result of the development of ultrasonic agitation systems which markedly reduce the steeping time. It is anticipated, however, that future developments will largely concern liquid CO_2 extraction, which offers the potential for production of flavourings with a taste very close to that of the natural product.

In addition to flavourings used on a general basis, a number of flavour compounds are used only in restricted product groups. Examples include quinine, originally intended as a prophylactic against malaria and restricted to Indian tonic water, but now used also in bitter lemon and orange drinks, and ginger root used in ginger beer and ginger ale. Mention should also be made of drinks such as 'shandy' and 'lime and lager', which contain ale- or lager-type beer. Although alcoholic beverages are excluded from the definition of soft drinks, beverages containing less than 2% alcohol are not subject to excise duty in the UK and are not therefore classified as alcoholic. Such beverages may be sold without restriction despite the fact that the alcohol content can be as high as 1.9%, compared with that of ale which can be as low as 2.9%.

Sweeteners
Sweetness is an important aspect of the character of soft drinks and in many countries minimum sugar contents are stipulated. In the UK, for example, minimum added sugar of soft drinks consumed without dilution is 4.5% (w/v), with the exception of 'dry' ginger ale where the lower level of 3% is permitted in accordance with its use as a 'mixer' in alcoholic beverages such as whisky. Stipulations concerning minimum sugar contents may, however, change in future.

In continental Europe, traditional soft drinks are sweetened with sucrose derived from sugar beet, while in the UK cane sucrose is also used. UK regulations also permit the use of the intense sweetener saccharin and for this reason tend to have a lower sucrose content than equivalent drinks produced in continental Europe. Sucrose may be added in dry (granular) form or as a 67% (w/v) aqueous syrup.

Glucose syrup, manufactured by acid and enzyme hydrolysis of starch, may partially, or wholly, replace sucrose, and in the UK is used exclusively in health drinks. The major sugar replacement for sucrose, however, is high fructose corn syrup, which is used in very large quantities in the US as a direct replacement. Manufacture involves enzymic hydrolysis of starch to glucose, which is then converted to fructose by glucose isomerase. After purification this yields a 42% fructose syrup which is enriched to 55% by a selective absorption procedure. High fructose corn syrup is not produced in Europe due to restrictions imposed by the Common Agricultural Policy of the EEC.

In recent years the high sugar content of soft drinks has led to their being considered 'unhealthy' by the dietary conscious and the consequent development of low-calorie (diet) soft drinks in which all, or most of the added sugar is replaced by high intensity sweeteners. High intensity sweeteners are not restricted to low-calorie drinks, but are used more generally where sugar is an expensive commodity.

Low-calorie soft drinks were originally sweetened by a mixture of saccharin and cyclamate, saccharin alone being unsuitable. Fears over the safety of cyclamate led to withdrawal of approval for use in many countries, including the UK and the US, and stimulated the development of alternative intense sweeteners. Currently four types are in widespread use - saccharin, cyclamate, which remains in use in a significant number of countries, aspartame (NutraSweet™, Sanecta™) and acesulfame K (Sunnett™). A fifth, more recently developed, product, sucralose, seems likely to be adopted on a large scale in soft drinks. Widespread use of other types, such as neohesperidin dihydrochalcone, stevioside/stevia and thaumatin (Talin™), currently seems unlikely except in Japan. The use of intense sweeteners in soft drinks is summarized in Table 3.4 and the chemistry discussed briefly in pages 104–8.

Intense sweeteners cannot be directly substituted for sugars without loss of product characteristics and quality and a complete reformulation is required. A major difficulty is that there is no mouthfeel associated with intense sweeteners. This may require the addition of gums, or small quantities of sugars to provide mouth-feel. An alternative possibility is to increase the level of carbonation to provide an illusion of mouthfeel. This procedure has the additional advantage of masking undesirable taste side effects. Some intense sweeteners, including aspartame, are unstable at low pH values and formulation may include the use of buffers and reduced acidity.

Acidulants

Acidulants are of considerable importance in determining the sensory quality of soft drinks and care must be taken during formulation to obtain the correct sugar–acid balance. Carbonated soft drinks differ from non-carbonated in containing carbonic acid, but since this is not added *per se*, but formed from CO_2, its importance is often under estimated. Carbonic acid, however, is responsible for the extra sparkle in the mouth-feel, flavour and 'bite' which distinguishes carbonated soft drinks from their non-carbonated counterparts.

A number of acidulants are permitted in soft drinks, of which citric acid is most common. Each has its own characteristics and some, such as phosphoric and acetic acids, are limited to application in

* Sweetness synergy occurs between many blends of intense and bulk sweeteners. This results in higher perceived sweetness and improved taste. The components of two-sweetener blends should be used in inverse ratio of their relative sweetness to each other so that each contributes 50% of the perceived sweetness. Optimum ratios for multicomponent blends cannot, however, be predicted.

Table 3.4 Use of intense sweeteners in soft drink formulations

Sweetener	*Properties and limitations*
Acesulfame K	Relative sweetness 110-200. Synergy with aspartame, cyclamate and sucrose, but not saccharin. Bitter-astringent aftertaste. Stable during storage and to heat treatment, does not react with other ingredients. Contributes potassium ions which influences choice of emulsifiers and cloud agents.
Alitame	Relative sweetness 2000-2900. Good taste quality. Superior stability to aspartame, but does degrade to tasteless products. Incompatible with some other ingredients including sodium metabisulphite. May be unsuitable for use in colas. Not yet approved for use.
Aspartame	Relative sweetness 120-215. Good taste quality. Synergistic with acesulfame K, cyclamates, saccharin and sugars. Enhances fruit flavours. Stable to heat treatment, but degrades during storage. Maximum stability at pH 4.3, but up to 40% may be lost before drink considered unacceptable. Must be avoided by persons with the metabolic disorder phenylketonuria.
Cyclamates	Relative sweetness 30-140. Taste generally poor, but improved by combination with other sweeteners. Free acid may have flavour enhancing properties. Highly stable during processing and storage. Doubts exist over safety.
Neohesperidin dihydrochalcone	Relative sweetness 250-1800. Has menthol aftertaste which restricts application in soft drinks, although synergism and improved taste with most other sweeteners. Masks bitterness in citrus fruit juices. Good stability.
Saccharin	Relative sweetness 300-700. Has harsh metallic aftertaste totally unacceptable to some persons, although masking is possible. Synergy with most other sweeteners

Table 3.4 continued

Sweetener	*Properties and limitations*
	and sugars. Good stability. Persistent doubts over safety.
Stevioside/stevia	Relative sweetness 140–280. Has bitter, liquorice-like aftertaste which limits application. Enhances some fruit flavours. Some degradation occurs during storage especially in presence of phosphoric acid. Some doubts over safety and use not generally permitted.
Sucralose	Relative sweetness 400–800, being higher at low pH values. Sweetness quality resembles sucrose. Quality enhanced by aspartame in colas, but has negative synergy with aspartame in bipartite blends. Synergy with other intense sweeteners, but not sucrose. Stability very good, some interaction with iron salts.
Thaumatin	Relative sweetness 1300–2000. Liquorice aftertaste. Synergism with acesulfame K, saccharin and stevioside, but not cyclamate or aspartame. Sweetness increased two-fold by aluminium ions. Relatively stable but is denatured by high temperatures with loss of sweetness. Interacts with some gums and stabilizers and may co-precipitate with artificial colours.

Note: Relative sweetness is defined by reference to sucrose which has a value of 1. Relative sweetness varies according to concentration and other factors and the value may also vary according to the method of determination. Values quoted for relative sweetness should be used only as a guide.

* Citric acid is also a chelating agent for metal ions. This property is valuable in lipid-containing foods since it inactivates the pro-oxidants copper and iron. Other chelating agents such as EDTA have no antioxidant activity as such, but enhance the activity of phenolic antioxidants. In this context EDTA may be referred to as a synergist.

specific drinks. The use of acidulants in soft drinks is summarized in Table 3.5 and the chemistry discussed briefly in pages 108–9.

Colours

Colouring has no direct effect on the sensory properties of soft drinks but, where permitted, additional colouring is used to reinforce the consumers perception of flavour In some cases the colour is actually of greater importance than taste in the overall impression made on the consumer: reds invoke the flavours associated with berry fruits such as blackcurrant and raspberry, orange and yellow invoke citrus flavours, greens and blues are associated with peppermint and herbal flavours, while browns complement the 'heaviness' of colas, root beer and dandelion and burdock. The wide use of azo dyes in countries such as the UK has led to consumers expecting bright and even (especially in the case of fruit-ades) lurid colours, in contrast to the situation in France where colours are not permitted and where the colour of soft drinks is much more muted. Overall, a wide range of colours is permitted, although there is some variation from country to country. Colours must have good performance in the presence of light, fruit acids (especially ascorbic), flavouring compounds and preservatives. Colours in use have a high tinctorial strength and thus the concentrations required are very low (20–70 mg/l).

Artificial colours, largely azo dyes, are most widely used and most suitable from a technological viewpoint due to their stability in the final product and their very high tinctorial strength. The safety of some artificial colours has, however, been in question for many years, and the number in use has been reduced both by legislation and by manufacturers sensitive to particular concerns over colours such as tartrazine.

Natural colours are an attractive alternative to artificial, and increasing use has been made of colourings such as curcumin,

---

* In a study concerning consumer perceptions of the thirst quenching properties of fruit-flavoured soft drinks, it was found that drinks coloured red, brown and orange were thought to be most effective at satisfying thirst. Unexpectedly, consumers also considered that the sweetest drinks would satisfy thirst the most. The study illustrated the key importance of colour as an influence of sensory characteristics other than appearance. It was concluded that as society faces the need for more non-traditional drinks to meet dietary recommendations, the importance of colour will increase further. (Clydesdale, F.M. *et al.* 1992. *Journal of Food Quality*, **15**, 19–38).

Table 3.5 Use of acidulants in soft drink formulation

Acidulant	*Properties and limitations*
Acetic acid	Used only where strong, vinegar character improves flavour balance. Not permitted in fruit drinks in the UK.
Ascorbic acid	Anti-oxidant properties may be of prime importance. Can initiate browning after heat treatment and destabilizes some colours.
Citric acid	Light, fruity character which is highly acceptable in fruit drinks. Most widely used acidulant.
Fumaric acid	May be used in place of citrate to obtain equivalent palate acidity at lower usage rate. Less soluble than citric acid requiring special solubilization processes. Not permitted in UK, but widely used elsewhere.
Lactic acid	Smooth flavour in comparison with other acids.
Malic acid	Slightly stronger than citric acid with a more pronounced fruitiness. In contrast to tartaric acid may be used without problem with hard water.
Phosphoric acid	'Flat, dry' flavour suited to non-fruit drinks. Especially effective in colas. Not permitted in fruit drinks in the UK. Highly corrosive requiring special care in handling.
Tartaric acid	More sharply flavoured than citric acid and may be used at lower rates. Calcium and magnesium salts are poorly soluble and tartaric acid is unsuitable for use in hard water areas.

chlorophyll and anthocyanins. Success has been limited by instability. Nature-identical synthetic carotenoids, such as β-carotene are, however, widely used as yellow–scarlet colouring. A development of considerable potential is represented by polymeric food colours. The underlying concept involves the binding of food colour to a polymer of sufficient size that it is neither affected by human metabolism nor absorbed into the bloodstream. Such colours are of lower tinctorial strength and more expensive than artificial colours, but have significantly better performance than natural colours.

Caramels used in soft drinks are usually the Class IV, ammonium sulphite type. This type of caramel has the additional advantage of strong emulsifying properties and is therefore of particular value where emulsions are present. The use of colours in carbonated soft drinks is summarized in Table 3.6 and the chemistry discussed briefly in pages 110–13.

Table 3.6 Use of colours in soft drinks formulation

Type of colour	*Properties and limitations*
Anthocyanins	Attractive red, blue and purple colours, of 'natural' appearance. Prone to decolourization, especially in light and interactions with sulphite and ascorbate.
Azo dyes	Bright colours, can be considered lurid, but popular with children. Stable during storage and of high tinctorial strength. Doubts exist over the safety of *all* azo dyes.
Caramel	Used only in dark, heavy drinks, such as colas and root beer. Stable during storage. Type used in soft drinks generally considered safe.
Carotenoids	Synthetic carotenoids have attractive 'natural' orange–red colour. Relatively stable. Fat-soluble and require special preparation for use in soft drinks.
Chlorophyll	'Soft, natural' colour, but unstable under acidic conditions and in light. Only permitted in a limited number of countries, although no proven link with public health concerns.
Polymeric dyes	Relatively expensive, but of better performance than natural colours. No known problems associated with consumption.

Preservatives

Carbonated soft drinks support the growth of only a limited range of micro-organisms (see page 119), but, despite this, preservatives are required to prevent spoilage during extended storage at ambient temperature. It should be appreciated that, in addition to substances such as benzoates which are specifically added for their preservative function, acidulants, including carbonic acid, also have a specific inhibitory function in the undissociated state. The extent varies according to the nature of the acidulant and is distinct from, but inter-related with, the lowering of the pH value. The inhibitory role of acidulants should, however, be regarded as an additional safeguard and not as the primary 'barrier' to microbial growth.

Four main types of preservatives are used in soft drinks, SO_2, usually in the form of a SO_2 generating salt, benzoic acid and benzoates, esters of *p*-hydroxybenzoic acid (parabens), and sorbic acid and sorbates. The use in soft drinks is summarized in Table 3.7 and the chemistry briefly discussed in pages 112–15.

Antioxidants

Although ascorbic acid is employed to protect sensitive compounds in the aqueous phase, the most vulnerable components are oil-based flavours. Oxidation may be initiated by the introduction of air during the emulsification process. Protection is conferred by the use of oil soluble antioxidants added before emulsification. Butylated hydroxy anisole and butylated hydroxy toluene were widely used in the past but are now the subject of increasing restriction and are being rapidly replaced by natural and nature-identical preservatives such as 'natural extracts rich in tocopherols', synthetic tocopherols and ascorbyl palmitate and its salts. Ascorbyl palmitate and tocopherols are synergistic and are therefore often used in combination.

Emulsifiers, stabilizers and clouding agents

Emulsions are used to impart cloud (neutral emulsions) and/or flavour (flavoured emulsions). The oil phase typically consists of a citrus essential oil containing an oil-soluble clouding (weighting) agent, while the aqueous phase consists of a solution of gum arabic, or a hydrocolloid of similar properties. An oil in water emulsion is formed using a two- stage homogenizer, optimal stability and cloud requiring the production of droplets 1–2 μm in diameter. The clouding agent must contribute to opacity without

Table 3.7 Use of preservatives in soft drinks formulation

Preservative	*Properties and limitations*
Benzoic acid and benzoates	Soluble sodium salt is normally used. Greatest activity at pH values below 3. Effective against a wide range of micro-organisms. Synergistic with SO_2. Free acid may precipitate if mixing inadequate. Allergenic.
Parabens	More effective than benzoic acid at pH values above 3. Anti-microbial activity increases with chain length (methyl → propyl), but solubility correspondingly decreases. Allergenic.
Sorbic acid and sorbates	Normally used as sodium, potassium or calcium salts. Most effective at low pH values but retains activity at pH values as high as 6–6.5. Free acid may precipitate if mixing inadequate. Allergenic, refused GRAS[1] status in the US and only permitted in drinks to be consumed after dilution in the UK.
Sulphur dioxide	Normally used as salt such as sodium metabisulphite. Most effective at pH values below 4 and against yeasts, moulds and Gram-negative bacteria. Activity lost by binding with fruit components to form sulphites. Produces taint detectable at low concentrations by some persons. Allergenic.

[1] GRAS = generally recognized as safe.

affecting stability by producing creaming, ringing or separation and must also have no effect on colour, taste or odour.

Brominated vegetable oil was used as an emulsifier system for many years, but while of very good technological performance was withdrawn due to safety fears. Many alternatives have been sought including sucrose esters, such as sucrose diacetate hexa-isobutyrate, rosin esters, protein clouds, benzoate esters of glycerol and propylene glycol, waxes and gum exudates. None have achieved universal acceptance and unacceptable background flavour is a common problem, especially with rosins and gum exudates. A new product based on modified soy protein may, however, prove satisfactory.

Stabilizers are used both to stabilize emulsions and maintain the dispersion of fruit solids. Stabilizers also increase viscosity and improve mouth feel. Alginates, carrageenan, pectins, various gums including guar and carboxy methyl cellulose are most widely used. Extract of Quillaia also has stabilizer properties, but is used primarily because of its foaming properties.

Foaming agents

A head of foam is considered desirable in carbonated soft drinks, such as shandy, ginger beer and colas. The most effective foaming agents are saponins, extracted either from the bark of Quillaia, or in the US, Yucca trees.

(c) Carbonation and filling

Carbonation may be considered as the impregnation of a liquid with CO_2 gas. In older plants, some of which are still in use, the pre-syruping method was employed, in which carbonated water and sugar syrup were metered separately into the bottle or other container. This method has been superseded in modern practice by pre-mix filling in which sugar syrup, water and CO_2 gas are combined in the correct ratio, before transfer to the filler as a complete beverage. The complete beverage thus comes into existence directly before filling and control of carbonation, and of the relative proportions of syrup and water is of critical importance. Various methods of proportioning are available, but of these the ratio control system using magnetic flowmeters is easy to clean and may be combined with a density control system for higher levels of accuracy if required.

The optimum level of carbonation varies according to the flavour and perceived character of the different drinks. In general terms, fruit drinks are carbonated to a low level (*ca.* 1 volume CO_2), colas, ginger beer, alcohol-containing drinks, etc., to a medium level (2–3 volumes CO_2) and mixer drinks such as tonic water and ginger ale to a high level (*ca.* 4.5 volumes), to allow for dilution in the non-carbonated liquor. Soda water filled into siphons, however, contains up to 6 volumes of CO_2, to maintain internal pressure during use. The use of large (2–3 l) polyethyleneterephthalate (PET) bottles requires a slightly higher level of carbonation compared with glass to compensate for the loss of CO_2 through the bottle walls during storage and on each successive opening during consumption.

BOX 3.3 **Beyond reasonable doubt**

Excessive pressure inside carbonated beverage bottles can make opening a hazardous procedure. The bottle can contain sufficient pressure energy to propel the cap from the bottle with sonic velocity, the related phenomena of 'missiling' and 'tailing', with risk of severe facial injury. Following a large number of incidents, a UK local authority, Hampshire County Council, instituted legal action against a cola bottler under the 1987 Consumer Protection Act. This resulted from an eye injury suffered by a 15 year old girl, who attempted to remove a recalcitrant cap with nutcrackers. It was alleged that the head space in the bottle, which determines the pressure energy and which is controlled by the bottler, was excessive at 7% and that the situation was exacerbated by the design of cap, which was of insufficient thread depth and of a double sealing design which delayed gas release. The prosecution was dismissed by the magistrate following a lengthy and complex technical defence by the bottler. In the opinion of an expert commentator, the prosecution should have succeeded and its failure suggests that the 1987 Consumer Protection Act is failing both to protect the consumer and to counter powerful vested interests. (Wilhoft, T. 1992. *British Food Journal*, **94 (6)**, 29-35).

CO_2 is a colourless gas of slightly pungent odour which, in part, forms carbonic acid on dissolving in water.

$$H_2O + CO_2 \rightarrow H_2CO_3$$

The acid is unstable and has never been isolated, but two series of salts, the carbonates and the bicarbonates are formed.

In practice CO_2 is the only gas suitable for producing the 'sparkle' in soft drinks. The solubility is such as to allow retention in solution at ambient temperature and yet also allow the release of an attractive swirl of bubbles from the body of the drink when slightly agitated. The gas is also inert, non-toxic and virtually tasteless, and is available in liquefied form at moderate cost.

Carbonated soft drinks in a sealed container are in an equilibrium condition where gas in the headspace provides the necessary equilibrium pressure to maintain the remainder of the gas in solution. The equilibrium pressure varies according to the amount of CO_2 in solution and the liquid temperature. Increase in temperature, or decrease in pressure, results in a metastable (supersaturated) condition in which gas is spontaneously released. This phenomenon, which may be referred to as 'foaming' or 'fobbing' occurs when a container is opened and continues during pouring into tumblers and during consumption. The release of CO_2 into the mouth is of particular importance in providing the 'hit' for those drinks designed to be consumed direct from can or bottle.

The fundamental role of the carbonator is to obtain close contact between CO_2 gas and the liquid being carbonated. Factors determining the degree of carbonation are:

1. The pressure in the system;
2. The temperature of the liquid;
3. The contact time between the liquid and CO_2;
4. The area of the interface between the liquid and CO_2;
5. The affinity of the liquid for CO_2 (affinity decreases as the sugar content increases);
6. The presence of other gases.

Pressure, contact time and contact surface are process variables in all types of carbonator. In many cases carbonators are equipped with an internal or external cooling facility (carbo-coolers), which permit the temperature to be controlled. Chilling to 2–6°C avoids the use of very high pressures during carbonation to high levels and also has advantages during bottling.

Operation of a carbo-cooler may be described by reference to the widely used Mojonnier™ system which is of the 'waterfall' type. The system consists of a main vessel containing CO_2 at pressures up to 6 bar. The vessel contains a variable number of heat exchange modules through which a refrigerant is circulated. Incoming product is directed to flow in films down the heat exchange surfaces. The large surface area of the film and the tendency to turbulent flow permits effectively simultaneous carbonation and cooling. Carbo-coolers of this and similar types may, however, be unable to obtain the desired level of carbonation under some circumstances. It is then necessary to pre-carbonate the product. This is normally achieved by injecting CO_2 into the pipeline carrying the product to the carbo-cooler.

Carbonated soft drinks are filled into either bottles or cans. Thick-walled, reusable, glass bottles were used for many years, but are being replaced by thin-walled, non-reusable glass and, increasingly, PET bottles. PET bottles were originally used only for large 2–3 l sizes, but are now also used for smaller, individual sizes and thus also compete with cans. Cans are of the ring-pull type; resealable cans have been introduced, but found little application.

In recent years filling equipment has achieved a high level of sophistication in terms of throughput and automation and plant capable of flow rates of 50 000 l/h is not uncommon. Lightweight cans and PET bottles present problems not encountered with robust glass bottles, and a number of modifications to filling procedures have been required. A further, and major innovation, has been the development of ambient temperature filling.

Product refrigeration has been widely used to overcome design problems in filling machines and to permit maximum output while maintaining a high standard of fill. Cooling to 3–4°C renders highly carbonated product quiescent, overcoming such problems as excess foaming and possible liquid loss. Under conditions of high humidity, however, cold filling results in condensation on the container surfaces and while this is of no importance where absorbent wooden crates were used for bottles, cans are subject to corrosion. In later years the use of non-reusable bottles packed in cardboard outers led to problems of damage to the outer by moisture impregnation. The use of warming tunnels, the solution adopted during canning, was considered technically unacceptable and cold filling abandoned. The reversion to ambient temperature filling initially required the use of filling speeds between 35 and 50% slower. Improvement to filling valve design now permits ambient temperature (15–20°C) at a speed only 15% less than that attainable with the product cooled to 2°C and further performance improvement is likely. The high energy costs associated with cooling and can warming now means that ambient filling is used for cans and beverages may be filled at up to 3.7 volumes carbonation at 20°C. Cooling remains necessary, however, during periods of high ambient temperatures.

3.2.2 Postmix dispensing

Postmix dispensing is very widely used in catering establishments in the major industrialized countries. Postmix dispensing involves installation, at each vending outlet, of what is effectively a miniature carbonated drink manufacturing unit, which mixes a concentrated syrup with carbonated water at point of dispensing. Water is obtained from the mains supply and usually receives no treatment other than straining through a scale trap, although in some areas a carbon prefilter is also fitted. The water is cooled to *ca*. 3°C and carbonated in a chamber containing CO_2 at a pressure of 3 bars. Syrup is supplied either in small stainless steel tanks or in a 'bag-in-box'. The syrup is either pumped or propelled by CO_2 through a cooling coil and mixed with the carbonated water at the dispense head. Mixing must be carefully controlled to keep the CO_2 in solution and the final product should leave the dispense head at 5–6°C and 4 volumes carbonation. Postmix dispense units have also been adapted to mixing of non-carbonated drinks.

3.2.3 Non-carbonated soft drinks

In the past, non-carbonated soft drinks have been typified by products such as squashes and cordials, intended for consumption after dilution. More recently the consumption of ready-to-drink non-carbonates has been stimulated by the development of a wide range of heat-treated, aseptically packed fruit-based drinks.

The ingredients and technology involved in manufacture of non-carbonated soft drinks is similar to that of the carbonates, with the obvious exception of carbonation. With the exception of some cheaper cordials, non-carbonated soft drinks are based on fruit juice or comminutes, although artificial flavouring may also be present. In formulating squashes and cordials it is necessary to take account of the effects of dilution and flavour and colour systems should be as insensitive as practical to the degree of dilution. Ingredients such as preservatives have a function only during storage and are therefore used at the same concentrations as in carbonates. This means the level in the drink at point of consumption is lower than in carbonates and permits greater flexibility. SO_2, for example, is relatively little used in carbonates due to off-flavours, but is a common preservative in citrus squashes. The use of SO_2 in this context does, however, introduce complications with respect to water treatment resulting in smell and odour likened to that of drains. This occurs as a result of a combination of superchlorination of water, use of SO_2 as preservative and citrus juice and for this reason water for squashes is usually taken direct from municipal supplies.

Non-carbonated soft drinks, consumed without dilution have previously commanded only a limited market. In the UK, orange juice (effectively a pre-diluted squash) was produced in waxed cardboard, or plastic, containers for sale in cinemas, etc. and a similar product, packed in foil capped milk bottles, is still distributed in parallel with home milk deliveries. Such products are of mediocre quality and limited further by a short shelf life. In-bottle sterilization has been used in continental Europe, but has never been adopted in the UK to a significant extent.

The situation has changed with the development of heat-treated, aseptically packed fruit-based drinks, which are stable at room temperature. This requires inactivation of ascospores of *Byssochlamys*. Blends of two or more fruit juices in water are common, together with sugars or intense sweeteners, flavouring and acidulant. Colouring is added in a minority of products and ascorbic acid is used both to control browning and act as antioxidant and to enhance vitamin C levels. Other water soluble vitamins may also be added. Drinks of this type are intermediate between soft drinks and fruit juices.

Fruit drinks of this type are aseptically packaged after heat treatment. Single-serving (250 ml) Tetra Brik™ cartons supplied with a drinking straw are most commonly used and particularly popular with children.

BOX 3.4 **Childish, but very natural**

Some fruit drinks are targeted directly at children and package design is sometimes based on popular cartoon characters. Such products are often to be brightly coloured and heavily sweetened. The association of fruit drinks with childhood tastes tends to diminish sales to adults. Ring-pull cans, which are thought to have greater appeal to adults, have been introduced in an attempt to broaden the market base.

Despite the use of heat treatment and aseptic packaging, a minority of products contain benzoate as preservative. Preservatives are redundant in products of this type and their presence highly undesirable.

3.2.4 Powdered drink mixes

Powdered drink mixes may either be soft drinks (as legally defined) in powder form or analogues of soft drinks. Products of this type, such as lemonade powder, have been available for many years but, until recently, were of poor quality. Powdered drink mixes offer advantages in terms of product stability and low bulk, and are used in large-scale catering. There is also a limited domestic market especially in niche areas such as hill walking.

The formulation of powdered drink mixes generally corresponds to the liquid counterparts. Mixes may either be complete, requiring no addition except water, or may also require the addition of sugar. Flavourings and clouds are prepared by spray drying and anti-caking agents such as silicates are incorporated. Where required sodium or potassium carbonate is added to provide a degree of carbonation.

Manufacture of powdered drink mixes is essentially a weighing and mixing operation, and a high level of automation is possible. Special attention should be paid to the choice of packaging, which should have barrier properties as well as physical strength. A triple-laminate of paper–polyethylene–aluminium foil is widely used.

3.2.5 Sports drinks, enriched drinks and neutraceuticals

In recent years a significant, if fragmented, market has developed for drinks which are consumed for specific purposes other than thirst quenching or pleasure. Such drinks are either supportive of a healthy life style or are claimed to have health-promoting, or even medicinal, properties. There is a considerable overlap between these products and conventional soft drinks, but the rationale of formulation is entirely different.

(a) Sports drinks

Sports drinks and their derivatives, 'lifestyle' drinks, have become widely available in recent years. A number of types are available to meet different exercise-related needs. The most common are probably fluid replacement drinks, formulated to facilitate rehydration after or during heavy exercise. Such drinks are also referred to as isotonic, electrolyte balance and electrolyte replacement. Electrolytes are present, however, primarily to facilitate water absorption, since restoring electrolytes after exercise is not usually a priority and is achieved by normal post-exercise feeding. Sodium and chloride are the principal electrolytes, others which are often present are potassium, magnesium, calcium, iron, phosphates and carbonates. Electrolytes are usually stabilized in ionic form by co-addition of citric acid or malic acid. Fluid replacement drinks also contain carbohydrates as an energy source. Small quantities of simple carbohydrates, such as glucose, aid water absorption, but larger quantities interfere. The amount of energy which can be supplied by simple carbohydrates is, therefore, limited. This problem can be at least partly resolved by use of maltodextrins, which can be added at higher levels without interfering with water absorption. Fructose is sometimes added, since this sugar is thought to enhance athletic performance. There is no evidence for this, however, unless the athlete has a hangover; fructose slightly increases the rate of alcohol metabolism. Vitamin mixtures, especially C, B-complex and E, are present in some fluid replacement drinks formulated for post-exercise use. More recent formulations, especially in the US, place emphasis on enhancing performance. Chromium and, controversially, creatine may be added for this purpose.

BOX 3.5 **Sickness to health**

The glucose-containing drink Lucozade™ is a well established British brand. For many years the drink was popular amongst convalescent persons and a well remembered feature of childhood illness. For this reason the brand became associated with ill health. More recently, however, the makers have been highly successful in launching a range of sports drinks, which have resulted in an association with health.

Manufacture of fluid replacement drinks is primarily a matter of blending ingredients. Typical formulations are illustrated in Table 3.8. Most fluid replacement drinks are supplied as powders for rehydration by the user. One type is also available as a liquid concentrate, which must be consumed with large quantities of water, while others are manufactured in ready-to-drink form and packed either in cans or in flexible foil packs incorporating a drinking nozzle. In this case the product is either heat-treated and aseptically packed or a preservative is added. The basic taste of electrolytes is unpleasant and it is usual practice to add fruit flavouring. A derivative is sold in deep-frozen form.

'Lifestyle' drinks are intended for persons leading a generally healthy life, rather than participating in specific sporting activities.

Table 3.8 Typical formulations of fluid replacement drinks

A. Powder	
electrolytes:	sodium, chloride
carbohydrate:	fructose, glucose
others:	citric acid, flavouring
B. Powder	
electrolytes:	sodium, potassium, magnesium, chloride, sulphate
carbohydrate:	maltodextrin
others:	malic acid, flavouring
C. Powder	
electrolytes:	none added[1]
carbohydrate:	maltodextrin
others:	citric acid, flavouring, saccharin
D. Concentrate (individual sachet)	
electrolytes:	sodium, potassium, calcium, magnesium, iron, chloride, phosphate, carbonate
carbohydrate:	fructose, maltodextrins
E. Ready-to-drink	
electrolytes:	sodium, potassium, magnesium, phosphate, sulphate, chloride
carbohydrate:	glucose, sucrose
others:	citric acid, vitamin C, flavour, colour, preservative

[1] Electrolytes are obtained from the maltodextrins

Many drinks of this type were derived directly from fluid replacement drinks and intended for consumption after aerobics, etc. In general, the carbohydrate content is lower and more emphasis is placed on flavour. Such drinks are also lightly carbonated. The potential market amongst those largely mythical creatures who spend all night dancing was recognized with the production of basically similar drinks, brightly coloured and flavoured with esoteric combinations. A second type of lifestyle drink has no connections with the more serious sports drinks and is based on combinations of fruit juices with various herbal ingredients, notably ginseng.

Other types of sports drinks are available, but less popular than fluid replacement drinks. Pre-competition (carbohydrate loading) drinks are intended to provide a high level of carbohydrate as readily available energy, either before strenuous exercise or as a means of enhancing recovery. Such drinks contain 20–25% carbohydrates, usually as a mixture of maltodextrins fructose and/or glucose. Vitamin B-complex is often added to enhance carbohydrate metabolism and electrolytes may also be present.

Nutritional supplements are also produced for the sports drinks market and are intended to build long term strength through provision of an easily digested, concentrated nutrient source. As such they are derived directly from products used in hospital nutrition and which pre-date sports drinks by many years. The benefit of such supplements in healthy persons is doubted by many nutritionalists, although they can be of value where pre-competition stress makes digestion difficult. The major component of nutritional supplements is a mixture of protein and amino acids. This can be obtained from a number of sources, including meat, milk, soya, casein, whey and gluten. Supplements may also contain carbohydrate, unsaturated fat, vitamins and minerals, especially calcium, zinc, magnesium and iron.

(b) Enriched drinks and neutraceuticals

These two categories embrace a wide range of soft drinks, and overlap with each other and with sports drinks. Enriched drinks are considered as soft drinks which resemble their conventional equivalent organoleptically, but which contain enhanced levels of a nutrient or group of nutrients. Fruit drinks are the most common base and may be enriched with a wide range of nutrients, including proteins, minerals, especially calcium, vitamins and fibre. Some are aimed at specific markets, including a childrens' cola, which contains no caffeine, is sweetened with glucose and contains a high level of calcium.

Neutraceutical is a new term which, at its broadest, may be applied to any substance which is a food, or part of a food, and

---

* Until the mid-1980s, it was considered that carbohydrates should be avoided before hard physical exercise to avoid triggering an insulin reaction. This involves an increase in insulin release in response to the high level of carbohydrate and a consequent fall in blood sugar. Studies of endurance athletes, however, showed that while carbohydrate intake directly before exercise did initially lead to depressed blood sugar levels, the situation was reversed after *ca.* 60 min exercise. It appears that exercise itself changes the normal insulin response, leading to lower insulin levels and stable blood sugar.

which provides medical or health benefits. Most drinks described as neutraceutical, however, are formulated for a specific purpose, rather than to enhance general health, and have no conventional equivalent. Neutraceutical soft drinks are particularly popular in Japan, where claims for medicinal properties are made which would almost certainly not be permitted in Europe and the US. Formulation for specific purposes means that there is very considerable market segmentation. Examples include a kiwi fruit-flavoured drink for singers and karaoke-performers, which contains ginseng and karin extracts to soothe the throat and improve quality of performance, and 'After alcohol', a concentrated mineral water which is intended for consumption after heavy drinking to prevent hangovers. In some cases the market is defined by blood group, Kio-Tsukou soda, a lemon-flavoured drink of high calcium and vitamin C content is designed for persons of blood group A, while the apple-flavoured Jori-Atsui soda contains high levels of vitamins C, B_1, B_2 and B_{12}, and is designed for persons of blood group O.

BOX 3.6 **In the eye of the beholder**

Among the more bizarre neutraceutical drinks are Japanese products which, it is claimed, beautify the skin and add lustre to hair. In the UK neutraceutical properties were ascribed to a spring water which, it was claimed, was an effective treatment for nappy rash when sprayed onto the affected skin.

3.2.6 Quality assurance and control

The extent of quality assurance and control, at factory level, varies according to the size of the operation and its nature. Franchise and other operations obtaining premixed syrup from a central source must obviously rely on that source to ensure correct composition to a far greater extent than an operation responsible for its own formulations. In all cases, however, basic procedures must be

* Soft drinks containing high levels of calcium are among the most common enriched drinks and are particularly popular in Japan. Calcium is obtained from bonemeal, dolomite (molluscs and shellfish) and oyster shells. Concern has been expressed at the high concentration of lead in some sources of calcium. This results from an accumulation of the metal in the bones of older animals and pollution of waters from which dolomite and oyster shell is obtained.

followed. These are: general plant operations, including training, designation of responsibility, medical policies, etc.; control of the filling operation including level of fill and, where applicable carbonation and headspace air; ensuring the product is correctly formulated and plant hygiene. The chemical composition and microbiological status of the water supply should also be monitored on a regular basis. It is essential to understand that high speed filling lines mean that the economic consequences of control failure are very severe. For this reason control *must* be aimed at prevention of problems or, at the very least, early detection. End product testing is required, but is usually for verification purposes.

The needs of the soft drink manufacturer have led to development of on-line control equipment and computers may be used for real-time control. Total solids, acidity, colour, clarity and degree of carbonation may all be determined on-line by various instruments. On-line microbiological analysis is not yet available, although methods such as ATP determination, mean that plant contamination can be detected at a very early stage. On-line monitoring of the bottling operation, especially level of fill is also widely applied.

Manufacturers who formulate their own syrups must also institute procedures for monitoring ingredients and the mixing operation. Small manufacturers may rely to a large extent on ingredient suppliers, but in this case specifications must be very carefully drawn up. Mixing of the syrup ingredients can be highly automated, but control equipment is not infallible and verification of the operation is required. Special monitoring is also applied to ensure the maintenance of hygiene standards in the syrup room.

Postmix dispensers present a special problem in quality control and assurance, in that operation is by technically unskilled persons. Quality control is the responsibility of the field engineer, who periodically monitors total solids, carbonation, taste and temperature. The field engineer is also responsible for hygiene, an area which is the cause of some concern.

3.3 CHEMISTRY

3.3.1 Chemical constitution of water used in soft drinks

Examples of manufacturers' requirements for water are listed in Table 3.9.

Table 3.9 Chemical standards for water for soft drink manufacture

Parameter	*Maximum permitted level* (mg/l)
Total dissolved solids	500–850
Alkalinity (as $CaCO_3$)	50
Chloride	250–300
Sulphate	250–300
Iron	0–0.3
Aluminium	0–0.2

Note: Data from Houghton and McDonald (1978) and based on the standards applied by five UK soft drink manufacturers.

3.3.2 Chemistry of intense sweeteners used in soft drinks

(a) Acesulfame K

Acesulfame (Figure 3.2) is the trademarked common name of 6-methyl-1,2,3-oxathiazine-4(3H)-one-2,2-dioxide. The common name is derived from the structural relationship of the molecule with acetoacetic acid and sulphamic acid and its potassium salt nature. Acesulfame K is not metabolized by man and is excreted intact, primarily in urine. It is metabolized by only a small number of micro-organisms and is considered to be non-cariogenic. Thin-layer chromatography (TLC) may be used for qualitative analysis and high performance liquid chromatography (HPLC) is the method of choice for quantitative analysis. Isotachophoresis may be used for the simultaneous determination of acesulfame K, cyclamate and saccharin.

(b) Alitame

Alitame (Figure 3.3) is the generic name for L-α-aspartyl-*N*-2,2,4,4-tetramethyl-3-thetanyl-D-alaninamide hydrate. Alitame is metabolized by man and excreted as a mixture of metabolites and surviving

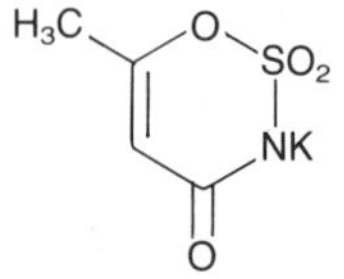

Figure 3.2 Structure of acesulfame K. $C_4H_4N\ SO_4K$: molecular weight 201.2; solubility in H_2O 270 g/l ('room temperature').

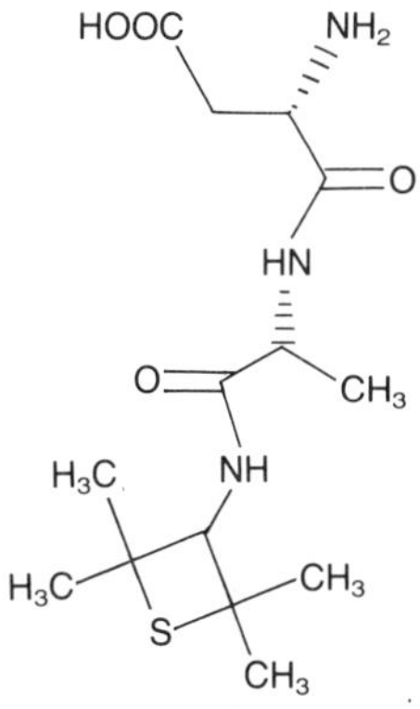

Figure 3.3 Structure of alitame. $C_{14}H_{25}N_3O_4S$ $2.5H_2O$: molecular weight 376; solubility in H_2O 131 g/l (25°C).

intact sweetener. In solution the aspartylalanine dipeptide bond is hydrolysed to yield aspartic acid and alanyl-2,2,4,4-teramethyl-thietane amide, which are tasteless at low concentrations.

(c) Aspartame

Aspartame (Figure 3.4) is the generic name for the methyl ester of *N*-α-aspartyl-L-phenylalanine. In man, aspartame is metabolized to aspartic acid, phenylalanine and methanol, and for this reason must be avoided by persons suffering phenylketonuria. Aspartame is non-carcinogenic. The compound undergoes degradation in solution, being most stable in the pH range 3–5 (maximum stability pH 4.3). Degradation involves hydrolysis of the ester bond to yield the dipeptide aspartyl-L-phenylalanine with the elimination of methanol. At pH values above 5, aspartame undergoes an intramolecular con-

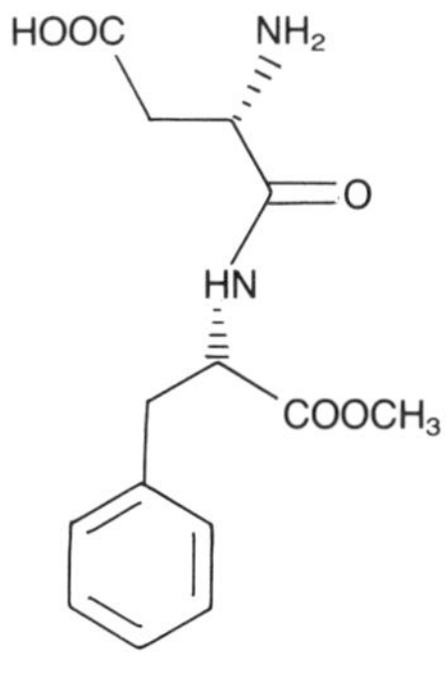

Figure 3.4 Structure of aspartame. $C_{14}H_{18}N_2O_5$: molecular weight 294.3; solubility in H_2O 1 g/l (20°C).

densation to yield the diketopiperazine, 5-benzyl-3,6-dioxo-2-piperazine acetic acid, also with the elimination of methanol, The diketopiperazine may itself be hydrolysed to aspartyl-L-phenylalanine and subsequently to the constituent amino acids. Aspartame may be analysed by spectrophotometric methods following reaction with ninhydrin, but for quantitative determinations HPLC is recommended.

(d) Cyclamates

Cyclamate, as normally used (Figure 3.5), is the sodium salt of cyclohexylsulphamate. Cyclamate is not hydrolysed by enzymes present in the gastrointestinal tract of man, but some common intestinal micro-organisms mediate hydrolysis of the sulphamate ester with consequent formation of the known carcinogen, cyclohexylamine (Figure 3.5). Urinary excretion is the primary route of cyclamate and cyclohexylamine with consequent exposure of the bladder to the carcinogen. Cyclamates are non-cariogenic. Analysis is by spectrophotometry or liquid chromatography.

(e) Neohesperidin dihydrochalcone

Neohesperidin dihydrochalcone is one of a group of phenolic compounds derived from the bitter flavonones of citrus fruit. Neohesperidin dihydrochalcone is prepared by hydrogenation of neohesperidin.

In man, metabolism of neohesperidin dihydrochalcone is mediated by intestinal microflora and yields isoferrulic acid, *m*-hydroxy-

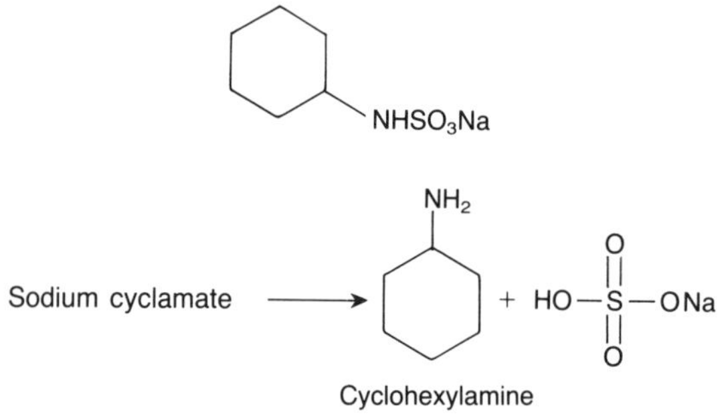

$NHSO_3Na$

NH_2

Sodium cyclamate → + HO—S(=O)(=O)—ONa

Cyclohexylamine

Figure 3.5 Structure of sodium cyclamate and formation of cyclohexylamine. $C_6H_{12}N\ NaO_3S$: molecular weight 201.2; solubility in H_2O 200 g/l (20°C).

phenylpropionic acid and *m*-hydroxycinnamic acid. Excretion is via urine. In solution, limited hydrolysis occurs to the aglycone, hesperitin dihydrochalcone and neohesperidose. The sweetener is non-cariogenic. Analysis is by chromatography.

(f) Saccharin

Saccharin (Figure 3.6) is the generic name for either the calcium or sodium salt of 1,2-benzisothiazolin-3-one-1,1- dioxide. It is absorbed by man, but not metabolized, and is subsequently excreted intact. Saccharin is stable in soft drinks. HPLC is the method of choice for determination, but spectrophotometric methods are also available.

(g) Stevioside/stevia

Stevioside (stevia) comprises a group of diterpene glycosides extracted from the leaves of *Stevia rebaudiana*. Extracts are variable in composition, but all contain several diterpene glycosides, of which stevioside and rebaudioside A are most important. Data concerning the fate of stevioside after consumption is both limited and conflicting. It appears likely that stevioside is not metabolized by man, but there is some evidence that microbial activity in the gut yields the aglycone portion, steviol, which is known to be mutagenic when metabolically activated. Degradation of both stevioside and rebaudioside A occurs in soft drink systems, but is only significant at storage temperatures of *ca*. 37°C. Degradation is also faster when phosphoric acid is used as acidulant. HPLC is most suitable for analysis of the components of extracts, but gas–liquid chromatography may also be used.

(h) Sucralose

Sucralose (Figure 3.7) is a chlorinated derivative of sucrose and the generic name for 4,1′,6′-trichloro-4,1′,6′-trideoxygalactosucrose

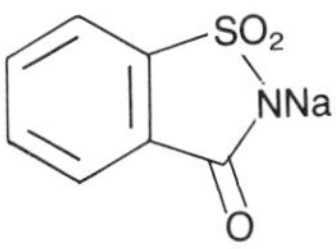

SO_2
NNa
O

Figure 3.6 Structure of sodium saccharin. $C_7H_4N\ NaO_3S$: molecular weight 205.16; solubility in H_2O 700 g/l (20°C).

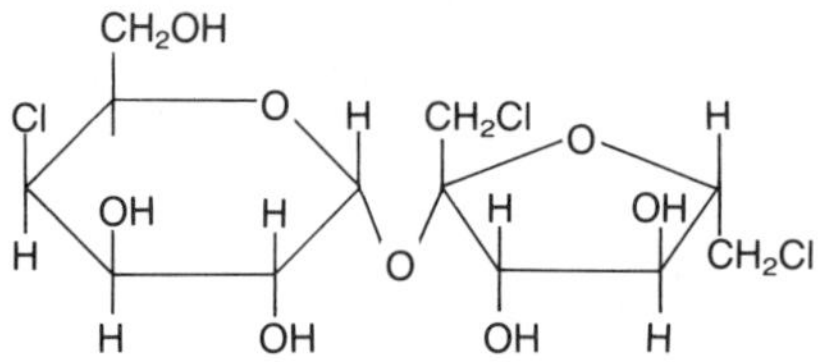

Figure 3.7 Structure of sucralose. Molecular weight 397.6; solubility in H_2O 280 g/l (20°C).

(trichlorogalactosucrose, TGS). Sucralose is not metabolized by man and is only poorly absorbed. The sweetener is stable in soft drinks under most conditions, although limited hydrolysis yielding 4-chloro-D-galactosucrose and 1,6-dichloro-D-fructose occurs at high storage temperatures and at extremes of pH value. Interaction with iron salts, possibly chelation, has been reported.

(i) Thaumatin

Thaumatins comprise a group of extremely sweet basic proteins isolated from the African fruit Katemfe (*Thaumatococcus danielli*). Five components have been isolated; thaumatins a, b, c, I and II. Thaumatins I and II, both of which have a molecular weight in the order of 20 000 Da, are of highest sweetness and present in the largest quantities. Thaumatins form an aluminium salt of twofold greater sweetness and the salt is used in the commercial sweetener Talin™

As a protein, thaumatin is metabolized by man by the same mechanisms as other proteins. No adverse toxicological effects are known. The sweetener is subject to denaturation by heat and extremes of pH value, but is more resistant than most soluble proteins. Thaumatin carries a net positive charge and will form salts with negatively-charged compounds, including gums and stabilizers.

3.3.3 Chemistry of acidulants used in soft drinks.

The chemical structure and properties of added acidulants are summarized in Table 3.10.

Table 3.10 Chemical structure and properties of acidulants used in soft drink manufacture

Acidulants		*Molecular weight*
Acetic acid	Ethanoic acid $CH_3CO.OH$ pK_1 4.75 palate acidity 1.0 g/l	60
citric acid	2-hydroxy-1,2,3-propane tricarboxylic acid $HO.OC.CH_2.C(OH), (CO.OH).CH_2.CO.OH$ pK_1 3.08, pK_2 4.74, pK_3 5.40 palate acidity 1.22 g/l	192.1
Fumaric acid	*trans*-buten-dioic acid $HO.OC.CH{=}CH.CO.OH$ pK_1 3.03, pK_2 4.44	116.1
Lactic acid	2-hydroxy propanoic acid $CH_3CH(OH).CO.OH$ pK_1 3.08 palate acidity 1.36 g/l	90.1
Malic acid	2-hydroxy butan-dioic acid $HO.OC.CH(OH).CH_2.CO,OH$ pK_1 3.4, pK_2 5.11 palate acidity 1.12 g/l	134.1
Phosphoric acid	Orphophosphoric acid CH_3PO_4 pK_1 2.12, pK_2 7.21, pK_3 12.67 palate acidity 0.85 g/l	98
Tartaric acid	2,3-dihydroxy butan-dioic acid $HO.OC.CH(OH).CH(OH).CO.OH$ pK_1 2.98, pK_2 4.34 palate acidity 1.0 g/l	150.1

[1] Palate acidity is a subjective measurement, but of value as a guide to the relative sourness of each acid. The concentrations (in water) quoted were considered to give the same perceived acidity in a series of taste trials. [2] pK_x = dissociation constant(s). Based on data from Taylor (1990)

* The presence of acidulants, such as phosphoric acid, means that soft drinks can be corrosive and contact materials must be carefully chosen. An unusual case of copper toxicity, involving vomiting and headaches, followed consumption of cola in a restaurant at Niagara Falls during 1977. The problem was due to the use of copper piping during repairs to the dispenser.

3.3.4 Chemistry of colours used in soft drinks

(a) Natural and nature-identical colours

The most commonly used colours in this category are the carotenoids, which include the hydrocarbons, carotenes and their oxygenated derivatives, xanthophylls. Carotenoids may be extracted from natural sources such as annatto (*Bixa orellana*) but, while annatto itself finds limited application in soft drinks, most carotenoids are prepared by synthesis. Use of synthetic carotenoids has significant advantages in terms of control of purity and colour properties.

Three synthetic carotenoids are currently in use (Figure 3.8), β-carotene (yellow to orange), β-apo-8-carotenal (orange to red) and canthaxanthin (red). Of these, β-carotene finds widest application, but canthaxanthin is also widely used in place of Amaranth (FD & C Red No. 2), which is no longer permitted in the US.

Carotenoids are fat-soluble compounds and require special preparation for use in water-based soft drinks. Carotenoids may be incorporated into emulsions or, in the case of synthetics, be prepared as gelatin or carbohydrate-coated beadlets.

The search for an alternative red pigment to amaranth also led to renewed interest in anthocyanins and their derivatives (Figure 3.9), although these pigments may also be employed to impart blue and purple colouration. All anthocyanins are derivatives of the basic flavylium cation structure. The flavylium nucleus is electron deficient and consequently highly reactive. Reactions in food systems normally involve decolourization and interaction with sulphites and ascorbic acid is of particular significance in relation to soft drinks. Anthocyanins are also of only moderate stability to light.

Anthocyanins are widely distributed in the plant kingdom, grapes being the common source for food use. Anthocyanins with a

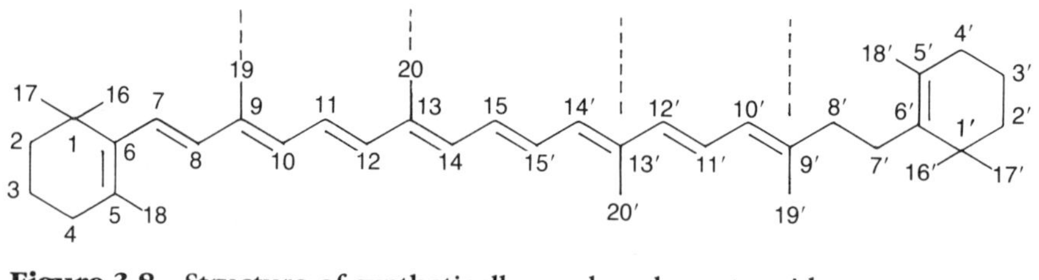

Figure 3.8 Structure of synthetically produced carotenoids.

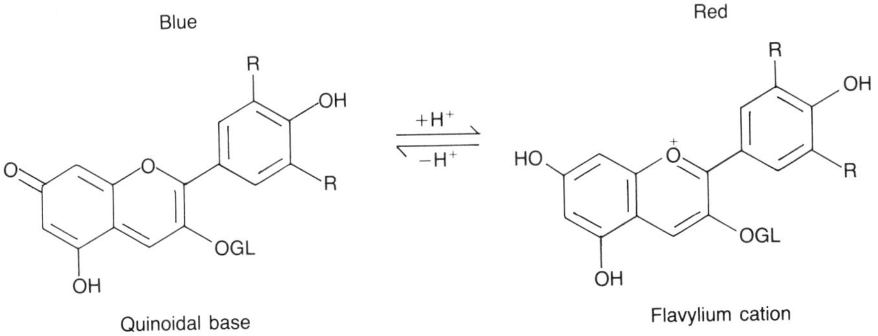

Blue

Red

Quinoidal base

Flavylium cation

Figure 3.9 Structure of anthocyanins.

methyl or a phenyl group at position 4 have been prepared, which have stability as great as or greater than that of artificial red colours, but these are not currently permitted in foods.

Chlorophylls (Figure 3.10) have found limited application as a green colouring, although use is not permitted in the US. Stability is only moderate in the presence of fruit acids and light, but may be improved by use of the potassium salt. 'Copper chlorophyll' is produced by replacement of the magnesium ion by copper in

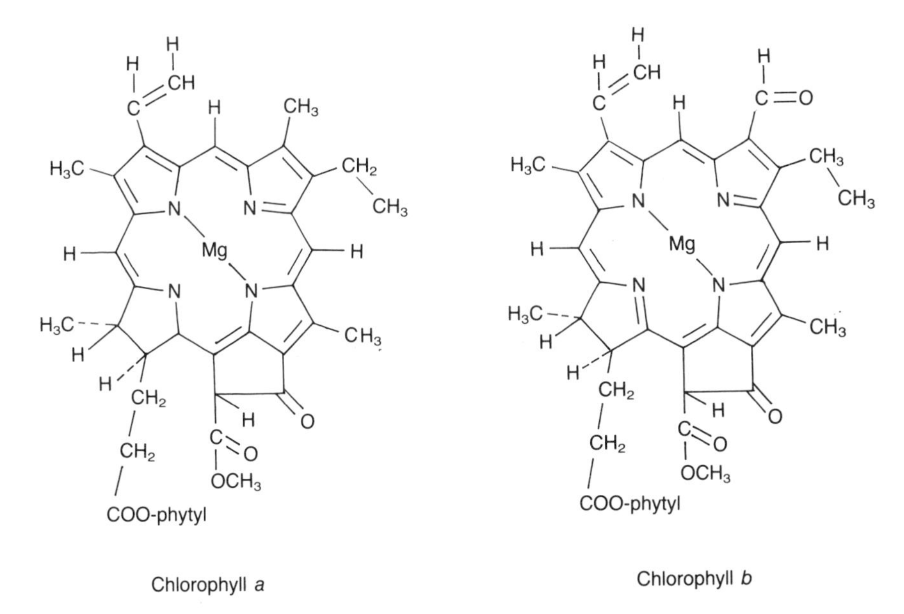

Figure 3.10 Structure of chlorophylls.

chlorophyll, chlorophyllin, pheophytin or pheophorbide, and is olive green in colour and oil-soluble. The colour is permitted in Europe, but not the US. Commercial chlorophyll is usually obtained from nettles, grass and alfalfa, and is relatively impure, containing other plant pigments, fatty acids and phosphatides.

Caramel is widely used in colas as well as root beer, dandelion and burdock, etc. The Class IV ammonium sulphite type used in soft drinks is produced by controlled heat treatment of carbohydrates in the presence of ammonium and sulphite containing compounds.

(b) Artificial colours

The most important group of artificial colours are the azo dyes (Figure 3.11), a number of which are permitted for use in soft drinks (Table 3.11). Carcinogenicity has been associated with structure in the azo dyes in that only those that are polar and contain two azo bonds are carcinogenic. This has been attributed to benzidine formation as a result of *in vivo* reductive cleavage. However, aramanth, which contains only one azo linkage, was withdrawn from use in the US in 1976 and more recent work, using novel methodology, suggests that some currently used azo dyes including carmoisine and sunset yellow may be potentially more hazardous than previously considered.

(c) Polymeric food colours

Polymeric food colours consist of a high molecular weight polymer backbone with appropriate colouring and solubility components. A wide variety of structures are possible, but the backbone is a linear polymer of 20 000 to 1 000 000 Da made from materials such as polyacrylic acid, polyvinyl alcohol and polyvinylamine. Anthraquinone-type colourings are preferred, attached to the backbone by a carbon–nitrogen bond, although use of other colours, including azo dyes is feasible. Three solubilizing groups may be used: $-SO^{-}_{3}-RSO_{3}^{-}$, $SO_2-R-O-SO_3^-$ and $-SO_2-R'-SO_3^-$ (where R is a 2–4 carbon alkyl group and R′ a 2 carbon alkyl group).

3.3.5 Chemistry of preservatives used in soft drinks

(a) SO_2 and sulphites

In aqueous solution, SO_2 and sulphite forms sulphurous acid and ions of bisulphite and sulphite.

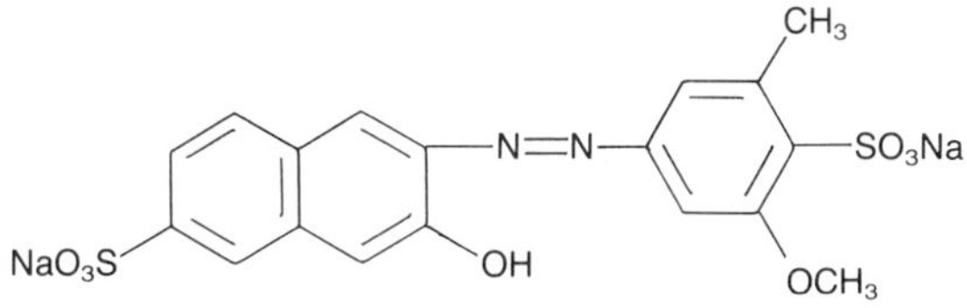

Allura red (FD & C No. 40)

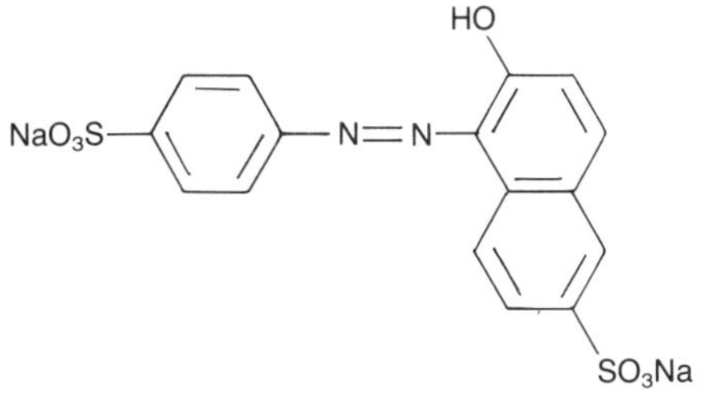

Sunset yellow (FD & C (yellow No. 6))

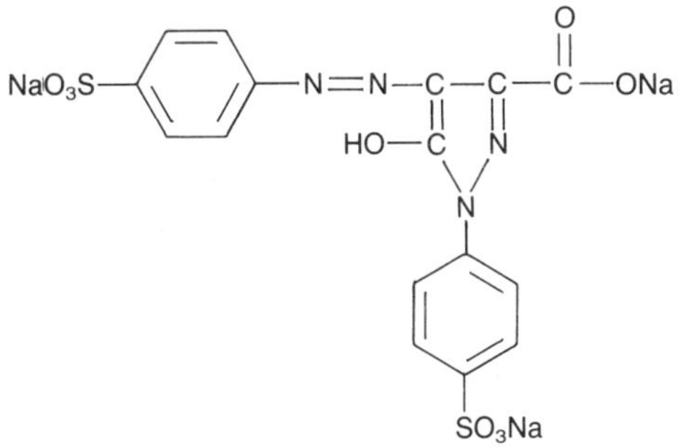

Tartrazine (FD & C (yellow No. 5))

Figure 3.11 Structure of azo dyes.

$$SO_2 + H_2O \rightleftharpoons H_2SO_3$$
$$2H_2SO_3 \rightleftharpoons H^+, HSO_3^- + 2H^+, SO_3^{2-}$$

Antimicrobial activity is dependent on the concentration of the undissociated sulphurous acid molecule which is highest at low pH values. Several mechanisms have been proposed for its antimicrobial activity including the intracellular reaction between bisulphite and acetaldehyde, reduction of disulphide linkages within

Table 3.11 Azo dyes currently used in soft drinks

Allura red (FD & C Red No. 40)[1]
Aramanth[2]
Carmoisine[2]
Ponceau4R[2]
Sunset yellow FCF (FD & C Yellow No. 6)
Tartrazine (FD & C Yellow No. 5)

[1] Approved in the US, not in the UK.
[2] Approved in the UK, not in the US.

enzymes and blocking respiratory reactions involving nicotinamide dinucleotide. SO_2 is also effective in preventing both enzymic and non-enzymic browning reactions.

(b) Benzoic acid and benzoates

The sodium salt of benzoic acid is more soluble than the free acid (Figure 3.12) and is most commonly used. The undissociated acid, formed from the salt in solution, is responsible for antimicrobial activity, which is optimum in the pH range 2.5–4.0.

(c) Sorbic acid and sorbates

In common with many other preservatives, the undissociated acid is responsible for antimicrobial activity. Sorbic acid (Figure 3.12) is most effective at low pH values, although the preservative remains effective at pH 6.5. Sorbic acid is effective against yeasts and moulds, and some bacteria. Inhibition is probably due to interference with cellular dehydrogenases by the α-unsaturated diene

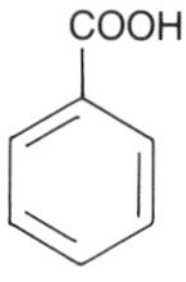

COOH

Benzoic acid

$CH_3 . CH{=}CH_2{-}CH_2{=}CH . COOH$

Sorbic acid

Figure 3.12 Structure of benzoic and sorbic acids.

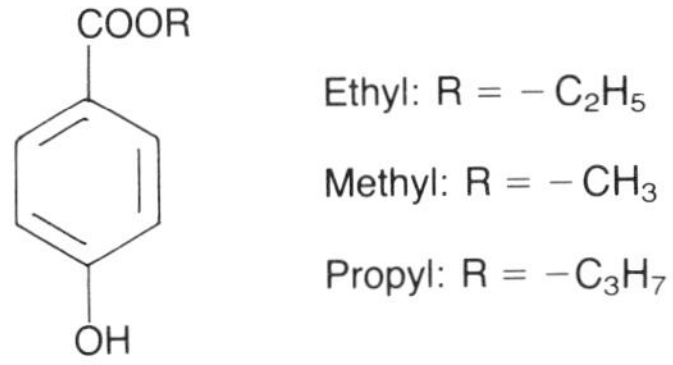

COOR

OH

Ethyl: R = $-C_2H_5$

Methyl: R = $-CH_3$

Propyl: R = $-C_3H_7$

Figure 3.13 Structure of parabens.

system of the aliphatic chain, which most yeasts and moulds are unable to metabolize.

(d) Esters of p-*hydroxybenzoic acid (parabens)*

Methyl (Figure 3.13), propyl and heptyl esters are most widely used, especially in the US. Use may also be made of ethyl and butyl esters. The antimicrobial activity of parabens increases with chain length, but the solubility decreases and for this reason short chain members are preferred. Parabens remain undissociated and thus retain antimicrobial activity at pH value 7.0 and above, although this is of little practical consequence in soft drink manufacture.

The antimicrobial spectrum of preservatives is discussed in greater detail in pages 119–120.

3.3.6 Chemistry of antioxidants used in soft drinks

All antioxidants are compounds which react readily with oxygen and thus serve as oxygen scavengers. Synergism occurs between some antioxidants including ascorbic acid and tocopherols. The fat-insoluble nature of ascorbic acid prevents this being of practical benefit in protecting soft drink emulsions, but this problem may be overcome by esterification to a fatty acid such as ascorbyl palmitate.

---

* Most lactic acid bacteria are resistant to sorbic acid, although sensitive strains do occur. Resistance is associated with metabolism of the acid to an ethoxylated hexadiene, which has a pronounced aroma of geranium leaves. This results in a distinctive form of spoilage of soft drinks and other beverages preserved by sorbic acid. Ethoxylated hexadienes have been implicated as a cause of mild food poisoning in spoiled soft drinks, although direct evidence is lacking.

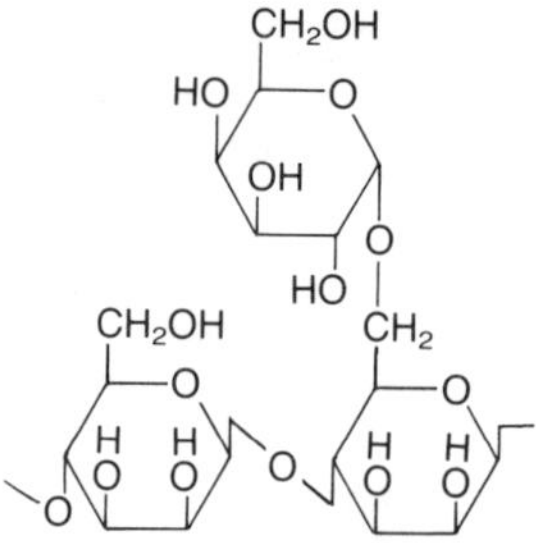

Figure 3.14 Structure of the repeating unit of guar gum. Molecular weight *ca*. 220 000 Da.

3.3.7 Stabilizers

Stabilizers used in soft drinks are hydrocolloids such as guar gum (guaran), a galactomannan containing a backbone of (1-4)-β-D-mannopyranosyl units of which every second unit bears a (1-6)-α-D-galactopyranosyl unit (Figure 3.14). Hydrocolloids of this type are not true emulsifiers but stabilize emulsions and suspensions of fine fruit particles by increasing viscosity of the liquid phase.

3.3.8 Emulsifiers

Emulsifiers are used to enhance the initial formation, and then to maintain, a uniform dispersion of oil droplets in the aqueous phase. The small size of the droplets (1-2 μm) means that the system tends to instability since the potential energy increases with total interfacial energy, which is the driving force of coalescence. Emulsifiers are used to minimize the total interfacial energy and act by adsorbing at the oil-water interface in an orientated manner. To function as an emulsifier a molecule must contain both hydrophobic and hydrophilic regions.

There are three mechanisms of emulsion stabilization, electrostatic, steric and particulate adsorption. Of stabilizers used in soft drinks, proteins act by electrostatic stabilization. The emulsifier is adsorbed at the interface and the droplet becomes surrounded by a cloud of ions of opposite charge to that carried by the emulsifier head group.

* According to the DLVO theory (named for its originators Derjaguin, Landau, Verwey and Overbeek), emulsion stability depends on the combined potential of the attractive van der Waal's forces and the repulsive electrostatic forces being greater than the thermal energy of the droplets. If this condition is met, the emulsion is stable unless heated or subject to increased gravitational forces by centrifugation.

The chance of close encounters between neighbouring droplets is considerably reduced by mutual repulsion by the ion clouds, which counter the van der Waal's forces favouring droplet coalescence.

In contrast to proteins, non-ionic emulsifiers, such as sucrose esters, act by steric stabilization. Such emulsifiers have highly hydrophilic head groups which, when the molecule is adsorbed to an oil droplet, holds a layer of water molecules. The water molecules act as a 'buffer' and prevent close contact between adjacent molecules. Where the hydrophilic group is a polymer, there may also be an entropic barrier to coalescence. In this situation close approach is unfavourable due to the restriction on the configuration of the polymer chain resulting from the proximity of another polymer chain.

3.3.9 Heading agents

Heading agents used in soft drinks are saponins, which are a group of glycosides, characterized by bitter taste, foaming in aqueous solutions and haemolysis of red blood cells. Saponins are usually regarded as undesirable food constituents, but extract of Quillaia is permitted in the UK, as both a stabilizer and heading agent, while extract of yucca is used in the US.

BOX 3.7 **Travellers to the promised land**

Two species of *Yucca* are used as a source of yucca extract, the Mohave yucca (*Y. mohavensis*) and the Joshua tree (*Y. brevifolia*). Both are native to the Mojave desert of California and are today of economic importance in the area. The Joshua tree was named after the prophet Joshua by Mormon migrants, the crooked branches appearing to resemble the upraised arms of the prophet, directing the migrants to the promised land.

The mechanism of heading agents is analogous to that of emulsifiers in that the interfacial tension between two phases, in the case of foam gas and liquid, is reduced. Heading agents also form an elastic protective barrier between entrapped air bubbles.

3.3.10 The chemical analysis of soft drinks

Total solids (or °Brix) of the soft drink, or ingredients, such as liquid sugar is determined manually by refractometry. Refractometry may also be used for in-line control, although measurement of density is often preferred. Special instruments are available for simultaneous determination of carbonation and headspace air in the end-product, but infrared spectroscopy is used for on-line determination of carbonation, as well as acidity. Colour of the end-product is traditionally assessed against standards, although comparators of various levels of sophistication are also used. Clarity is assessed visually or by use of a sensitive turbidometer. A variety of methods have been applied to analysis of artificial sweeteners, but HPLC is the method of choice in most circumstances. Analysis of artificial sweeteners is not yet possible on an on-line basis.

3.4 MICROBIOLOGICAL PROBLEMS ASSOCIATED WITH SOFT DRINKS

Microbiological problems associated with soft drinks are almost entirely those of spoilage. Occasional allegations of foodborne disease associated with soft drinks are, however, made occasionally. The vast majority of these are without substance, although a small number of cases are known where mild sickness has followed consumption of soft drink containing large numbers of yeast or visible mould pellicles.

3.5.1 Source of micro-organisms in soft drinks

The vast majority of soft drinks are made from heat-treated sugar syrup and disinfected water. The immediate source of contamination, with the exception of endospore-forming bacteria, is usually poorly cleaned plant, although aerial contamination at the point of filling is a possibility. In many cases, however, the ultimate source of contamination is ingredients, and in the past both sugars and fruit concentrates have been notorious as the source of yeasts. The application of strict standards has, however, much reduced such problems.

* Sickness following consumption of drinks containing mould pellicles is probably due to revulsion. The problem was fairly common during summer months when use of returnable bottles was widespread. This resulted from mould growth in the dregs before the bottle was washed and filled. The extent of the problem is considerably less following the adoption of non-returnable bottles, but can result, where hygiene is poor, from mould colonization of process plant.

3.5.2 Soft drinks as an environment for micro-organisms

Soft drinks, in general, support the growth of only limited types of micro-organisms. This arises from the presence of a number of inhibitory factors and, in consequence, micro-organisms able to develop are those tolerant of the different stresses. Soft drinks vary in the degree of stress placed on micro-organisms according to their formulation and this affects not only the rate of growth, but the composition of the microflora.

In carbonated soft drinks, the elevated concentration of dissolved CO_2, resulting from an over-pressure greater than 0.1 MPa is an important factor in controlling the growth of micro-organisms. Carbonation becomes fully effective at carbonation levels above 2.5-3 volumes, acetic acid bacteria being most sensitive. Yeasts are the most resistant members of the spoilage microflora, *Brettanomyces* spp. and *Dekkera anomala* being able to grow in a medium containing 4.45 volumes CO_2, although in practice only *Brettanomyces* spp. has been isolated from spoiled drinks with carbonation levels greater than 4 volumes.

Acidity is an important stress factor in all types of soft drinks, although the extent of stress varies from the relatively high pH value fruit drinks to low pH value, highly acidic colas. pH value also interacts with stress factors such as acidulants and preservatives, all of which have greatest antimicrobial activity at low pH values. Acidulants have a specific antimicrobial effect distinct from the lowering of pH value which varies according to the nature of the acidulant. Acetic acid appears to be the most effective inhibitor, of the acids used in soft drinks, but there appear to have been no recent detailed studies.

The spectrum of antimicrobial activity of preservatives varies. SO_2 is considered effective against all micro-organisms, although in fruit drinks effectivity may be reduced by binding. Resistant strains of lactic acid and yeasts may also arise. In contrast to SO_2 both benzoic acid and sorbic acid have been considered to be effective against yeasts and moulds but not against bacteria. This is an oversimplification and both acids have activity against at least some bacteria, although there appears to be considerable variation. Resistance may arise amongst otherwise sensitive populations. Sorbic acid is most effective when used in conjunction with SO_2

and when initial microbial numbers are low. Parabens are of broad spectrum activity, the extent of resistance is not known.

A number of other ingredients of soft drinks have a degree of antimicrobial activity which can be of importance in some circumstances. Citrus oil, for example, can have a significant effect in highly carbonated drinks, while herbal oil is effective in high acid, highly carbonated colas. Synthetic flavours and essences have slight antimicrobial activity which may contribute to the overall inhibitory effect of the preservative system.

In some drinks microbial growth is limited by lack of available nutrients. Soda water, for example, has a very low level of nutrients together with a high (above 5 volumes) level of carbonation and is highly stable. Synthetic ingredients tend not to support growth and are resistant to spoilage, while in contrast natural materials, nature-identical materials and stabilizers support growth and are susceptible to spoilage.

3.5.3 Effect of heat treatment

In addition to problems due to post-process contamination, ingredients and products heat treated to eliminate vegetative spoilage micro-organisms may be spoiled by mould genera such as *Byssochlamys*, which produce heat-resistant ascospores. Ascospores of *Byssochlamys* are of greater heat resistance at low pH values and a temperature of 100°C is required to ensure control.

3.5.4 Micro-organisms responsible for spoilage of soft drinks

(a) Acetic acid bacteria (Acetobacteriaceae)

Acetic acid bacteria of the genera *Acetobacter* and *Gluconobacter* are occasionally involved in spoilage of soft drinks, the sugar-loving *Gluconobacter* being generally more important. Products containing benzoate and/or sorbate and packed in plastic containers appear to be most vulnerable. The spoilage potential of *Gluconobacter* varies from strain to strain, some strains causing off-flavours while others may be present in high numbers but have no detectable effect on flavour or aroma. Spoilage by both *Acetobacter* and *Gluconobacter* may also involve slime and turbidity.

(b) Lactic acid bacteria

Three genera of lactic acid bacteria, *Lactobacillus*, *Leuconostoc* and *Pediococcus*, are involved in the spoilage of soft drinks. Spoilage involves fermentation, tainting and, in the case of *Leuconostoc*, slime production. *Pediococcus* and some strains of *Leuconostoc* produce diacetyl in fruit drinks resulting in a characteristic 'buttery' spoilage.

(c) Moulds

Mould spoilage involves visible growth together with bitter flavours and discolouration. A number of genera have been isolated from spoiled drinks including *Penicillium*, *Alternaria*, *Aureobasidium* and *Fusarium*.

(d) Yeasts

Yeasts are the dominant spoilage organisms in soft drinks. Spoilage patterns typically involve film formation, fermentation with gas production, turbidity and sediment and 'fruity' flavours. A large number of genera have been isolated from soft drinks and ingredients (Table 3.12). Amongst the most important spoilage yeasts are *Candida*, *Brettanomyces*, *Saccharomyces* and *Zygosaccharomyces*. *Zygosaccharomyces* has a particularly high level of resistance to preservatives and can be very difficult to eradicate from plant, while *Brettanomyces* can be a problem in highly carbonated drinks. Other yeasts including *Rhodotorula* and *Cryptococcus* are also common isolates, but appear to be of low spoilage potential.

(e) Zymomonas

Zymomonas is only rarely involved in spoilage of soft drinks but can be difficult to eliminate from plant (*cf. Zymomonas* in brew-

Table 3.12 Genera of yeast isolated from soft drinks

Brettanomyces	*Candida*	*Cryptococcus*
Dekkera	*Pichia*	*Rhodotorula*
Saccharomyces[1]	*Torulaspora*[2]	*Yarrowia*
Zygosaccharomyces		

[1] *Saccharomyces cerevisiae* is the most common spoilage yeast.
[2] *Torulaspora delbrueckii* is common in raw materials and on equipment, but is rare in the end product.

eries, page 358). Spoilage involves fermentation and is characterized by gas production, taints and a heavy sediment.

(f) Other bacteria

A large number of other bacteria have been isolated from soft drinks and ingredients. Sugar, for example, has been recognized as the source of endospores of *Bacillus*, which persist in the final drink. There have been anecdotal accounts of soft drink spoilage by *Bacillus* species, but in no case has the incident been fully investigated. Water is a potential source of a wide range of organisms, some of which are of significance as index organisms of poor hygiene. Growth of bacteria in water distribution systems may lead to tainting in the final product and environmental members of the Enterobacteriaceae have been implicated in the spoilage of high pH value soft drinks. As with *Bacillus*, however, reports were anecdotal and the incidents not fully investigated.

3.5.5 Microbiological methods

In general methods applied to other foods can be used for soft drinks and ingredients, including cultural techniques, microscopy and 'rapid' methods such as impediometry. Microbiological examination of the final product may not be necessary on a routine basis where incubation of the product at 25°C and examination for spoilage is often adequate. In such cases bottles showing gas formation should be removed from incubation and the pressure released to avoid risk of explosion.

(a) Cultural methods

Classical cultural methods may be used without modification, although the technique of pipetting highly carbonated liquid is one of life's little challenges. Numbers of micro-organisms are likely to be small and membrane filtration techniques permit greater sensi-

* Various genera of bacteria may occur in the final drink in relatively large numbers due to colonization of the process plant. *Staphylococcus warnerii*, for example, has been isolated from lemonade at levels of *ca.* 10^3 cfu/ml. The organism is unable to grow in the lemonade and has no spoilage or, as far as is known, public health significance. The persistence in the plant does, however, suggest inadequate sanitization.

tivity. Membrane filtration can be difficult with viscous material or material containing fruit particles.

Commonly used microbiological media are suitable for use (Table 3.13), in the case of water, standard media and methods should be used (see Chapter 1, page 11).

(b) Microscopic methods

Simple direct microscopic examination is valuable under many circumstances. More sophisticated methods such as the direct epifluorescent technique (DEFT) may also be used and can distinguish live from dead cells.

(c) 'Rapid' methods

Impediometry has been adapted for use with soft drinks, although application to date is limited. Considerable interest has also been shown in the determination of ATP by the luciferin-luciferase reaction. This method permits a result to be obtained in as little as 30 min, depending on the pretreatment required.

Table 3.13 Culture media used for microbiological examination of soft drinks and ingredients

Micro-organism	Media
Bacteria (general)	plate count agar[1]
	nutrient agar[1]
Acetic acid bacteria	WL nutrient agar[1]
	Cirigliano agar[2]
Lactic acid bacteria	WL nutrient agar[1]
	MRS agar[1]
Moulds[3]	buffered yeast agar[1]
Yeast	WL nutrient agar[1]
	OGYE agar[1]
	Rose-Bengal-chloramphenicol agar[1]
Zymomonas	WL nutrient agar[1,4]

[1] Details of this medium and its usage may be obtained from the *Oxoid Manual*, Oxoid Ltd, Basingstoke, UK.
[2] Cirigliano, M.C. (1982) A selective medium for the isolation and differentiation of *Gluconobacter* and *Acetobacter*. *Journal of Food Science*, **47**, 1038-9.
[3] Cultural methods are rarely necessary, but may be useful for assessing efficiency of washing of reusable bottles.
[4] Incubated under 95% H_2:5% CO_2.

EXERCISE 3.1.

Although colas are traditionally dark coloured drinks, clear colas have been successfully launched in the US and the UK, while an amber coloured cola has been launched in the UK. Discuss the effect of these changes on consumer perception of colas. Do you consider that these 'new' colas extend the overall market or merely compete with existing products. Discuss the possibility of creating new markets by changing the colours of traditional carbonates, including dandelion and burdock, cream soda and ginger beer (non-alcoholic).

EXERCISE 3.2.

You have been appointed to the Scientific Advisory Secretariat of the government of a small Pacific-rim country with a developing 'tiger' economy. Major emphasis in the past has been concerned with development of an innovative electronics industry, but there has also been a small input into biotechnology and food science. One possible development is the extraction of naturally occurring intense sweeteners and preservatives from indigenous plants. You are asked to draw up a research programme, suitable for presentation to the government, for screening up to 100 compounds considered to have commercial potential as food sweeteners or preservatives. Your considerations should include assessment of functional properties, toxicity and safety testing and the possible use of biotechnology to enhance production of promising compounds from slow growing shrubs.

EXERCISE 3.3.

You have been appointed microbiologist to a small company manufacturing a range of non-carbonated, concentrated soft drinks. Formulations and most of the plant is old fashioned but, by making a virtue of necessity, the company have exploited the 'tradition factor' and successfully transformed a downmarket product into a premium brand. The image is reinforced by the use of water abstracted from an adjacent artesian bore, which receives no treatment other than filtration. The laboratory is well equipped, but under-utilized and the plant, as a whole, has a history of poor control resulting in lost production and a high rate of complaints due to various factors including spoilage. To your horror you find you have inherited two, potentially serious situations, which require immediate resolution. Assess these situations, suggesting possible causes and indicating what further information must be obtained. What recommendations are you able to make concerning overall quality management within the plant based on the limited information immediately available to you.

1. Freshly bottled drinks can contain up to 5×10^5 cfu/ml bacteria (based on counts on plate count agar incubated at 25°C for 3 days). Numbers fall to less than 10^3 cfu/ml within a 'few' days. The predominant organisms are Gram-negative rods. (Note that microbiological analysis is only made on the finished products).
2. One product, raspberry and blackberry drink, is particularly prone to spoilage due to sediment, 'beery taste' and, usually, gas production. The Customer Relation department 'thinks' that most complaints are of product made at the end of the working week, but record keeping is inadequate. Microscopic examination shows large numbers of cells resembling yeast, but viable counts (on malt extract agar, incubated at 25°C for 3 days), never exceed 5×10^2 cfu/ml.

4

TEA

OBJECTIVES

After reading this chapter you should understand

- The nature of the tea plant, *Camellia sinensis*
- Cultivation and harvesting of the tea plant
- The relationship between plucking and quality of the made tea
- The processing of black tea
- The processing of green and semi-fermented teas
- The processing of pickled tea
- The manufacture of instant and decaffeinated teas
- Herbal and fruit 'teas'
- The physiological effects of tea drinking
- The chemical and biochemical changes during processing of the various types of tea
- The nature of the flavour and aroma compounds present in the various types of tea

4.1 INTRODUCTION

Tea, in the context used by most consumers, is a beverage consisting of an infusion of the processed and dried leaves of the tea plant, *Camellia sinensis*. Tea has been consumed since antiquity and, while the origin of the beverage remains unknown, the Chinese are recognized as being regular consumers by the 5th century AD. Tea was first drunk for its supposed medicinal properties, but subsequently became accepted merely as a refreshing beverage. The tea drinking habit gradually spread along the trade routes of Asia minor and was introduced to Europe by Dutch traders during the 17th century. In England, tea overtook coffee in popularity during the 18th century and has been established as the 'national drink' since that time.

For over 200 years, China held a monopoly of tea imported into Europe, but this was broken by the establishment of tea plantations in India and later Sri Lanka (Ceylon). In the latter case, the industry developed largely because of the failure of the coffee crop due to leaf rust disease. More recently African countries, especially Kenya and Malawi, have become established as significant tea producers.

BOX 4.1 **Tea and scandal**

Adulteration of tea was widespread during the 19th century. In some cases the 'tea' was entirely spurious, black tea being found to consist of tea dust bound with gum to sand and dirt and 'faced' with black lead (graphite). In London and other large cities an entire industry existed to manufacture 'tea' from used tea leaves, sloe and bay leaves, gums and colourings including copper carbonate and lead chromate. Unadulterated tea was introduced by the Central Co-operative agency, but initially met with little success since purchasers preferred the brightly coloured, highly adulterated product.

In the UK, the majority of tea is retailed as branded blends. This product, although remaining dominant, is likely to be partially replaced by speciality teas. These comprise teas of specific origin, such as Darjeeling, specific occasion teas, such as Breakfast tea, and flavoured teas, such as Earl Grey (see page 157).

In recent years herbal 'teas', properly described as tisanes, have gained small, but significant, market share. The major attraction of such products is the lack of caffeine, although sales also benefit from a general image of purity and from an undoubted 'nostalgia factor'. The origin of herbal teas lies with medical herbalism and with the historic rural practice of preparing a wide range of beverages from plants. As such, herbal teas are more closely related to traditional soft drinks, such as dandelion and burdock, than to *Camellia* tea. Concern over caffeine, however, has also led to the development of decaffeinated *Camellia* tea and it is likely that this product will take an increasing market share.

4.2 AGRONOMY

4.2.1 The genus *Camellia*

The genus *Camellia* comprises evergreen shrubs which, in commercial cultivation, are maintained as a low bush in continuous vegetative growth. The genus contains a large number (*ca*. 82) of species, but only the 'tea Camellia', *Cam. sinensis* is grown commercially. Other species, especially *Cam. irrawadiensis* and *Cam. taliensis*, are of importance as a source of genetic material.

Camellia is a cross-pollinated plant and for this reason is highly heterogeneous. A number of varieties and sub-varieties have been described some of which have, in the past, been incorrectly assigned species status. Three principal taxa of cultivated *Camellia* are recognized; China, Assam and Cambod (Figure 4.1), which in

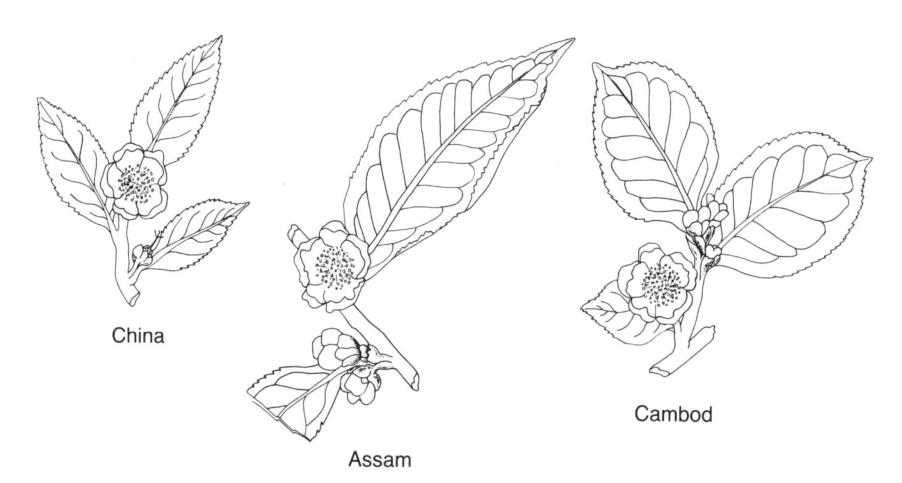

Figure 4.1 The principle taxa of *Camellia sinensis*. Reproduced with permission from Willson, K.C. and Clifford, M.N. (eds) 1992. *Tea. Cultivation to Consumption*. Copyright 1992, Chapman & Hall, London.

* For many years there was considerable controversy relating to the relationship between *Camellia* and *Thea*, the genus to which the tea plant was originally assigned. The prevailing wisdom until the second half of the 20th century was that *Camellia* and *Thea* were separate genera, the presence of eugenol glycoside in the essential oil of *Camellia*, but not *Thea*, being considered to be of major significance. Subsequently, however, it was recognized that actual differences between the two 'genera', based on morphological, anatomical and biochemical criteria, were minimal and that *Camellia* and *Thea* are synonymous.

some taxonomic schemes are classified as *Cam. sinensis*, *Cam. assamica* and *Cam. assamica* var. *lasiocalyx*. Interbreeding between the taxa occurs freely and the extent of hybridization is such that it is doubtful if archetypes of the China, Assam and Cambod varieties still exist. It is also likely that existing tea populations contain genes derived from other *Camellia* species.

Breeding programmes to improve the quality of commercial tea plants have existed for many years. Recent efforts have concentrated on improving yield, and have been greatly assisted by the adoption of improved methods of selection, hybridization and clonal propagation. The gene pool available has also been enlarged by use of indigenous tea varieties. Improvements in tea yield, however, are constrained by the necessity to maintain a standard of quality acceptable to the ultimate consumer. This constraint limits the use of *Camellia* species other than *Cam. sinensis*.

4.2.2 Cultivation of *Camellia*

(a) Climate and soil type

Tea is grown commercially in a large number of countries, the northernmost being Georgia in the former Soviet Union, and the southernmost being South Africa and Argentina. India, China and Ceylon are the three major producers, but significant quantities are also grown in Java, Sumatra, Japan as well as in parts of Africa.

It is generally accepted that air temperatures in the range 18–30°C are optimal for shoot growth. The minimum air temperature for shoot growth appears to be 13–14°C, while net photosynthesis and growth are both markedly reduced at temperatures in excess of 30°C. Soil temperature is also important and optimum growth occurs between 20 and 25°C. Maximum yield requires long hours of sunshine and *Camellia* becomes dormant when day length falls significantly below 12 h. Temperatures must be within the normal range for at least 6 months of each year, but lower temperatures and shorter days are not significant outside the growing period. A high rainfall and high air humidity are required, and in most areas an annual rainfall of *ca*. 1800 mm, distributed more or less evenly over the year, is necessary for continuous crop production. Cultivation of *Camellia* is not usually possible where the annual rainfall is less than 1150 mm unless irrigation is used.

Camellia is grown in a wide variety of soil types, developed from different parent rock under different climatic conditions. Types which support the commercial cultivation of tea include alluvial soils, drained peat, sedimentary soils derived from gneiss and granite, and soils derived from volcanic ash. Growth of *Camellia* is favoured by acidic conditions, a pH value of 5.0–5.6 being considered optimum. Tea will grow in a soil with a pH value as low as 4.0, but soils of pH value only marginally above 5.6 are considered unsuitable unless pH adjustment is employed. Soils of pH values above 6.5 are not amenable to treatment and cannot be used for commercial tea growing.

Tea has the basic nutrient requirements common to all plants. The nature of tea cultivation and harvesting, with frequent pruning and removal of the young shoots, however, means an exceptionally high nitrogen requirement which cannot be met by soil. Application of nitrogenous fertilizers is therefore necessary, especially for high-yielding clones, although high levels of use are thought to reduce the quality of the made tea.

Commercially grown *Camellia* has a low requirement for phosphorus, but potassium is important as a consequence of its intimate relationship with nitrogen metabolism. There are also relatively high requirements for calcium and magnesium. Tea has only relatively small requirements for sulphur, which is usually a constituent of nitrogenous fertilisers, but deficiencies occasionally occur due to the use of unsuitable soils in propagation nurseries.

Camellia has a relatively high requirement for zinc and deficiencies can occur. Localized deficiencies have also been reported for copper (Malawi) and boron (north-east India). Copper deficiency also has significance in the post-harvest behaviour of tea, low levels

---

* An exploration of the relationship between the tea plant and calcium reveals an apparent paradox. The preference for acid soils means that *Camellia* has often been considered to be a calcifuge (a plant which 'dislikes' calcium), yet the plant can be successfully grown in soils containing over 0.5% calcium. Further, it has been suggested that *Camellia* is particularly sensitive to calcium carbonate, although this salt is used as a fertilizer in parts of India. At present, the paradox remains unresolved, although there is general agreement that excess calcium, in any form, is deleterious both to the growth of the young plant and yield of the mature tree. This has been attributed to disturbance of the balance of bases in the soil. (Othieno, C.O. 1991. In *Tea. Cultivation to Consumption* (eds Willson, K.C. and Clifford, M.N.). Chapman & Hall, London, pp. 137–199).

of a copper-protein enzyme resulting in failure of the fermentation process. Aluminium is also an essential element for the tea plant.

The physical properties of soil are of importance in determining suitability for tea cultivation. In general, soil should be deep and well drained and, while tea can be grown in shallow soils with a high water table, management is difficult. The soil should be of good water-holding capacity to minimize problems due to low rainfall and absence of irrigation. Texture *per se*, however, appears to be relatively unimportant.

(b) Agricultural practice

Choosing a site for tea planting requires care even in established growing areas. Local variation in soil type means that assessment of suitability must be made. The presence of certain types of indigenous vegetation (indicator plants), which are known to grow in soils successfully used for tea cultivation, often forms the basis of assessment. It is also valuable to obtain a knowledge of the history of the site and any previous agricultural uses. Some previous crops can result in persistent problems with fungal diseases, nematodes or termites, while land previously used for housing or animal enclosures (hutsites) often has a high pH value. In most cases remedial action is possible, but the resulting cost must be included in economic considerations. The location of the site must also be carefully assessed with respect to requirements for drainage, control of soil erosion and access. Local climatic variations, such as excessive exposure to high winds, should be identified at an early stage.

Land clearance is required before tea planting. Planting should follow as quickly as practical after clearing, but precautions must be taken against disease originating with the cleared plant population. This requires killing of large trees by ring barking ahead of clearing. The accumulation of large quantities of decaying vegetation can lead to a significant rise in pH value and cleared material should be removed from the site. Burning is not recommended as the ash will also raise the soil pH value.

Both drainage and erosion control measures should be commenced as soon as possible after land clearing. Drains are required in flat land, especially that of a swampy nature, while all but very gentle slopes require erosion control. Erosion control requires a

BOX 4.2 Into the forest dim

Considerable concern has been expressed over the adverse ecological effects of clearing rain forest for tea planting, or other agricultural purposes. There is, however, some disagreement over the extent of change. It is generally accepted that the plant species and associated fauna change markedly following tea planting, but streamflows and evapotranspiration are claimed to be similar to that of rain forest, once the tea plants have matured.

combination of earthworks, usually comprising terracing with associated drainage, and cover crops. Some cover crops have an additional function where the soil is of poor nutrient status due to previous cropping, of aiding rehabilitation. Oats are widely used as a cover crop at high altitudes and buckwheat at lower altitudes. Some leguminous plants such as lupins are also suitable, but pasture legumes should be avoided. Cover crops need not be removed before planting, but may be retained to continue erosion control and provide protection for the young tea plants. The cover crop is then normally removed from the line of planting when the tea is first weeded, but is retained between the lines for 2 years. Cover crops may also be used as a source of mulch and in some countries, including Malawi, ground for young tea is mulched before or immediately after planting.

Tea plants are normally raised in a nursery before being planted out in fields. Exceptionally, tea is propagated by planting seeds directly into the field, the 'seed at stake' system. This system was widely used in the early days of commercial tea planting, but later largely abandoned due to difficulties of obtaining consistent high quality amongst the plants and problems of weed control. The plants do, however, come into production early and the system is economical in terms of labour costs. For this reason a seed at stake system has been reintroduced in Australia, where labour costs are significantly higher than other major tea producing countries.

Traditionally all tea was raised as seedlings grown from seeds, which were obtained from a seed orchard, the *bari*. Tea of a particular type and character is referred to as a *jat* and this term is also used to describe tea obtained from a specific *bari*. Numerous attempts have been made to develop grafting, budding and vegetative propagation for commercial use. This has been successful only in the case of vegetative propagation, which in many areas is now more commonly used than propagation from seed. The development of vegetative propagation was parallelled by the production of a series of clones of high yield and a quality superior to that of seedling tea. Such clones have comprised a high proportion of *Camellia* planted over the last 25-30 years.

Planting policy in many areas has involved the use of a single clone. This policy is economically advantageous since the high level of uniformity over a wide geographical area simplifies cultural operations and allows full benefit to be obtained from mechanization. A high quality tea may also be obtained by processing one clone separately. At the same time monoclonal planting can be hazardous, if the clone chosen is adversely affected by local weather conditions, or if susceptible to a pest or disease. For this reason, at least six clones should be employed in each area.

Tea seeds are produced by selected plants, which are allowed to grow with only minimum pruning. A relatively high level of superphosphate or organic fertilizer is required. The viability of *Camellia* seeds rapidly diminishes during storage and ideally seeds should be used when fresh. Viability may be prolonged to some extent by packing the seeds in moist charcoal.

The quality of seeds within a single batch can vary considerably and a simple grading by flotation is applied immediately before planting. Flotation involves nothing more than placing seeds in a bath of water; seeds which sink within 24 h are high quality, while those which sink between 24 and 72 h are usable, but likely to produce a relatively high proportion of substandard plants. Seeds which remain floating after 72 h are unsuitable for use. Seeds require cracking before use to permit water entry and ensure a

* In lowland Sri Lanka, yields were initially significantly increased by the replacement of seedling tea with clones. Subsequently, however, yield has fallen back. This has been attributed to the fact that 90% of planting has involved only two clones, one of which is unsuited to the prevalent soil type and susceptible to drought stress. Under these conditions, resistance to *Macrophoma* infection (stem canker) is markedly reduced and the plant also becomes prone to termite attack and die-back after pruning. (Sivapalan, P. 1986. *Sri Lanka Journal of Tea Science*, **55 (2)**, 53-7).

high and even rate of germination. Traditional methods remain in use by which seeds, maintained moist by watering, are placed in the sun and inspected at frequent intervals. Cracked seeds are then removed for planting. In some countries, especially Sri Lanka, pregermination is common practice.

Nursery beds should be good quality topsoil which is sometimes treated by a pre-emergent weedkiller to reduce weed competition. Fumigation, or heating to *ca.* 62°C, may be required in some soils as a precaution against nematodes. An alternative means of planting is to place cracked seeds in sleeves covered by *ca.* 2.5 cm of soil. Shade is necessary for young plants, the shade density being progressively reduced until the plant is exposed to full sunlight and fertilizer is usually applied as a foliar spray.

A number of methods of vegetative propagation have been investigated, but use of single-leaf cuttings is most widely used. Bimodal and trimodal cuttings give a higher initial number of branches in the field but have a high transpiration rate in the nursery. Under some management systems, cuttings for propagation are taken from producing bushes, but results are often unsatisfactory and it is preferred to use special 'mother' bushes maintained in separate nursery plots. It is common practice to apply fertilizer at a high rate and special pruning is applied to the bushes.

Cuttings must be kept moist during preparation and stored in water until planting in the nursery. The soil in the nursery is critical to the success of vegetative propagation, and must be of pH value below 5.0 and of low humus content. Cuttings are planted individually in soil-filled polythene sleeves, with the leaf and bud slightly above soil level. Growth regulators may be used to improve rooting, triacontanol being most effective. Planted sleeves are supported in frames, giving a bed *ca.* 1.5 m in width, and placed on raked, well drained soil. Newly planted beds must be shaded, and kept cool and moist. This usually requires frequent watering or installation of a misting system. A highly effective, although expensive, method is to use a polythene tunnel, which is sealed after watering and which need not be opened for several weeks. A temperature of 25–30°C is considered optimal for development of cuttings, but before planting-out in the field hardening-off is required. This involves gradual removal of shade and, where polythene tunnels are used, opening of the tunnel to expose the young plants to temperatures in the ranges 5–15°C and 35–40°C. Fertilization of the young plants is by foliar spray and commences as soon as roots are established.

In the field, bushes should be planted as closely as possible, to give complete ground coverage without overcrowding. Optimum spacings are dictated by the branching patterns of different clones, climate and soil type, and the requirements of mechanical harvesters. Tea plants are most commonly planted as hedges, plants within the hedge are usually planted 0.6–0.8 m apart, with a space of *ca.* 1.2 m between hedges.

Sites for each bush are staked out in advance of planting. A hole is then dug alongside the stake and the new bush planted immediately, the hole being filled with a mixture of soil and fertilizer. *Camellia* bushes grown from seed are planted-out as stumps, which are ready for planting when at least 1 cm in diameter at the collar. Stumps must also have adequate starch reserves and this necessitates unshaded growth for 3–6 months before planting-out. The stump is carefully removed from the nursery soil, the stem cut off 10 cm above the soil line and the root cut off 45 cm below the soil line. Lateral roots are also removed. Stumps must be kept moist before planting-out in holes *ca.* 60 cm deep and 20–30 cm in diameter. Shading is required for a short period during sunny weather.

Vegetatively propagated clones may also be planted-out as stumps, but have several spreading roots, rather than the single taproot of seedlings This makes handling very difficult and clones are usually planted-out in the sleeves used for establishing the cuttings. Plants in sleeves are ready for planting-out when the roots have reached the bottom of the sleeve and there is 20–25 cm of top-growth. Sleeves are watered before removal from the nursery beds and kept moist until planting in the field. Care must be taken to avoid disturbing soil within the sleeves or subsequent root development could be adversely affected. Sleeved plants do not require shading if adequately hardened-off.

New plants are required not only for the establishment of new tea fields, but also for replacing dead plants (infilling). Before replacement, however, it is essential to ensure that death was not due to plant disease or to unsuitable soil. It is usually considered advisable to rehabilitate the soil before replanting.

Plants used for infilling must be sufficiently vigorous to withstand competition from established neighbouring bushes. This requires the use of seedling stumps which have been retained in the nursery for an additional year or, where vegetative propagation is used, a large sleeved plant from a vigorous clone. As far as possible, conditions should favour the rapid establishment of the new plant. This requires preparing a large planting hole to ensure the absence of the roots of neighbouring bushes in the proximity of the new plant and use of fertilizer at a very high level. Shading should be at a minimum and infilling is best done immediately after pruning of the established plants. Neighbouring bushes should be cut back.

In addition to pruning, an essential part of tea plantation management (see pages 145–7), established *Camellia* bushes require continuing attention. Weed growth is always a potential problem and control is a prerequisite for successful tea growing. Manual weeding can markedly reduce yield due to root damage and has now been almost entirely replaced by no-cultivation regimens using herbicides.

No-cultivation weed control systems have been a major factor in increasing tea yields. This results not only from the avoidance of damage to root systems, but also from the continuous mulching which results from pruning and leaf-fall.

Irrigation is an absolute necessity in countries such as Zimbabwe, where annual rainfall may be less than 700 mm. In other areas irrigation, although not a necessity, can be justified on the basis of increased yield. In southern Tanzania, for example, an annual application of *ca*. 700 mm of irrigation water consistently increases yield from *ca*. 1100 to *ca*. 1650 kg/hectare. Under similar conditions in Malawi, however, response to irrigation varies widely from year to year. This has been attributed to high temperatures (up to 36°C) in Malawi during September to November, restricting growth irrespective of the water status of the soil. For this, and other reasons, it is necessary to assess the commercial viability of irrigation on an individual site-by-site basis. Many factors must be considered, some of which cannot be directly quantified (Table 4.1). In general terms, however, irrigation should be considered where regular dry seasons of 3–6 months may be expected and where potential soil water deficits exceed 300 mm annually.

Table 4.1 Factors affecting the commercial viability of irrigation

Factors
Direct financial benefits
higher yield
Indirect financial benefits
improved crop distribution
improved utilization of resources
Physical constraints
availability of water
Financial constraints
capital cost of reservoirs
capital cost of water distribution and irrigation equipment
operating and maintenance costs

The role of shade and shelter in tea husbandry has been the subject of discussion for many years. For many years shade was almost invariably provided, but subsequently the finding that yields were often higher without shade led to the removal of shade trees. Shade trees are, however, valuable in some low-altitude tea growing areas, particularly Assam, due to the maintenance of air temperature within the optimum range. Leaf-fall from shade trees can also improve the nutrient status and physical condition of soil.

High winds have an adverse effect on tea through physical damage, reduction of the leaf temperature and increase of the transpiration rate. These effects can be mitigated by belts of shelter trees. It is again necessary to assess the individual site, since shelter trees can reduce yield through competition for nutrients and water and through shading. Belts of shelter trees are usually only justified where very high winds are common.

(c) Pest and disease control

Tea is a perennial crop, which is maintained in the vegetative phase by continual pruning and plucking. This, together with the monocultural nature of its production, means that tea bushes

* Hail is a major problem in some tea growing areas, including Kenya. The current crop is usually totally destroyed, while damage to the stems reduces future yield and increases susceptibility to pests and disease. Considerable attention has been given to minimizing the long-term effects of hail damage, the most effective approach being the controlled application of growth promoters to assist recovery.

provide a constant supply of food for pests as well as a series of stable ecological niches. Tea can, therefore, be prone to a number of pests and diseases, control of which can be difficult. In general problems are greater in Asia, where the plantations are much older, than in Africa, where there are surprisingly few problems. This situation may well change with increasing age of the plantation and increasing use of clonal material.

A large number of pests are known to infest tea plants, although some are of only minor economic significance. Each geographic area of tea production has its own fauna, although many of the more important genera occur in more than one country.

Pests which attack the leaves and shoots are often the greatest cause of problems and lead both to immediate loss of yield and to a debilitation of the bushes. Important pests of this type include the Mosquito bug, thrips, aphids, caterpillars, various types of mite and cicadellids. In contrast, stems and branches of *Camellia* are attacked by relatively few pests. Young plants are most susceptible to attack and may require special protective measures when planted near mature bushes. Pests of potential economic importance include the larvae of the carpenter moth, various types of boring insects and termites. Nematodes are the most important root pests. A large number of species which parasitize the roots of tea plants have been identified, but in many cases no pathogenic role has been established. Various grubs, bugs, weevils and the cockchafer beetle can also damage roots and stems at ground level.

The increasing use of vegetative propagation has reduced the importance of flower and seed pests. Thrips have been implicated in damage to flowers, although there may also be a role in pollination. Weevils and larvae of the false codling moth can cause extensive damage to seeds.

Pest control is best undertaken through management of the plantation rather than through extensive use of pesticides. The breeding of pest resistant cultivars is an obvious possibility, but has received relatively little attention. Exceptions are hard-wooded clones of relatively high resistance to termites and nematode-resistant clones for planting in infested areas.

Pest infestations can be minimized by correct timing of operations such as pruning and plucking, while removal of dead, damaged or diseased wood is effective in minimizing access of termites. Pest control measures should be extended to shade trees, which often act as secondary hosts. *Grevillea robusta*, however, which is widely grown in south India is relatively pest-resistant, while in the case of termites, shade trees can play a positive role by diverting pests away from the tea crop. Where cover crops are grown, it is necessary to select species which are unsuitable as hosts of tea pests.

Plant pests are themselves subject to attack by predators and attempts have been made to enhance the activity of native predators. Attempts to introduce predators have not been successful, with the exception of biocontrol of the leaf feeding caterpillar, *Homona coffearia*, by the parasite *Macrocentrus homonae*.

A large number of insecticides are available for control of tea pests. In some cases use is restricted to soil-dwelling pests in nurseries, but others are applied to the tea itself. Pesticides that taint, or leave high levels of undesirable residues, can only be used under restricted and carefully controlled conditions. Pesticides are usually applied after harvesting to minimize residues in the tea at point of consumption. Maximum residue levels of commonly used pesticides have been defined (Table 4.2) and most importing countries have continuous monitoring programmes.

Camellia is susceptible to a number of diseases, most of which are caused by fungal infections. In Asia, blister blight caused by the fungus *Exobasidium vexans* is now endemic and the cause of considerable economic loss, but the disease is not currently present in Africa. Only young stems and leaves are affected, but the disease is particularly serious in the recovery period after pruning. Grey blight (*Pestalotia theae*) and brown blight (*Colletotrichum camilleae*) are common in all areas and usually appear together on old or damaged leaves. Neither of the fungi is of high virulence and usually only weakened or injured bushes are affected. Brown blight may, however, cause defoilation and death of healthy young plants

* High levels of application of nitrogenous fertilizers can lead to a significant build up of pests and this factor must be taken into account in fertility management. In contrast, potassium-based fertilizers at high levels of application can reduce the incidence of pests in the soil. A similar effect can be obtained by soil addition of potassium, or zinc, acetate. The mechanism appears to involve both inhibition of the pests in the soil and an increase in the resistance of the tea plant.

Table 4.2 Maximum permitted levels of pesticides in tea

Pesticide	Maximum residue limit[1] (mg/kg tea)
Brompropylate	5.0
Cartap	20.0
Chloropyrifos-methyl	0.1
Cyhexatin	2.0
Cypermethrin	20.0
Deltamethrin	10.0
Dicofol	5.0
Endosulfan	30.0
Ethion	7.0
Fenitrothion	0.5
Melthidathion	0.1
Parathion-methyl	0.2
Permethrin	20.0
Propargite	10.0

[1] Recommendations of the United Nations Food and Agriculture Organisation and the World Health Organisation.
Note: Data from Muraleedharan, N. (1992). In *Tea: Cultivation to Consumption*, (eds. K.C. Willson and M.N. Clifford), Chapman & Hall, London.

under warm humid conditions. In most circumstances blights are significant only as indicators of an underlying weakness, for which remedial action should be taken.

Two economically significant foliar diseases of tea have been reported only in Japan and Taiwan. Anthracnose caused by *Colletotrichum theae-sinensis* is one of the most important diseases of tea in these two countries. The disease is most prevalent in mountain areas in years of high rainfall. Affected leaves fall off the plant and, in severe autumn cases, yield is reduced the following year. Net blister blight (Japanese exobasidium blight), caused by *Exobasidium reticulum*, is also prevalent in cool, damp mountain regions. In severe cases, infected leaves fall off and shoots die back reducing yield during the following year.

Two fungal diseases, eye spot (*Pseudocercospora ocellata*) and brown spot (*Calonectria* spp.) occur under damp conditions in parts of Africa (Malawi, Tanzania and Zimbabwe) and Mauritius, respectively. Eye spot is not usually of economic significance, but brown spot can result in a major loss of yield.

Bacterial shoot blight is one of the few bacterial diseases of importance in tea cultivation. The causative organism is *Pseudomonas syringae* pathovar *theae*, which invades dormant shoots during autumn and early spring. The disease is most serious amongst young tea plants exposed to high winds and results in the death of leaves and shoot tips.

Camellia is subject to a relatively large number of stem diseases. Wood rot is the most important and results from the invasion of pruning wounds by largely saprophytic fungi, especially species of *Hypoxylon*. Selective pruning of dead and dying branches and application of fungicidal wound dressing can form the basis of effective control measures. Stem cankers, of varying severity, are common, especially in young bushes during, or immediately after, prolonged periods of water stress. Other economically important causes of stem cankers are *Phomopsis theae*, which primarily affects susceptible clonal plants, *Macrophoma theicola*, *Nectria cinnabarina* and *Poria hyperbrunnea*. *Hypoxylon serpens*, a causative organism of wood rot, is also a cause of stem canker and gains entry to the plant through tissue damaged by sunscorch.

Root diseases are most common where tea has been planted in fields recently cleared from virgin forest or where shade trees have been removed without ring barking to reduce root starch reserves. Disease leads to destruction of the roots and a corresponding reduction in supplies of water and nutrients. The lack of visible signs of infection means that recognition of root rot may be delayed. This can result in infection becoming established in an area and severe problems in eradication.

Armillaria root rot, caused by *Armillaria mellea*, is common in Africa, but is much less common in other areas. The infection spreads from root to root, infected roots having longitudinal splits in the bark and wood, which contain a compact fungal growth. Charcoal root disease (*Ustulina deusta*) is widespread in India, Sri

* Algal plant diseases are relatively rare, but red rust disease of tea is caused by the alga *Cephaleuros parasiticus*. The organism, which is transmitted by windborne spores, is an opportunistic pathogen and weak plants, whether old or young, are most seriously affected. The alga is visible as circular, brick red to orange patches, which appear during April to July when fructifications are produced. Red rust disease is common in the plains of north-east India and in low-altitude areas of Sri Lanka, but is rare in other tea-producing areas.

Lanka and Indonesia, but occurs only sporadically in Africa. Infection is spread both by root contact and windborne spores and damaged, or nutrient deficient, trees are usually involved. Red root disease (*Poria hypolateritia*) is of major economic importance in Sri Lanka, Indonesia and south India, but has not been reported in Africa. In advanced stages of the disease, infected root tissue collapses completely. Infection is spread from diseased to healthy roots by mycelia. Brown root disease is common in India, Sri Lanka and Indonesia. The causative organism is *Fomes noxius* and spread of infection is by root contact. Violet root disease is important only in Assam and even in that area occurs only in heavy, clay soils which are prone to waterlogging. Visible growth of the causative organism, *Sphaerostilbe repens*, is seen as thick, purple–black strands under the bark of affected roots. It is not known, however, whether the fungus is responsible for death or develops after the root dies as a consequence of waterlogging.

Control of disease, like control of pests, is best achieved by use of resistant plants and by attention to cultural procedures. In the case of seedling tea, attempts to select or breed highly resistant varieties have been unsuccessful despite the large amount of genetic material available. It has, however, been possible to select clones which show a high level of tolerance to blister blight as well as to cankers caused by *Macrophoma theicola* and *Phomopsis theae*. Despite these successes, selection or breeding for resistance to root infections currently remains an unattainable goal.

Disease control through cultural practice depends basically on the growth of healthy and vigorous plants.

Maintaining *Camellia* is a vigorous state requires attention to a number of factors. Climatic stress, for example, can be minimized by regulating the shade according to the season and by preservation of soil moisture in the dry season by mulching. Cultural operations must be controlled so as to minimize damage to bushes. The introduction of mechanical pluckers in Japan has led to a significant increase in the incidence of grey blight. This is a con-

---

* Generalizations can always be misleading and there are important exceptions to the rule that healthy and vigorous tea plants have the highest level of disease resistance. *Exobasidium vexans*, the cause of blister blight, causes most severe damage to vigorously growing plants. Conversely, recovery from blister blight by previously vigorous plants is more rapid than recovery by weaker plants.

sequence of the higher level of leaf damage leading to greater susceptibility to invasion by *Pestalotia theae*. Pruning can also create entry points for invasive pathogens and pruning cuts should be covered with a suitable fungicidal wound dressing. It is good management to use the regular maintenance pruning as an opportunity to examine the bush and to remove all dead, diseased and damaged wood.

Root pathogens are not fully adapted to life outside a host and cannot persist indefinitely in soil. It is possible to apply effective, but not absolute, control of root pathogens by removal of live and dead host material. This involves as complete as possible removal of all roots of both tea and shade plants, and necessitates culling of apparently healthy bushes as well as those obviously infected. In practice, it is not possible to remove all infected host material, but the planting of non-hosts, such as grasses for a period of at least 2 years reduces the amount of infective material in the soil and also improves the nutritional and physical condition. The incidence of root disease usually increases following felling of shade trees, but this may be controlled by ring barking before felling.

Changes in agricultural practice can lead to a change in the pattern of disease and, possibly, an increase in the overall incidence. The effects of planned changes, such as introduction of new clones, removal of shade trees and alterations to the pruning cycle should be carefully considered, taking account of uncontrollable factors including climatic extremes. Where possible, the economic consequences of changes in disease patterns should be quantified and set against estimated benefits arising from changes in practice. In Sri Lanka, for example, removal of shade trees and planting of susceptible clones, partially coincided with changes in climatic patterns leading, in the absence of adequate water conservation measures, to a high level of drought stress. The overall consequences included the emergence of *Macrophoma theicola*, a causative organism of canker, as an economically important pathogen. In south India, alterations to the pruning cycle and the introduction of pruning during the dry season led to a high incidence of sunscorch. This, in turn created highly favourable conditions for the growth of *Hypoxylon serpens*.

Fungicides should be used only when other approaches to disease control have failed. Two types of fungicide are used, protectant multi-site inhibitors and systemic single-site inhibitors. Protectant fungicides are largely based on copper and have been in use for many years. Such fungicides are highly effective in control of foliar diseases even during wet weather, but application is necessary at 7–10 day intervals. Protectant fungicides are also ineffective in control of diseases such as stem cankers, where the fungus is protected from contact by the bark.

Systemic fungicides have been replacing protectants since the late 1960s, although copper fungicides are still the most widely used agent of foliar disease control. Systemic fungicides are effective at low concentrations and require less frequent application, typically every 14 days. Systemic fungicides are also active against fungi located inside the bark. Among the more widely used are pyracarbolid, bitertanol and tridemorph. Considerable care must be taken in the use of systemic fungicides to avoid development of resistance.

Soil fumigants have been used in attempts to control root diseases. Methyl bromide has been found to be most effective and has the additional beneficial effects of killing nematodes and weed seeds. Growth of *Trichoderma viride* and some species of *Penicillium*, which are antagonistic to *Armillaria* and *Poria*, is enhanced. Root disease may also be controlled by watering plants with systemic fungicides at the time of planting and repeating the treatment three to four times over the first year of growth.

(d) Bringing into bearing, pruning and harvesting

Tea is harvested by removing the top section, usually two or three shoots and a bud (the flush) from an actively growing stem. New stems develop along the top and sides of the bushes, being produced by the underlying foliage (maintenance foliage). It is necessary to shape new bushes so that they meet at the sides and develop an upper surface with an even density of new shoots (plucking points). This is achieved by a combination of pruning and bending branches down and pegging into position and is commonly referred to as 'bringing into bearing' or 'frame formation'. Bringing into bearing by pruning must be carefully timed so that the plant has good starch reserves and to minimize risk of damage to the exposed stems by sunscorch or hail. A time towards the end of the main growth period is usually chosen. The nature of the prune is dictated by climatic conditions and type of tea, and varies according to locality. Specific clones may have special requirements.

After the bush is brought into bearing, but before regular plucking has commenced, a procedure commonly known as 'tipping-in' is applied. This procedure is intended to level the plucking surface and to increase the number of plucking points. Tipping-in involves a very coarse plucking from the taller shoots and must normally be repeated three to five times. Tipping-in is also applied to mature bushes before regular plucking recommences after maintenance pruning.

Continuing growth of tea plants is necessary to permit the formation of new maintenance foliage to produce the actively growing stems for harvesting. This means that the bushes inevitably grow larger and periodic maintenance pruning is required to remove all stems and leaves above the basic frame of the bush. This is usually achieved by making a straight cut across the bush at the desired height.

The frequency of pruning, the pruning cycle, varies between 1 and 3 years. The bush must be in good physiological condition before pruning, with high starch reserves. If necessary fertilizer may be applied, but sufficient time should elapse to allow absorption by the plant. For economic reasons, it is desirable to prune during the dry season when yield is low, although precautions against sun-scorch are required and it is prudent to avoid pruning during periods of highest temperature. Pruning should also be completed before the bushes suffer drought stress. In some regions it is customary to leave one or two unpruned stems ('lungs' or 'breathers') to continue photosynthesis after pruning and to assist recovery. These should be removed as soon as possible after the pruned stems have regrown.

The severity of the prune varies according to circumstances and requirements. The bush is out of production for a number of months, the length of time depending on the severity and the speed of recovery. Recovery is largely dependent on nutrient status, and prunings must be left in the field to provide nutrients and prevent mineral deficiencies. Although crop is lost in the aftermath of pruning, the yield after recovery is increased.

A very light pruning ('skiffing') is sometimes possible as an alternative to full maintenance pruning. Only the overcrowded upper layer is removed and bushes are out of production for only a short period. On the other hand, maintenance pruning cannot maintain a bush in satisfactory condition indefinitely and it is then necessary to prune very heavily, below the lowest normal prune. This is variously referred to as 'down' or 'collar' pruning. Bushes must be in very good physiological condition before down pruning and are out of production for significantly longer periods.

Harvesting of tea is traditionally a manual operation, which involves *ca*. 80% of the labour employed. This is expensive, even in the low wage economies operative in tea-growing areas and there has been an increasing trend to various levels of mechanization. Manual pluckers can, however, be trained to be selective and to pluck only at the optimal level, which is not currently possible with mechanized pluckers. This relative non-selectivity leads to a lower quality crop due to the inclusion of immature and overgrown material. The yield is also often thought to be lower when mechanical plucking is used, but evidence is rather contradictory. Mechanical pluckers range from motor-powered, hand-held shears to sophisticated self-propelled harvesters, which may be adopted for use in other aspects of tea cultivation. Choice depends largely on the scale and economics of the operation as well as factors such as access and type of bush.

Irrespective of whether manual or mechanical plucking is used, it is usual to maintain a flat plucking 'table', parallel with the ground. In a few regions, a cylindrical row is maintained, although such a table has a lower yield, and is more difficult and expensive to pluck. Manual pluckers are able to maintain the correct level by use of a light pole, while wheeled harvesting equipment maintains a preset level. Maintaining a flat plucking table is, however, difficult where hand held mechanical shears are used.

In the case of black tea, an important factor in determination of the quality of the end product (the made tea), is the number of leaves plucked during harvesting. The actively growing bud gives the highest quality tea, quality deteriorating with the position of each leaf down the stem. The lowest quality tea is obtained from the stem itself (Figure 4.2). The lower leaves, however, are larger and plucking more leaves markedly increases yields. Picking the first two leaves and the unopened bud is generally considered to provide the best compromise between the conflicting demand of high quality and high yield. In contrast plucking three, or more, leaves and a bud is the system employed where the demand for high yield outweighs considerations of quality.

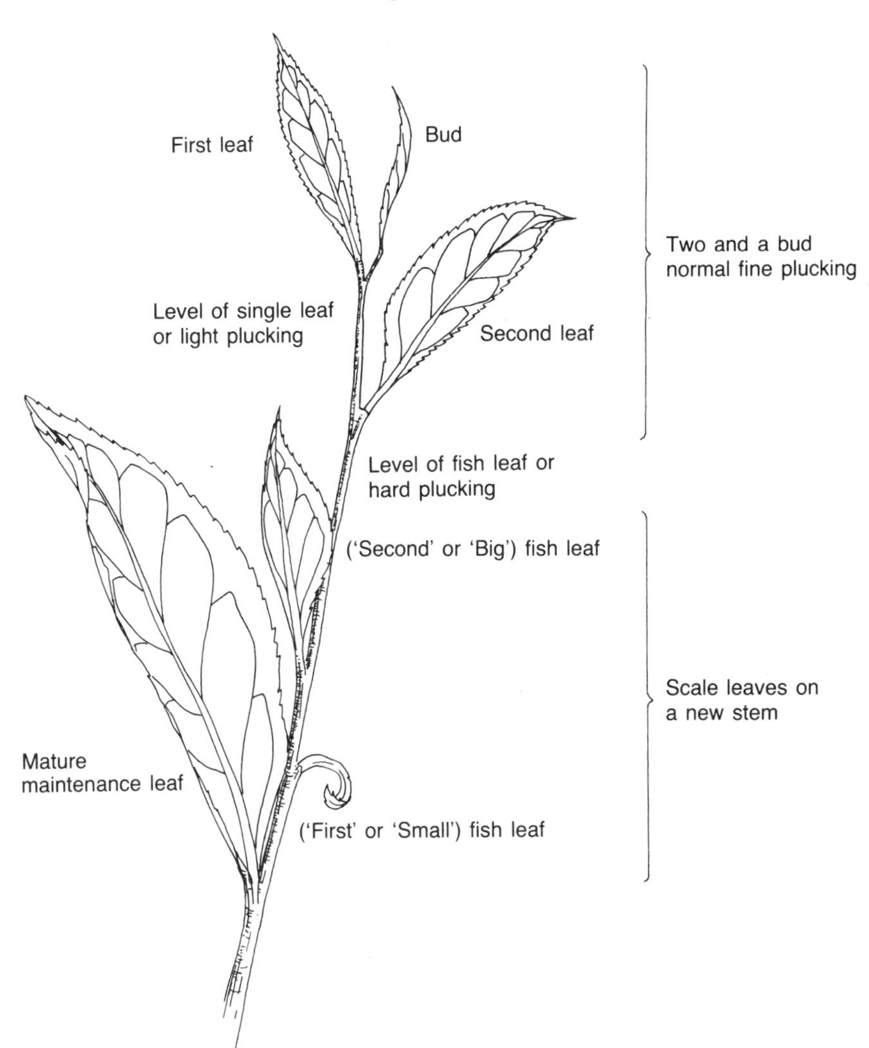

Figure 4.2 Plucking of tea. Reproduced with permission from Willson, K.C. and Clifford, M.N. (eds) 1992. *Tea. Cultivation to Consumption*. Copyright 1992, Chapman & Hall, London.

Following harvest, new shoots develop to form the next 'flush'. For various reasons, including dormancy, the rate of development of the new material is not constant. At any given time, therefore, each bush will bear shoots at all developmental stages from the immature to the highly overmature. Over the field as a whole, variation will be greater with seedling tea than with clonal tea. At some stage, however, most of the shoots will be of the correct size

BOX 4.3 Take some more tea

Plucking two or less leaves and a bud is known as 'fine', while plucking three or less leaves and a bud is known as 'coarse'. At times buds only have been picked to provide the highest quality teas, but this is not economic and buds, or particles of high grade tea from the buds, are now separated from the rest of the tea at the factory. Tea made from buds only was previously referred to as 'tips', but this term is now used in a more general sense to describe high quality tea produced by separation from lower quality material in the factory.

for harvesting. In practice, the time between pluckings, the 'plucking round', is fairly constant providing that weather conditions are stable. The plucking round may be as short as 4–5 days at times of peak growth, but in excess of a month when conditions are unfavourable. The high cost of harvesting, even when mechanical equipment is used, means that it is necessary to use the plucking round which provides optimal balance between quality, yield and the relative cost of more, or less, frequent harvesting. There has been some disagreement concerning the optimum plucking round, but there is now a consensus which suggests that frequent plucking improves both quality and yield. Long plucking intervals lead to a poor leaf standard with a large number of mature leaves. The chemical composition also changes as a result of the large number of coarse leaves and this is reflected in a lowering of perceived quality of the made tea.

The requirements for green tea are considerably less rigorous than for black and usual practice is to remove shoots, carrying several leaves, on an infrequent basis. Four pluckings a year is common and highest yield is obtained by removing all the material which develops between successive pluckings.

Planning and management of tea harvesting must take account of the need of the bush to add new leaves to the maintenance foliage. This requirement is met by varying the depth of plucking and allowing an additional leaf to remain above the previous plucking level. This is often known as 'light' plucking, as opposed to 'heavy' plucking in which leaves are removed to the previous plucking level. Practice varies on a regional basis, but in all cases it is necessary to balance light and hard plucking. This results in a vigorous maintenance layer, capable of supporting formation of the crop, while preventing excessive growth of the bush.

Withering commences as soon as the tea is plucked, but is uncontrolled until the tea reaches the factory. It is therefore important that the plucked tea should reach the factory as quickly as possible and provision of efficient transport is an important aspect of plantation management. It is also important to minimize damage to the leaf.

4.3 TECHNOLOGY

The technology of tea processing evolved over a number of years. The processing stages are relatively unsophisticated, but bear directly on the quality of the made tea. For this reason considerable skill is involved in ensuring that the end product is of a quality which meets the demands and expectations of the consumer.

4.3.1 Black tea

Black (fermented) tea is technically the most complicated of the teas, the processing stages being summarized in Figure 4.3.

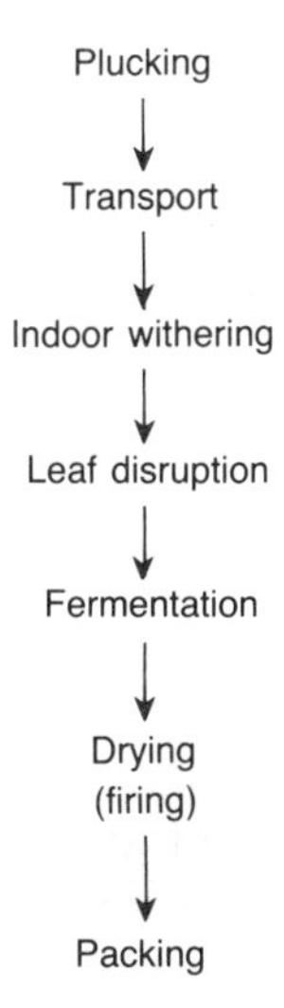

Figure 4.3 Processing of black tea.

(a) Withering

During withering, leaves lose moisture and, as a result, changes occur in the permeability of the cell membranes. This preconditions the leaf for the subsequent maceration and fermentation stages and is known as the physical wither. At the same time a number of biochemical changes occur - the chemical wither. For a number of years, it was thought that the chemical wither might only be of benefit to 'flavoury' teas, but it is now also known to improve the quality of 'plain' and 'medium' teas.

During withering it is necessary that air should circulate as freely as possible around the leaf. This is usually achieved by use of a withering trough, a long, low, narrow rectangular box fitted with a perforated floor through which a fan blows air upward through a shallow bed of leaves. An air heater may be fitted, but it is important to avoid over-heating the leaf and the leaf temperature is usually maintained close to ambient.

Economics dictate that the withering trough should be loaded as fully as is compatible with adequate withering and higher loadings are possible where air is of lower humidity. A problem with the simple withering trough is the relatively slow rate of withering in the upper leaves of the bed. The most effective solution is to enclose the withering trough and fit baffles to enable the direction of the air flow to be varied without reversing the fan. The use of such equipment is recommended at all times in areas of high air humidity.

Although withering is of considerable importance with respect to the quality of the made tea, the process is often poorly controlled and subject to considerable variation. Further, little effort has been made to determine the optimum withering time, which in many factories is determined purely on the basis of operational convenience. Withering times of 12-16 h are most common, although some factories operate on a 24 h withering cycle. Systematic studies, however, have shown that a withering time of not longer than 14 h results in a tea of the highest quality and that a marked decrease in quality occurs when withering is extended beyond 20 h. Time, rather than extent of withering, appears to be of importance since variations in chemical composition occur in green leaf when withering is to the same degree, but for different time periods.

In most cases physical withering occurs concurrently with chemical withering. Investigations in Kenya however, where increased tea growing has led to a shortage of factory space, have shown that a two-stage wither may be used without affecting the quality of the tea. In this process, the leaf is stored in a holding unit, where the chemical wither takes place, before transfer to undergo a short physical wither under conditions of high air flow.

(b) Leaf disruption

Leaf disruption is a physical process which may be achieved in a number of ways. The basic function, however, is a reduction in the size of the leaf material and some degree of cell disruption, which exposes new material to air during the subsequent fermentation process. Choice of method of leaf disruption is not merely a matter of choosing the most convenient or cost effective machine, since the extent of size reduction varies. Further different machines operate most effectively at different moisture contents, and thus the choice of method of leaf disruption affects the requirements for both withering and drying.

The orthodox roller is a batch machine, which has been used for many years. A batch of leaf is loaded into a cylindrical hopper, positioned above a circular table, bearing a series of ridges, or battens, on its upper surface. Downward pressure is placed on the cylinder of leaf in the hopper, forcing the leaf at the lower end into contact with the table. The table and hopper move eccentrically to each other, bruising, rolling and disrupting the leaves. The degree of disruption can be varied by varying the downward pressure on the cylinder. The orthodox roller requires a fairly dry leaf with a moisture content of *ca*. 60% and the output contains a relatively high proportion of large particles. These can be recycled back to the roller or, alternatively, the orthodox roller may be used in combination with a different type of machine.

The rotorvane machine was developed in an attempt to reproduce the physical action of the orthodox roller on a continuous basis. The rotorvane machine consists of a horizontally mounted rotor, fitted with pairs of opposed vanes, surrounded by a cylindrical outer jacket, the inner surface of which is ribbed. Leaves enter the rotorvane machine from a hopper and are propelled to the discharge end by the rotor. Disruption of the leaves occurs by rubbing against the ribs of the outer casing, an adjustable choke at the discharge end allowing pressure within the machine to be varied. The temperature of the leaf rises during passage through the rotorvane and there must be provision for cooling immediately after disruption. The rotorvane may be used alone, or as a preliminary treatment before the crush, tear and curl (CTC) process. To some extent, rotorvanes have been replaced by the more modern Boruah continuous roller machine. This machine has a conical rotor which oscillates through 180°. The rotor is placed in a stationary outer chamber fitted with ridges and battens to act as rubbing surfaces.

The CTC machine is very widely used and produces small particles, popular for use in tea bags. Detailed design varies, but the machine basically consists of one or more pairs of contra-rotating rollers. The surfaces of the rollers are machined in a pattern designed to give a tearing and cutting action and this is aided by a speed differential, usually 10:1, between the rollers. CTC machines produce a high value product, but require skilled maintenance. The machine is able to handle leaves with a relatively wide range of moisture contents, but operates most effectively at 68–70%. Cooling air is supplied to prevent over-heating of the leaves.

The Lawrie Tea Processor (LTP machine), is one of the simplest designs and consists of hinged knives and beaters mounted on a rotor which spins within a circular casing. Cutting and crushing takes place as the leaves move from a hopper through the machine to the discharge end. The LTP machine is ineffective at leaf moisture levels in excess of *ca*. 71% and over-heating can be a serious problem.

(c) Fermentation

Fermentation involves maintaining the disrupted leaves in contact with air for a period ranging from 40 min to 3 h. During this period a number of chemical and biochemical reactions occur (see pages

---

* The term 'fermentation' as applied to tea can mislead the unwary into thinking that micro-organisms are involved. There is, however, no evidence of any involvement in production of black or green tea, although micro-organisms play a role during manufacture of the various types of pickled teas. Growth of spoilage micro-organisms, including the moulds *Aspergillus* and *Penicillium*, during fermentation has been considered to be a cause of poor quality in the made tea. It would appear, however, that problems of this nature occur only if hygiene standards are very poor, temperature control is totally inadequate or the fermentation is of unusual length.

181–3), most of which are oxygen-dependent. An adequate supply of air to each tea particle is, therefore, a prerequisite. There is also a need to control temperature, to limit the rate of undesirable secondary reactions.

Originally, fermentation involved spreading thin layers of tea onto the floor. This method is effective, but involves a very large floor space and is now practised only on a very small scale. Attempts to employ thicker layers initially failed due to oxygen starvation, but systems were subsequently developed which permitted the aeration of deep layers. These evolved into complex systems, such as the Williamson system, involving the use of large, wheeled tubs as fermentation vessels, which could be coupled individually to an air supply manifold. Subsequently continuous fermenters were developed, in which air was blown upward through a moving bed of tea leaves. A continuous supply of oxygen is not necessary, and fermentation continues in deep beds providing air is replenished on a periodic basis. This has led to the development of modern continuous fermentation equipment, which involves the use of deep beds and intermittent replenishment of air. The limited air flow means that dehydration is not a problem and humidification of the air is not practised. The rate of evaporative cooling is thus unaffected and temperature control to within 1°C is possible.

(d) Drying (firing)

Drying of tea invariably involves the use of hot air. In the early stages of drying, enzymic reactions continue and some of the end-products are of importance in determining the character of the made tea. Ultimately, however, enzymes are inactivated or destroyed by heat and the reduction in water activity level.

The water content of tea at the end of fermentation is *ca.* 60–72%, which is reduced to *ca.* 2.5–3.5%. Dryers originally used were either of a batch or semi-continuous design, but these have been

---

* The quality of green tea deteriorates rapidly during storage. For this reason green tea over one year in age is known as 'stale tea' and is of low commercial value. Fermentation has been investigated as a means of converting 'stale' green tea into higher value black tea. The 'stale' tea is partially rehydrated, mixed with a fresh tea leaf homogenate and fermented for 10 h at 26°C. Sufficient aroma precursors remain in the 'stale' tea to produce an acceptable quality black tea (Gur, W. *et al.* 1992. *Bioscience, Biotechnology, Biochemistry*, **56**, 992–3).

almost entirely replaced by belt dryers or, more recently, fluidized bed dryers. Belt dryers are of the multi-pass type, in which tea travels along a series of perforated belts mounted in a hot-air chamber. The most common configuration is four to eight belts stacked vertically. Tea enters at the top of the dryer and air at the base (counter- current). As each pass is completed the tea falls onto a lower belt, finally leaving the dryer through a discharge valve.

Fluidized bed dryers are widely used in the food and other industries. The basic design consists of a drying chamber containing a perforated floor on which a layer of product is deposited. Hot air is blown upwards through the floor at a velocity controlled between that needed to expand the product bed and that at which individual particles float. The bed may either be stationery or vibratory and mechanical sweeps are often fitted to assist product movement through the dryer. Fluidized bed dryers are of two types, the well-mixed and the plug-flow. Well-mixed dryers achieve extremely good contact between the material being dried and the drying air and are highly suitable for dealing with tea of high moisture content direct from the fermentation stage. There is, however, a wide distribution of residence time in the dryer and variation in the moisture content of the end-product.

Plug flow fluidized bed dryers operate on a 'first in-first out basis' and the end-product has a flat moisture profile. Plug flow dryers, however, are less effective than well mixed when dealing with tea of high moisture content direct from fermentation. A combination of well mixed and plug flow types is effective, but problems remain due the variation in residence time in the well mixed stage.

(e) Cleaning, sorting and grading

Tea leaving the dryer is in a relatively crude state and consists of a mixture of different sized leaf particles together with a quantity of stalk and fibre. The tea is sorted into portions by size, during passage through a vertical stack of vibrating mesh screens. Stalk and fibre is removed by electrostatic attraction during passage close to electrically charged rollers. The particle size in each portion, the grade, is determined by the size of the mesh. There is, however, no internationally recognized nomenclature, although the grade names (Table 4.3) do give an indication of the method of processing and of leaf particle size.

Table 4.3 Grading of tea

	Orthodox teas	*CTC*[1] *teas*
Whole leaf tippy	golden flowery	none
	orange pekoe	
	golden flowery	
	orange pekoe	
	flowery orange	
	pekoe one	
	flowery orange pekoe	
	orange pekoe	
Brokens	tippy golden broken	
	orange pekoe	broken pekoe one
	golden broken	broken pekoe
	orange pekoe	
	flowery broken	
	orange pekoe	
	broken orange	
	pekoe one	
	broken orange pekoe	
Fannings	broken orange pekoe	fannings one
	pekoe fannings	pekoe fannings
	golden orange fannings	
	pekoe fannings	
Dusts	pekoe dust	pekoe dust
	dust one	dust one
	dust	dust

[1] Crush, tear and curl.

The humidity in the cleaning and sorting room should be as low as possible both to minimize moisture pickup by the tea and to ensure effective operation of the electrostatic cleaning equipment. In areas of high humidity, localized heating in the vicinity of the electrically charged rollers is required.

(f) Bulk packaging

The traditional wooden tea chest has remained in use for many years. Despite this, the chest has been recognized as less than ideal, moisture pickup during transportation leading to quality deterioration. Multiwall kraft paper sacks have been developed for tea and are increasingly used, especially for exports from African countries. A barrier layer can be incorporated and, in conjunction with heat sealing, offers a very effective means of controlling moisture pickup. Undesirable chemical reactions do continue even in the absence of significant moisture pickup and 'freshness' is undoubtably lost in the period between packaging and consumption. This may be very considerably reduced by vacuum packing tea, at point of production, in aluminium foil–polyester–polyethene laminates, which are of extremely low oxygen and moisture permeability. Protection from oxygen must, however, be continued to point of consumption and this involves the use of nitrogen flushed storage tanks at the blending stage and gas flushing of the final packaging.

(g) Packaging of black tea as a consumer product

Leaf tea is retailed in two basic forms, loose in paper containers within a cardboard outer or in tea bags. Small quantities are sold in metal caddies, although this form of packaging is now largely restricted to the novelty or gift market. The development of 'very fresh' tea has also necessitated the application of gas flushing and at present this involves packaging the blended tea in nitrogen flushed, two-ply metallized polyester pouches. Various types of tea bag are available, although in some cases differences are restricted to the presence, or absence, of a string and tag. There has, however, been some genuine innovation in construction of the bags to improve extraction during 'mashing'. This has been of significance in fast-food outlets and the vending machine sector which previously made very heavy use of instant tea. A further innovation has been the use of foil or plastic pouches as containers for tea bags, which offer a higher degree of flavour protection. In

> BOX 4.4 **Fresh from nature's mould**
>
> 'Very fresh' tea was first introduced into the UK as Premier Brand's Typhoo Extra Fresh™. Such tea is seen as representing a new market sector, since the technology can be applied to tea of any type. The tea, which is of milder but stronger aroma than conventional, is aimed at tea drinkers currently buying mainstream blends, but who are willing to pay rather more for improved quality. An alternative approach was taken by the chain store, Marks and Spencer Ltd, which imports Kenyan tea, as tea bags packed in plastic pouches, by air freight. The tea is very fresh and high quality and commands a premium price, which justifies the high transport costs.

recent years, there has been concern over the possible presence of dioxins in the paper used for tea bag manufacture and the use of alternatives to chlorine for bleaching the paper is now widespread.

(h) Flavoured teas

Flavoured teas have been produced for many years, the most famous being Earl Grey, which is flavoured with essence of the citrus fruit, bergamot. Flowers, such as jasmine, are used to flavour some China teas. Other fruit flavours have been introduced, including various exotic combinations, but lemon is dominant. The technology of flavoured tea manufacture is proprietary, but is basically simple. Flavouring is added either as an essential oil or, especially for tea bags, as a water-soluble concentrate. Tea should be chosen which has a natural flavour compatible with that of the fruit flavouring. Natural, synthetic or nature-identical flavouring may be used. Synthetic is, in many cases, most widely used and gives a long shelf life. Natural flavouring is increasingly preferred and tends to give a better quality, more delicate product but the shelf life can be short due to rapid loss of flavour. Nature-identical flavours, where available, probably offer the best compromise.

4.3.2 Green tea

Green tea differs technologically and chemically from black tea. The fermentation stage is completely omitted from processing and enzymic activity in the leaves is inhibited by heating (steaming or pan-firing) as soon as possible after plucking. The character of green tea is thus largely determined by the endogenous components of the leaves at the time of plucking, rather than by compounds formed in post-harvest reactions.

(a) The Sen-cha process

Sen-cha tea is produced from the leaves of unshaded plants, although shading is used for the variant Gyokuro and Ceremony teas. The basic manufacturing process (Figure 4.4) involves a series of heating, drying and curling operations.

Freshly plucked leaves are steamed for 45-60 s before being dried and curled in a cylinder (roller), in air at a temperature of *ca.* 100°C. This primary drying and rolling lasts 40-50 min and reduces the leaf moisture from *ca.* 76% to *ca.* 50%. The leaves are rolled for

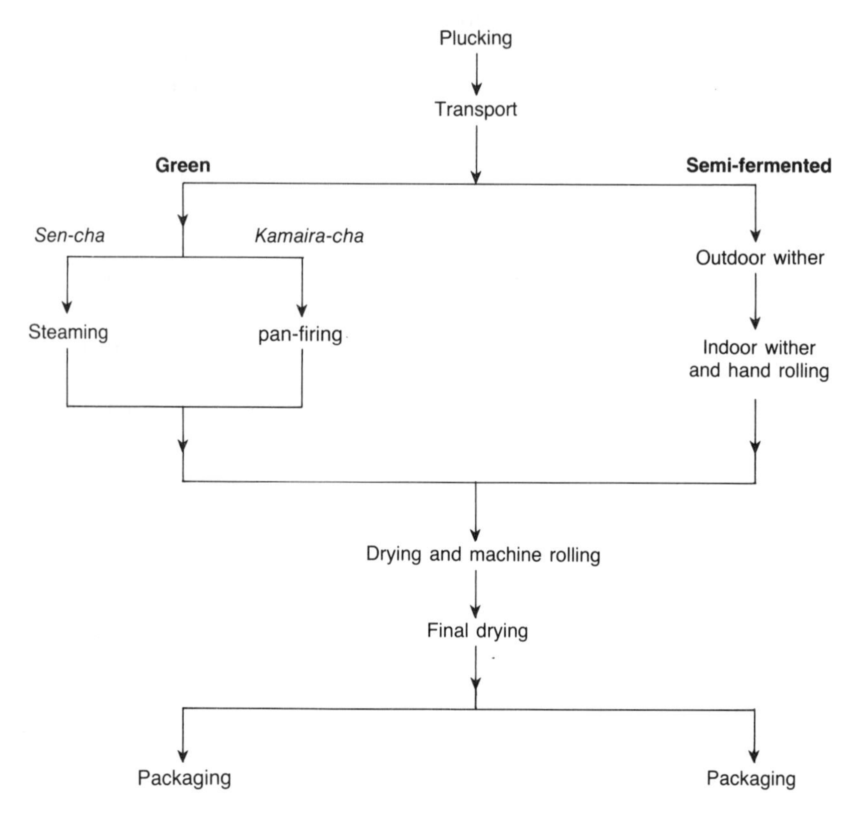

Figure 4.4 Processing of green and semi-fermented tea.

a further 15 min without application of heat, are pressed and then pass to a secondary drying stage, involving heating in air at 50–60°C for 30–40 min. This reduces the moisture content to *ca*. 30%. The leaves are then subject to further rolling before being dried by direct heating at 80–90°C. This process lasts for 40 min, during which the leaves are subject to twisting by pressing and rolling using a 'curling-hand'. Finally the moisture content is reduced to 6% by drying.

Tea produced by the factory requires 'refining', which is usually carried out at wholesale level. This involves repeated sifting of the tea to remove stalks, old leaves and dust. The refined green tea is usually redried to improve the aroma of the made tea.

(b) The pan firing process (kamairi-cha)

Pan firing is used in the production of 'Chinese' green tea (Figure 4.4). The process differs from the sen-cha process in the use of dry heat (parching) in place of steaming. Parching involves heating in a pan at temperatures of 250–300°C for 10–15 min. Slow agitation (*ca.* 5 cycles/min) must be employed to prevent burning. During the heating the tea develops the characteristic pan fired aroma. Pan fired leaves are then passed to a roller and processed, without heating, for 10–15 min before final drying in a pan at 100–150°C. The rolling stage determines the final shape of the tea and three end-product variants can be produced; gun tea (small pellets), chun-mee and sow-mee (fine, twisted) and pan fired (flat, polished leaves, pale white in colour).

4.3.3 Semi-fermented tea

Semi-fermented tea is characterized by oolong and pouchong teas, differences lying in the intensity of fermentation. Semi-fermented teas have a flowery characteristic and specially selected clones are used. The basic manufacturing process is illustrated in Figure 4.4.

The first stage involves outdoor withering in the sunlight. Sunlight is important in initiating biochemical reactions involved in producing the character of semi-fermented teas. Outdoor withering is a simple process in which fresh shoots are spread onto shallow bamboo baskets. During outdoor withering the temperature increases to 35–40°C, the length of the wither ranging from 30 to 60 min depending on the ambient temperature.

After the outdoor wither, the tea is transferred to the floor of a withering room for the indoor wither at room temperature. This lasts 6–8 h in the case of oolong tea, but only 3–4 h for pouchong. During the indoor wither the leaf layer is manually agitated every hour. The withered leaves are then pan-fired at 250–300°C for *ca.* 15 min. Pan firing inactivates enzymes and terminates biochemical changes. The remainder of the processing is the same as that applied during the production of pan fired green tea.

4.3.4 Pickled tea

Pickled tea is produced by various processes in different tea producing countries. In each case the pickling involves a fermentation in which indigenous micro-organisms play a significant role.

Pickled tea proper is made from fresh tea leaves, which have been steamed for up to 1.5 h. The steamed mass of leaves is then buried in a pit, or packed into bamboo cylinders or baskets and allowed to ferment for 2–4 months. In some cases NaCl is added to suppress undesirable micro-organisms. Little is known of the course of the fermentation, but it appears to follow a natural succession of initiation by members of the Enterobacteriaceae, followed by lactic acid bacteria, yeasts and moulds. Micro-organisms are responsible for some of the characteristic sour and flowery flavour, but biochemical reactions are also involved. Pickled tea to be consumed as a beverage, such as the Japanese Goishi-cha and Awa-cha, is dried after pickling. Some types, such as the Thai Miang (eating tea), are eaten rather than drunk.

A form of pickled tea may also be made from an infusion of dried tea leaves. Teekvass (tea fungus, wunderpilz), which is produced in south-east Asia and eastern Europe, is made by steeping dried tea leaves in water, adding sugar and boiling. After cooling, the infusion is inoculated from a previous fermentation and fermented for 8–12 days at 25–30°C. A mixed microflora develops which consists primarily of yeasts including species of *Saccharomyces*, *Torulopsis*, *Candida*, *Pichia* and *Kloeckera*. The yeasts are accompanied by *Acetobacter aceti* ssp. *xylinum*, which is important in producing a cellulose film up to 5 cm thick on the surface of the infusion. This finally breaks up and sinks to the bottom. A new film starts to develop, but at this stage the teekvass is ready for consumption.

4.3.5 Decaffeinated tea

Coffee has long been associated with caffeine and its adverse effects, but there is now increasing consumer appreciation that tea is also caffeine-containing, albeit at a lower level per cup than coffee. Decaffeinated tea is well established in the US, but represents a relatively new market in the UK.

The basic principle of decaffeination is simple. Tea is mixed with an organic solvent, which dissolves the caffeine. Solvent is removed from the tea as completely as possible, although trace amounts inevitably remain. Caffeine is recovered from the solvent for sale and the solvent cleaned for re-use.

Three solvents have been proposed for decaffeination, ethyl acetate, methylene chloride and supercritical CO_2. Ethyl acetate was used in early decaffeination processes, but is relatively non-selective and removes other compounds which results in some loss of quality. Ethyl acetate occurs naturally in tea leaves, although the level in decaffeinated leaves is higher due to solvent residues. Methylene chloride is very widely used at present and is highly effective. Methylene chloride, however, is a widely used industrial solvent which, at high concentrations, can have adverse effects on health. It is unlikely that the levels present in decaffeinated tea are of any risk to human health, but an alternative solvent is considered desirable, if only because of the association of methylene chloride with paint strippers and dry cleaning fluids.

Supercritical CO_2 is CO_2 under very high pressure. At such pressures the gas liquefies and acquires solvent properties. Under these conditions CO_2 is highly selective for caffeine and other components are unaffected. The process is essentially physical in nature, no chemical reactions occur and there are no residues.

In most cases, decaffeination involves treatment of the dried black leaf. Decaffeination of the green leaf before drying is used on a small scale and appears to improve product quality.

4.3.6 Instant tea

Although an 'instant' tea was produced in the late 19th century, large-scale manufacture was not feasible before the development of spray drying equipment capable of drying tea concentrates without significant damage to the organoleptic qualities of the product. Both black and green instant teas are now produced.

BOX 4.6 **A Boston tea party**

Instant tea is produced both in Europe and the US, but there are significant differences between the two markets. In Europe, the requirement is for a product which, when reconstituted with hot water and consumed while warm, closely resembles conventionally made tea. In contrast, in the US, instant tea is usually fruit flavoured, lemon being particularly common, and consumed as a cold beverage without the addition of milk or sugar.

Instant tea may be made from fully processed black, semi-fermented or green dried tea leaves. Use may also be made of partially fermented undried leaf. This is particularly common in producer countries, since undried leaf does not have to pass through the auction system and is therefore significantly cheaper. The manufacturing technology of instant tea varies and, in some cases, processes are specific to one manufacturer. The basic technology is. however, common to all processes and is summarized in Figure 4.5.

(a) Extraction

The key requirements for the extraction stage are a high total yield of tea solids and highly concentrated extract. In the case of instant green tea it is also necessary to ensure the leaf is heated to at least 70°C to inactivate enzymes. Modern extraction systems are based on counter-current flow and either batch or continuous systems may be used. Batch systems consist of a series of 10 or more tanks, the first of which contains fresh tea. Each of the other tanks

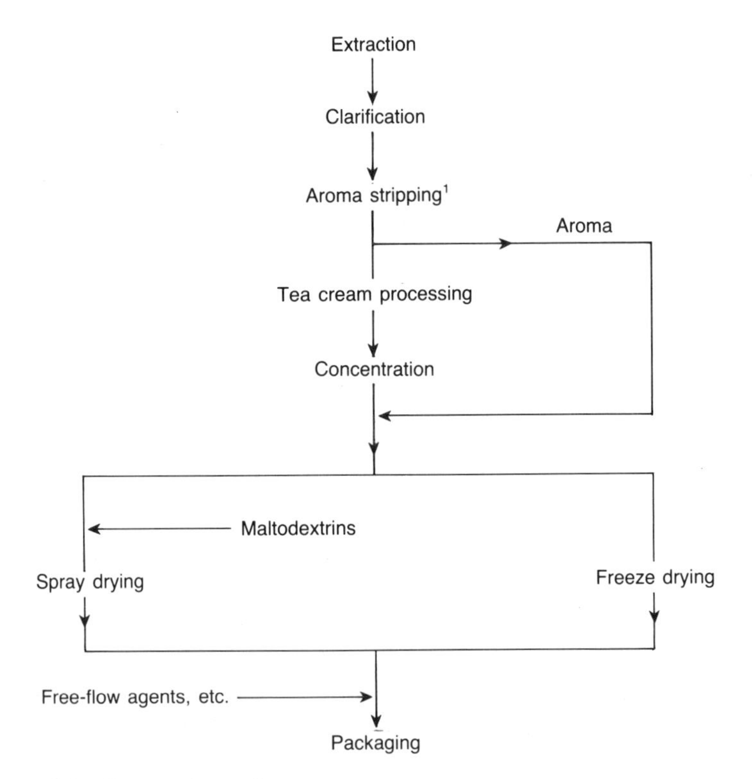

Figure 4.5 Processing of instant tea.

contains tea which has been extracted one, or more times. Thus the final tank of a ten tank series contains tea which has been extracted nine times. Hot water, at 80–90°C, is run into the last tank and flows forward through the other tanks to the first tank containing fresh tea. After this extraction, the last tank is emptied, loaded with fresh leaves, and becomes the first tank. The flow of hot water during extraction may be controlled by a series of valves. Alternatively the tanks can be mounted on a circular platform, which is turned to change the extraction sequence.

A large number of continuous extraction systems have been proposed. One of the simplest and most widely used consists of an inclined tube containing a spiral conveying blade. Tea leaves enter at the lower end and are conveyed upward through a falling stream of hot water, which passes through holes in the spiral blade. The whole extractor is maintained at high temperature by heating jackets.

Two-stage extraction systems are used, in which the first extraction is extracted at a relatively low temperature of 15–50°C followed by a second, separate, extraction at 50 to 100°C. Clarification is necessary after extraction. Centrifugation, decantation or filtration may be used, individually or in combination.

(b) Aroma stripping

Loss of aroma is a major factor leading to low perceived quality in instant tea. Most modern processes strip the aroma constituents, rather than attempting to retain the aroma with the tea during con-

---

* In the absence of a theoretical understanding of the extraction process, many extraction procedures were designed on a purely empirical basis. Theoretical studies have, however, shown that the extraction of solubles from leaf tea in a column extractor involves a system of three solubles. Each of these obeys a first-order solution law, one extracted simultaneously, one fairly rapidly extracted and one extracted only slowly. Diffusion of caffeine has been shown to be a greatly hindered process and the rate limiting step in extraction is diffusion through the leaf matrix and consequently the size and quality of the leaf. Caffeine extraction is unaffected by pH value, but is depressed by sodium, potassium and calcium chlorides. In contrast, theaflavin extraction proceeds at a higher rate at pH 8.0 and is depressed by calcium chloride, but unaffected by sodium and potassium chlorides. (Long, V.D. 1978. *Journal of Food Technology*, **13**, 195–210; Spiro, M. and Price, W.E. 1987. *Food Chemistry*, **25**, 49–59; Price, W.E. and Spitzer, J.C. 1993. *Food Chemistry*, **46**, 133–6).

centration. Aroma stripping involves passing the stripping gas through the extract, or spraying extract into a stream of gas. An inert gas, such as nitrogen, is preferred for stripping and gives a higher quality product, although steam is widely used. This is more economical and in modern systems quality differences can be slight. Aroma compounds are added directly back to concentrated extract.

(c) Tea cream processing

On cooling, a tea solution becomes opaque and lightens in colour. This is due to the formation of tea cream, a colloidal substance which contains the same components as the original extract. Tea cream can be removed by cooling followed by precipitation or centrifugation. This method is effective, but results in a poor quality product due to loss of flavour compounds with the tea cream. Solids yield is also reduced. An alternative method, which is widely used, is to maintain the temperature above 65°C. This is effective, but results in a 'stewed' flavour due to polymerization of theaflavins.

Tea cream causes particular problems in production of instant tea for the US market, which is required to be of clear, bright appearance when reconstituted with cold water. A number of possible approaches have been taken to the problem including chemical or enzymic solubilization of the components of tea cream, or their physical removal (Table 4.4).

(d) Concentration

Tea extract requires concentration before drying; a solids concentration of 40–45% is most common, although concentrations as low as *ca*. 20% and as high as 51% have been used. High solids concentrations in the dryer feed are desirable to improve reconstitution properties of the instant tea powder. Thermal evaporation under reduced pressure is most commonly used, evaporators usually being of the falling film or plate types (see Chapter 2, page 40). Such equipment permits evaporation at temperatures as low as 45°C, thermal damage being further controlled by the use of low temperature differentials between the heating medium and the tea extract and very short residence times. Aroma recovery equipment may be fitted to avoid the need for a separate aroma stripping stage.

Table 4.4 Means of minimizing problems due to tea cream in instant tea powders

Method	Comment
A *Removal*	
precipitation; centrifugation	removes flavour compounds
B. *Solubilization*	
maintain above 65°C	polymerizes theaflavins leading to 'stewed' taste
oxidize at 70°C, pH 9-10	requires bleaching with SO_2 to restore colour
oxidize with MnO_2 catalyst	may be bleached by mixing with untreated cream and equilibrating
treat with tannase	
remove high molecular weight components by ultrafiltration	
react cream with catechins from green tea	

More recently efforts have been made to reduce the damage suffered during concentration to a very low level by use of reverse osmosis or freeze-concentration. Reverse osmosis offers advantages in terms of reduction of thermal damage and avoidance of aroma stripping. There are, however, problems due to microbial growth and membrane fouling, which requires the application of high pressures to maintain flux. The degree of concentration possible is less than by thermal evaporation which necessitates removal of a large quantity of water in the dryer, or operating reverse osmosis in combination with an evaporator.

Freeze-concentration involves the removal of ice crystals formed when the tea extract is cooled beyond freezing point, There is no thermal damage or aroma stripping and concentration to *ca*. 35% solids is possible. The operating costs, however, are some three to four times greater than thermal evaporation or reverse osmosis. Instant tea is not normally regarded as a premium product and it is unlikely that the consumer would support the significantly higher cost of freeze concentration.

(e) Drying

The vast majority of instant tea is dried using a spray dryer. This consists of a chamber in which atomized tea extract, with a droplet diameter between 10 and 150 μm is mixed with hot air at an inlet temperature of *ca.* 240°C. Drying is almost instantaneous and the temperature of the product does not exceed 70°C. Large dried particles are removed from the base of the dryer, while small particles remain in the air stream and are removed by a series of cyclones and filters. Dryers used in instant tea production are of the co-current type in which air and product flow are in the same direction.

The method of atomizing the tea extract has consequences for product properties. Two types of atomizer are used in instant tea production, the rotary (disc) type and the pressure nozzle type. The rotary evaporator consists of a rotating disc driven by a high-speed electric motor. Product enters the centre of the disc and is directed to exit points on the periphery. The pressure nozzle evaporator consists of a nozzle, or a cluster of nozzles, through which the feed is forced under high pressure. Rotary atomizers produce a powder of lower bulk density, desirable in instant tea powder, but air inclusions within the droplet mean that the keeping quality of the end product may be low. Pressure nozzle atomizers produce a powder of higher bulk density, but which is generally of better keeping quality and better reconstitution properties. In either case CO_2 or nitrogen may be injected into the feed before atomization to lower the bulk density of the powder.

Spray dryers used in the production of instant tea usually complete the drying operation within the drying chamber, producing a powder with a moisture content of 3–5%. Such single-stage drying is considered obsolescent in the dairy industry where spray dryers are operated in series with one, or two, fluidized bed dryers. Two- and three-stage dryers are more economical to operate and produce a better quality powder, although initial capital outlay is higher. A plant of this type has been installed for instant tea production in the US and will probably become more common in future. Instantization (agglomeration) is rarely used for instant tea powders, although solubility is improved and the need for free-flow agents avoided.

A loss of aroma occurs during spray drying, but is minimized by using a feed of high solids concentration. Aroma loss can be further reduced by addition of carbohydrates such as maltodextrin to the feed.

Freeze dried instant tea has been introduced on a number of occasions, but appears to have been met with only limited marketing success. Freeze dried tea is granulated after drying and may be labelled as tea granules, rather than as instant tea.

(f) Additional ingredients

With the exception of flavouring, etc., used in instant lemon tea powder, most additional ingredients are used to improve convenience of use, especially in vending operations. The powder tends to clump, for example, and addition of a 'free- flow' agent such as tricalcium orthophosphate is common practice. Foaming is a further problem, especially in cold water-reconstituted tea powder for the US market, where a complete absence of foam is a necessity. Silicon powder is highly effective and acts both to prevent foam formation and to defoam. A compound derived from tea seed oil has also been proposed for this purpose.

Attempts have also been made to use additives to improve the colour and flavour of tea. A number of commercial products are available, and include a tea aroma distillate and a tea aroma oil. Commercially available products based on linalool (see Chapter 7, page 340) have also been developed. In production of instant green tea, flavour is improved by addition of freeze crushed unprocessed tea leaves to the concentrated tea extract before drying.

Instant lemon tea powder contains lemon flavouring, often with a carrier, together with various other ingredients including acidulants and anti-oxidants. Instant white tea is also available and is a blend of tea powder with spray-dried skim milk powder and a 'whitener', usually consisting of glucose syrup, vegetable oil and caseinates.

4.3.7 Other tea products

Although the tea plant may be utilized in production of by-products, such as caffeine and polyphenols, the development of a wide range of consumer products is difficult. Some years ago, a mixture of instant tea and coffee was marketed, but apparently failed to achieve success. More recently, processes have been proposed for the manufacture of ready-to-drink tea flavoured with lemon or other fruit. Technology is proprietary, major problems to be overcome being prevention of precipitate formation and development of oxidative off- flavours. It is likely that ready-to-drink tea would be stored under refrigeration and marketed in competition with existing soft drinks. There are also a number of ready-to-drink herbal infusions, which may be equated with herbal tea. These appear to be marketed primarily through the health food sector. A tea-flavoured mineral water is also available.

BOX 4.7 **Enter the dragon**

A number of herbal drinks have appeared which are claimed to have stamina and energy enhancing properties. These are usually fruit-based and contain in addition, ginseng, honey and various other herbal ingredients. One variant, formulated by 'Grand Masters of the Dragon Karate sect' contains ginseng, lychee and raspberry juice and honey in an 'exotic' base of wolfberry and *Camellia* extract - better known as cold tea!

4.3.8 'Tea' from other plants

A range of herbal teas, made from leaves, seeds, fruit and other plant parts is available, the most common varieties in Europe and the US being comfrey, camomile, peppermint, fennel, lemon verbena and rosehip. In South Africa as much as 15% of total tea consumption is the herbal rooibos tea, while matte is widely consumed in Brazil and some other South American countries. Wide use is also made of blending, some blends incorporating ginseng. Products made only from fruits and containing no other plant parts may be described as 'fruit cups' or 'fruit infusions'. The technology of herbal tea manufacture is simple, usually involving no more than drying, size reduction and removal of extraneous material. Flavouring with fruit essence is common and is considerably more popular with herbal than with true black tea. From a marketing viewpoint, fruit flavouring is seen to reinforce the healthy cottage life-style association, but more realistically its use may be to mask the unpleasant taste of some herbal teas.

A number of claims have been made concerning the health-promoting and even medicinal properties of herbal teas. Such claims stem, historically, from the association with medical herbalists and are largely unsupported by evidence. The toxic effects associated with widely consumed herbal teas are, however, well documented (page 172).

4.4 BIOLOGICAL ACTIVITY OF TEA

In many countries, tea is drunk, in relatively large quantities, by a high percentage of the population. It is, therefore, inevitable that the beverage should be associated with effects that are both beneficial and deleterious to health. Equally inevitable, although still regrettable, is the fact that many of the assertions made concerning the relationship between tea drinking and health have been made in the absence of any critical appraisal of evidence. For the vast majority of consumers, tea is an extremely safe beverage, with has the beneficial effects of being a source of dietary water, a mild stimulant and, perhaps most importantly, a source of pleasure. There is no evidence of serious medical side effects of consumption, although tea may have adverse effects on a small minority of individuals.

4.4.1 Tea and nutrition

Water is the most important dietary component of tea. Tea is perceived as an efficient thirst quencher, equalling tap water and being superior to mineral water, fruit juice, carbonated drinks and coffee. The use of soft water (low calcium content) results in a high oxalate content and slightly, but significantly increases the risk of kidney stone formation.

With the possible exception of some inorganic ions, the nutritional contribution of other constituents of tea is insignificant. Equally there is no conceivable risk from toxic trace elements, such as cadmium and fluoride, which have been detected in tea at very low levels.

4.4.2 Adverse effects of tea

Two major constituents of tea have been associated with adverse effects amongst consumers, caffeine and polyphenols.

(a) Caffeine

The physiological consequences of caffeine consumption are discussed in detail with respect to coffee (Chapter 5, pages 234-5).

Tea contains the same quantity of caffeine as coffee, on a weight by weight basis, but less tea is used per cup than coffee and thus the overall caffeine intake tends to be lower. This is probably one of the reasons why caffeinism is markedly less common amongst tea drinkers than among coffee drinkers. There may, however, be contributory substances to caffeinism which are present in coffee but not tea. Further, the link between caffeine consumption and coronary heart disease, established in coffee drinkers, seems unlikely to apply to tea drinkers. The reason may lie in the fact that tea, like filter-made coffee but unlike boiled coffee, does not increase levels of low density lipoprotein in the blood plasma and, in fact, may actually have a beneficial effect.

(b) Polyphenols

Polyphenols are an important quality determinant in tea (see page 173). Polyphenols, mistakenly referred to as 'tannins', have been associated with damage to the intestinal mucosa and with carcinogenicity. To a large extent these fears have arisen because of confusion between the chemical nature of tea polyphenols and condensed tannins including tannic acid. Tea consumption has been associated with various types of cancer, but evidence appears to be equivocal at best and may involve association in the absence of a true causal relationship. In general, it may be stated that, as far as is currently known, tea consumption is not related to an enhanced risk of cancer. Equally there appears to be no evidence that tea has a protective role against carcinogens, although green leaf aqueous extracts and green tea polyphenols have been shown to have strong scavenging activity against potentially harmful oxygen species.

(c) Tea and iron absorption

The observation has been made that tea, ingested alongside food, inhibits the absorption of inorganic and some forms of organic iron, although not that provided by heated haemoglobin. This interference is unlikely to be of significance to persons consuming meat as a significant part of the diet, since the haemoglobin provided by meat is almost invariably heated before consumption. However, tea consumption may contribute to iron deficiency in a sector of the population comprising mainly women on a vegetarian diet of low iron content.

4.4.3 Anti-microbial activity of tea

Historically, tea has been considered to have curative powers, but this aspect was not investigated systematically until the late 1940s, when black tea extracts were found to inhibit influenza virus. More recent work, primarily undertaken in Japan, has shown various extracts of black and green tea to have activity against enteropathogenic bacteria including *Campylobacter*, diarrhoeagenic strains of *Escherichia coli*, *Salmonella*, *Staphylococcus aureus* and *Vibrio* (including *V. cholerae* O1). There was also activity against influenza virus A and B, and against haemolysins (haemolytic toxins) of *Staph. aureus* and *V. parahaemolyticus*. All work was carried out *in vitro* and the significance of the findings to human health are not known.

4.4.4 Herbal teas

As noted earlier, claims have been made for the therapeutic properties of herbal teas. In general such claims are made in the absence of any supportive evidence, although some consumers may be susceptible to a placebo effect. In contrast, adverse effects, sometimes serious, are well documented and in the US at least four deaths and many cases of serious illness have resulted from herbal tea consumption. Claimed therapeutic properties and known adverse effects of various herbal teas are listed in Table 4.5.

4.5 CHEMISTRY OF TEA

The acceptability of tea as a beverage is largely dependent on the flavour of the product on consumption. This in turn is dependent on the original composition of the plant and chemical and biochemical changes during processing. Differences in processing are reflected in the character of the different types of tea.

The flavour of tea results broadly from taste and aroma, aroma being considered to be the more important. In general, non-volatile compounds are responsible for taste and volatile compounds for aroma, although the terms 'non-volatile' and 'volatile' should not be considered to be absolute.

Table 4.5 Claimed therapeutic and actual hazardous properties of herbal teas

Camomile	
therapeutic:	stress alleviation, relief of teething pain
hazardous:	can cause anaphylactic shock; contact dermatitis
Comfrey	
therapeutic:	general tonic
hazardous:	severe, and possibly fatal, liver damage. Possible carcinogen
Fennel	
therapeutic:	relief of throat infections, appetite stimulant, relief of 'wind' in babies
hazardous:	diarrhoea
Lemon verbena	
therapeutic:	relief of indigestion
Hazardous:	gastro-intestinal disturbance
Nettle	
therapeutic:	'blood purifier', enhances resistance to common cold
hazardous:	gastro-intestinal disturbance
Peppermint	
therapeutic:	relief of indigestion, 'improves' circulation
hazardous:	possible irritant in large quantities

4.5.1 Black tea

(a) Non-volatile compounds

The significant non-volatile compounds of black tea are pigmented, hot water-soluble polyphenolics, derived from precursors present in the shoots during fermentation and the early stages of drying. Polyphenols are directly related to the perceived quality of the made tea and are considered to contribute brightness, depth of colour, strength and mouthfeel.

Young tea shoots have a high content of polyphenols (Table 4.6) and as much as 30% of the dry weight consists of catechins (flavon-3-ols). Other phenolics are present, at low concentrations, but play no significant part in the formation of the characteristic polyphenolics of black tea. In addition to the high total content of polyphenolics, the range present in *Cam. sinensis* is one of the plants' unique features. No less than six catechin compounds are present

Table 4.6 Polyphenols of young tea shoots

Catechins (flavonols)	epigallocatechin gallate epigallocatechin epicatechin gallate gallocatechin catechin
Flavonol glycosides	
Leucoanthocyanins	
Phenolic acids	theogallin other compounds

at concentrations in excess of 1% (dry weight); (−)-epigallocatechin-3-gallate, (−)-epigallocatechin, (−)-epicatechin-3-gallate, (−)-epicatechin, (+)-gallocatechin and (+)-catechin (Figure 4.6). Smaller quantities of (−)-epicatechin-3,5-digallate and the 3-methyl gallates of (−)-epicatechin and (−)-epigallocatechin are also present in fresh shoots, but their significance in determining the character of the black tea is not known.

Polyphenols are situated in the cell vacuole. Highest levels are present in the bud and first leaf, and lowest levels in the internodes. Variations in climate and agronomic practice can affect the composition of the polyphenols. The proportion of epigallocatechin-3-gallate, for example, is lower in plants grown in colder climates and also falls during cold seasons elsewhere. Changes in the relative proportions of the different polyphenols also result from shading and different application rates of nitrogenous fertilizers.

Leaf disruption destroys the integrity of the cells and enhances contact between polyphenols and the enzyme polyphenol oxidase, which is situated in the cytoplasm of the intact cell.

In contrast to the polyphenols themselves, polyphenol oxidase activity is greatest at the internodes. The concentration of shoot polyphenols, however, is of greatest importance in determining the composition of the black tea.

* Polyphenol oxidase is an *o*-diphenol:oxygen oxidoreductase (EC 1.10.31). The enzyme is copper containing and has a molecular weight of *ca.* 150 000 Da. Pyrogallol and catechol are readily oxidized *in vitro* and the natural substrates in the tea plant appear to be catechins and their galloyl esters.

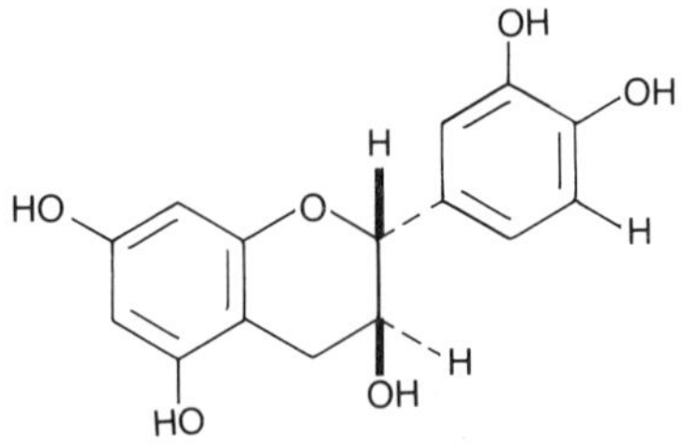

Catechin

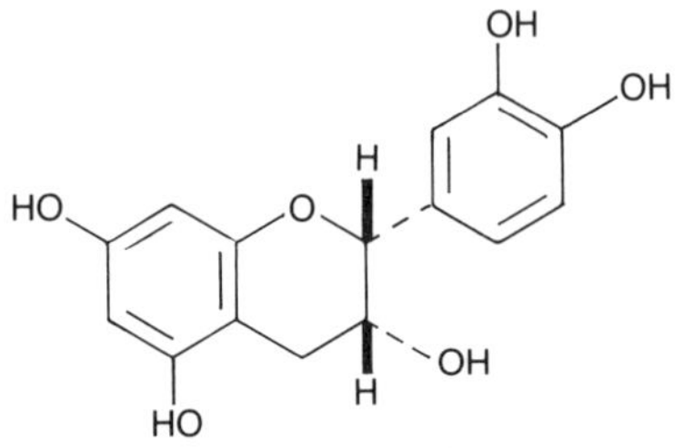

Epicatechin

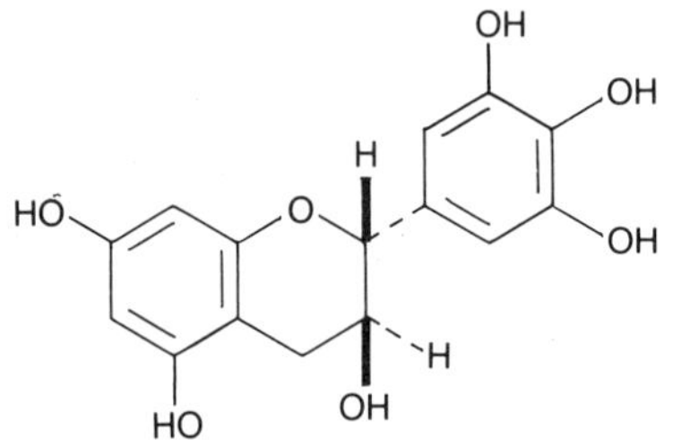

Gallocatechin

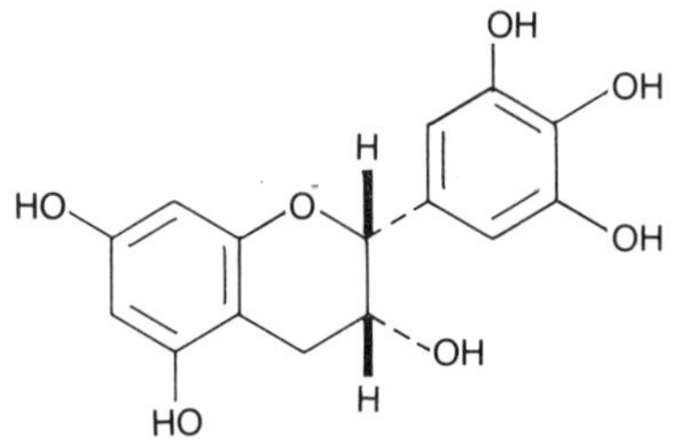

Epigallocatechin

Figure 4.6 Structures of catechin compounds of *Camellia sinensis*.

It has been known for many years that fermentation of tea involves the oxidation of catechins to form brown coloured pigments. After some difficulties it was shown that the pigments were of two types, which subsequently became known as the theaflavins and thearubigins.

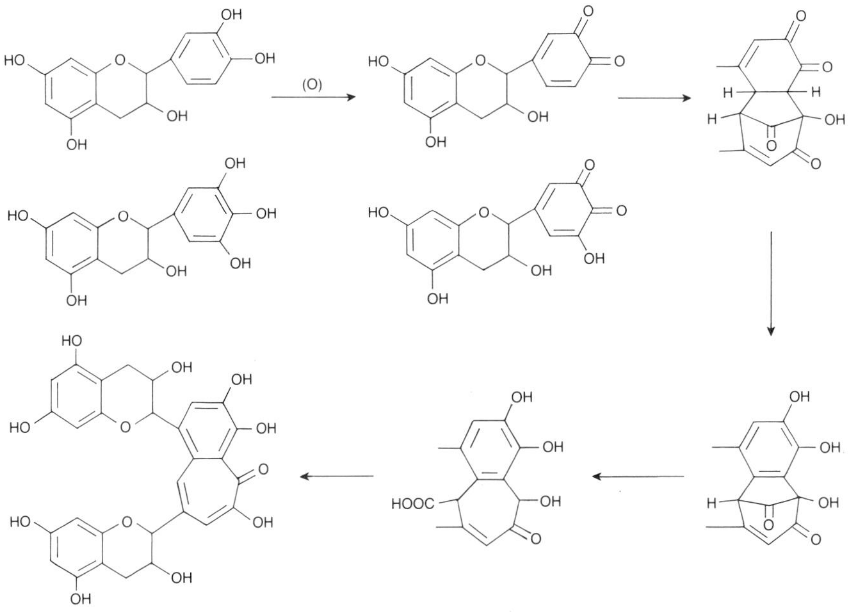

Figure 4.7 Formation of theaflavins. Reproduced with permission from Willson, K.C. and Clifford, M.N. (eds) 1992. *Tea. Cultivation to Consumption*. Copyright 1992, Chapman & Hall, London.

Theaflavins are formed by the enzymic oxidation and condensation of catechins with di- and trihydroxylated B rings (Figure 4.7). Theaflavins are recognized as providing the quality attributes 'brightness' and 'briskness'. The type of theaflavin formed is determined by the catechins involved in the reaction. Black tea has also been found to contain small quantities of epitheaflavic acid, theaflavic acid and flavic acid-3'-gallate. These compounds are formed by oxidation and condensation reactions of (+)-gallic acid with (−)-epicatechin, (+)-catechin and (−)-epicatechin gallate respectively (Figure 4.8). Gallic acid is present in the fresh leaf at very low concentrations, but accumulates during fermentation. The probable route is the breakdown of galloyl esters from catechins or theaflavin gallates. Oxidation mediated by polyphenol oxidase may be involved to a very limited extent, but the major pathway probably involves redox equilibration through electron transport by epicatechin, epicatechin gallate and possibly other catechins. This reaction does not occur to any significant degree in the early stages of fermentation, when the gallocatechins, epigallocatechin and epigallocatechin gallate are present in excess. The levels of theaflavic

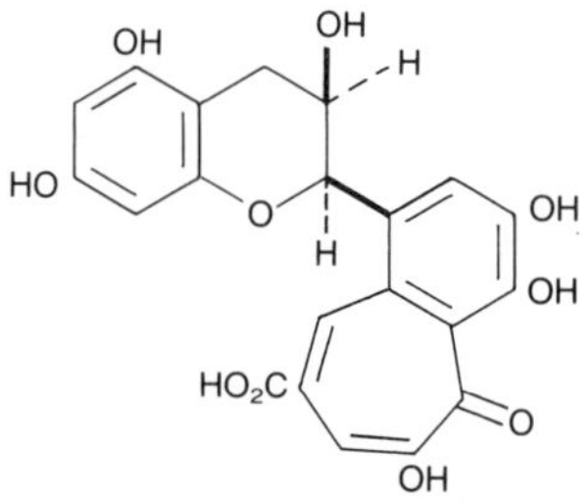

Theaflavic acid

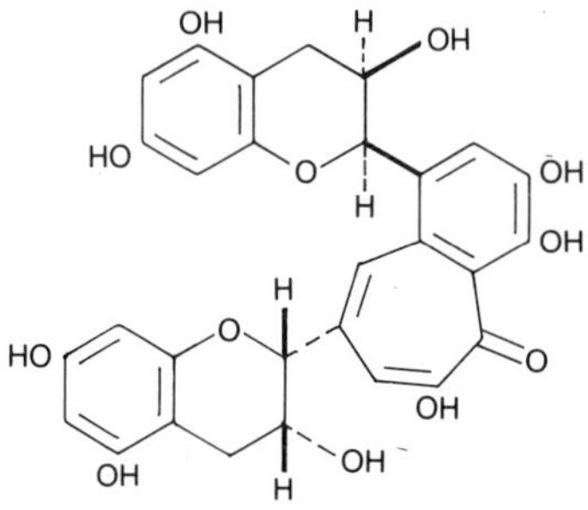

Theaflavin

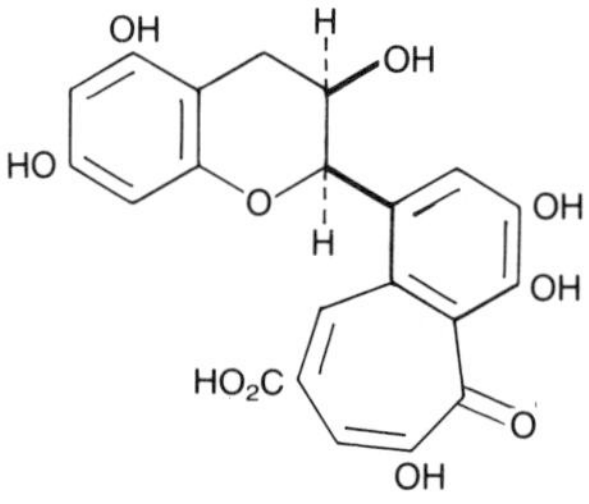

Epitheaflavic acid

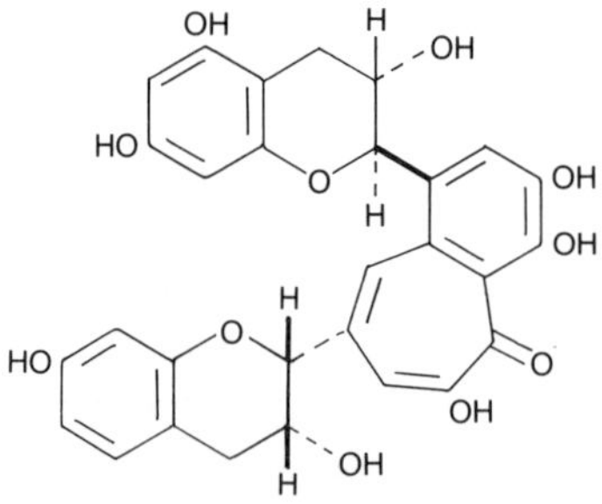

Neotheaflavin

Figure 4.8 Structure of theaflavic acid and related compounds.

acids and gallic acid will therefore depend on the extent to which fermentation proceeds.

Oxygen is required during the polyphenol oxidase-mediated oxidation of catechins to *o*-quinones and also during formation of the benztropolene ring of theaflavins. Oxygen deficiency during fermentation will thus lead to low levels of theaflavins in the black tea. A number of other factors also affect theaflavin formation. Formation of theaflavins cannot, however, be discussed in isolation to their degradation to thearubigins.

Thearubigins comprise 10-20% of the dry weight of black tea but, due to their relatively high solubility in hot water, comprise 30-60% of the solids in the made tea. In contrast to the theaflavins, relatively little is known of the chemical character of thearubigins. Thearubigins do, however, appear to be a structurally diverse group of compounds which result from various oxidative reactions of catechins. Oxidative degradation of theaflavins is generally recognized as being the major pathway of formation and it is also generally accepted that thearubigins can be formed directly by catechin-catechin interactions. A number of other pathways probably exist, involving the coupled oxidation of theaflavin intermediates. Interactions between catechol quinones with proteins, carbohydrates, nucleic acids and other macromolecules may also be involved.

Thearubigins are considered to be important in imparting 'strength' and 'colour' to the made tea. Thearubigin formation from theaflavins is considered to be deleterious to the quality of tea, because of the loss of theaflavins and a consequent 'softness'. In contrast, thearubigins formed by catechin-catechin interactions, without loss of theaflavins, are considered to have a positive effect on quality. The relative rates of formation and degradation of theaflavin and formation of thearubigins is largely dependent on oxygen availability, which in turn depends on the manufacturing processes. Fermentation is probably the most important stage, followed by leaf disintegration, but it is important not to overlook the high level of oxidative enzyme activity in the stages of the drying process before enzyme inactivation.

(b) Volatile (aroma) compounds

A very large number of volatile compounds, collectively referred to as the 'aroma complex', have been identified in tea. Those thought

Table 4.7 Significant types of aroma compounds of black tea

Acids
Alcohols
Carbonyls: Aldehydes
Esters
Hydrocarbons
Lactones
Miscellaneous oxygen compounds
Nitrogenous compounds
Phenols
Sulphur compounds

to be of greatest importance are listed in Table 4.7. It is likely that even the most comprehensive published lists are incomplete and the increasing sensitivity of instrumentation means that new components of the complex are frequently identified.

Components of the aroma complex may either be the primary products of plant biosynthesis, or secondary compounds produced during processing. Components can be classified into Group I compounds, responsible for 'green, grassy' aromas, desirable as part of the overall character of black tea, but undesirable at high concentrations and Group II compounds responsible for highly desirable sweet, flowery aromas. The ratio of total Group II to total Group I compounds, as determined by retention time during gas chromatography, has been proposed as a semi-quantitative means of determining quality, the flavour index (see page 184).

In black tea, the primary products of plant biosynthesis are less important than secondary compounds, but can still be significant. Most of the compounds which have been identified in fresh green leaves are alcohols and include linalool, oxides of linalool, nerol, geraniol, benzyl alcohol, 2-phenylethanol, nerolidol, *Z*-2-penten-1-ol, *Z*-3-hexen-1-ol, *E*-2-hexen-1-ol and *n*-hexanol. The relative quantities present depend on the clone and agronomic factors. There is considerable variation according to the country of origin which, at least partly, reflects the climate. Tea grown at higher altitudes usually has a higher concentration of aroma compounds and a superior flavour index, although an opposite effect is apparent with some clones. Seasonal changes are superimposed on climatic, especially where seasonal differences are greatest.

Shading tends to increase the quantity of aroma compounds and improve the flavour index, while high levels of nitrogenous fertilizer have an adverse effect on the flavour index, although the total quantity of aroma compounds increases. The effect of other types of fertilizer have been less fully investigated, but appear to have no significant effect.

The flavour index is lowest immediately after pruning, increasing throughout the cycle until the next prune. This results from a simultaneous decrease in the concentration of deleterious Group I components and an increase in the desirable Group II. A reverse effect occurs as individual leaves age and coarse plucking results in a marked decrease in the flavour index.

The fate of the primary products of metabolism during processing varies. Increases in concentration of some occurs as a result of enzyme activity during processing, while decreases result from glycosidation and formation of non-volatile products. These changes commence during withering, but fermentation is probably the most significant stage, the extent of changes being dependent on the extent to which fermentation is allowed to proceed.

Amino acids, carotenes, unsaturated fatty acids and other lipids and terpene glycosides are the most important sources of secondary products in the aroma complex. Amino acids increase in concentration during withering and, possibly, the initial stages of fermentation due to peptidase-mediated protein breakdown. As fermentation proceeds, the amino acids are converted to corresponding aldehydes, such as 2-methylbutanal (derived from leucine), pentanal (derived from isoleucine) and phenylacetaldehyde (derived from phenylalanine). The reactions are mediated by polyphenol oxidase, or peroxidase, the sequence being the oxidation of catechins to quinones and subsequent Strecker degradation of amino acids to aldehydes. Aldehydes accumulate during fermentation, but the concentration falls during firing due to reduction to primary alcohols, or oxidation to carboxylic acids. Reactions involving amino acids during tea processing are summarized in Figure 4.9.

Fresh leaves contain significant quantities of carotenes, which have been considered to be important precursors of components of the aroma complex and a high carotene content has been considered to be an index of quality. More recent work has suggested that,

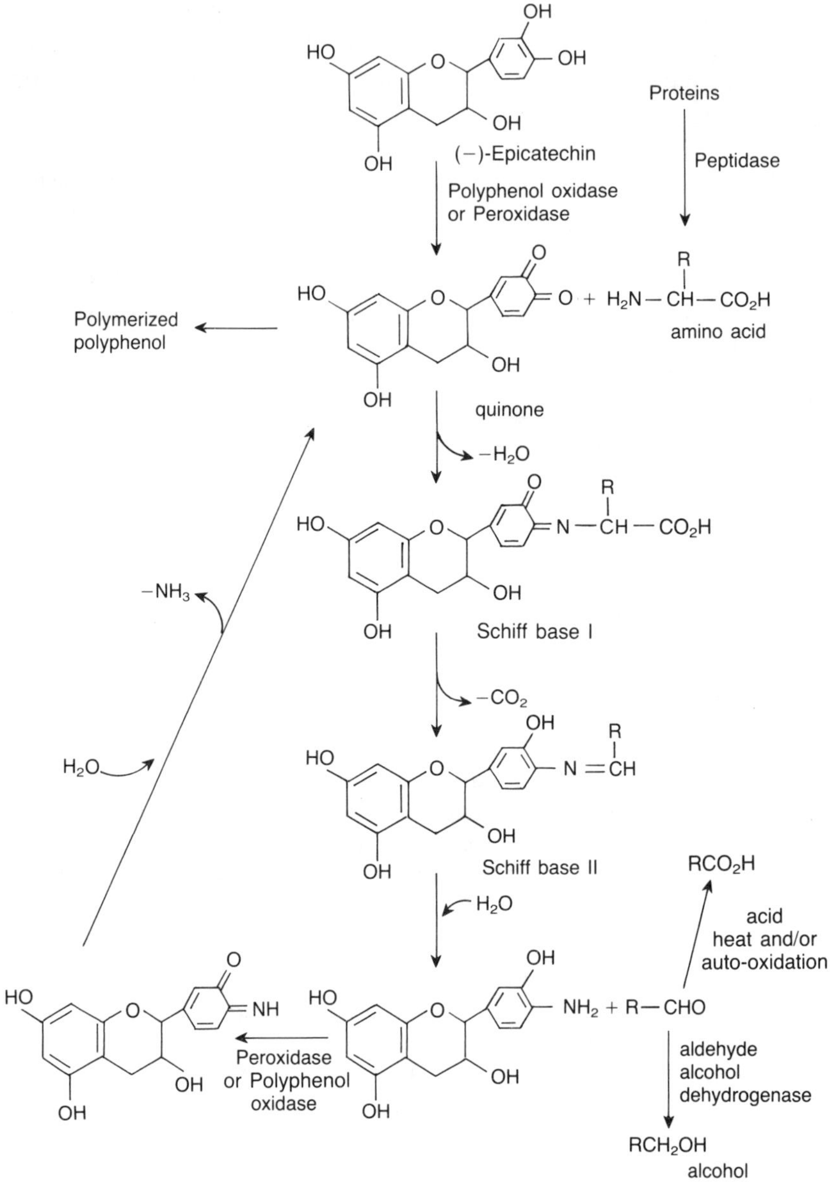

(−)-Epicatechin

Proteins

Peptidase

Polyphenol oxidase or Peroxidase

amino acid

Polymerized polyphenol

quinone

$-H_2O$

Schiff base I

$-NH_3$

$-CO_2$

H_2O

Schiff base II

RCO_2H

acid heat and/or auto-oxidation

Peroxidase or Polyphenol oxidase

aldehyde alcohol dehydrogenase

RCH_2OH alcohol

Figure 4.9 Production of aroma compounds from amino acids. Reproduced with permission from Willson, K.C. and Clifford, M.N. (eds) 1992. *Tea. Cultivation to Consumption*. Copyright 1992, Chapman & Hall, London.

while carotenoids are significant in determining quality, β-carotene and some other carotenes have a negative effect. This rather surprising finding may be due to transformation compounds produced during manufacture.

Carotene content decreases during the entire tea processing operation from withering to firing. Enzyme-mediated oxidation commences during withering and continues during fermentation. Beta-ionone is the primary oxidation product of β-carotene and is considered to be an important aroma compound. Degradation products of other carotenoids include β-ionone, α-ionone, 3-hydroxy-5,6-epoxyionone, 3,5-dihydroxy-4,5-dihydro-6,7-didehydro-α-ionone and terpenoid aldehydes and ketones, such as linalool. Further reactions result from pyrolysis during firing and, probably, from photo- and auto-oxidation. The extent of such reactions increases with increasing wither. Reactions involving carotenes and their degradation products are not fully understood, but primary oxidation products such as β-ionone are thought to be converted to dihydroactinidiolide, 2,2,6-trimethylcyclohexanone, 5,6-epoxyionone, 2,2,6-trimethyl-6-hydroxycyclohexanone and theaspirone. A number of other components of the flavour complex are also probably derived from carotenes. These include α-damascone, β-damascone, β-damascenone, 3-oxo-β-ionone, 1,2-epoxy-1′,2′-dihydro-β-ionone, 3,7-dimethyl-1,5-octadien-3,7-diol, dehydrovomifoliol and loliolide.

Lipids in the fresh leaf are largely composed of free fatty acids and fatty acid esters. Variety of tea and climatic conditions in the growing area determine the relative quantities of fatty acids present, but under most conditions linolenic acid is present in the largest quantities. Fatty acids are important precursors of aroma complex components, although the large number of reactions and interactions which can occur during processing means that direct relationships between the fatty acid content of the fresh leaf and the quality of black tea are rare.

Withering is an important stage in the production of aroma compounds, in which fatty acid esters undergo acylhydrolase-mediated hydrolysis to free fatty acids. Existing unsaturated fatty acids, together with those derived from fatty acid esters, are known to be degraded to aroma compounds, but the fate of saturated fatty acids, and any role in aroma development, remains obscure.

The initial stage in degradation of linolenic and linoleic acids involves oxidation to 13-hydroperoxylinolenic and 13-hydroperoxylinoleic acids respectively. In each case the reaction is mediated by lipoxygenase and molecular oxygen is required. The enzyme hydroperoxide lyase is then involved in formation of 12-OXO-(9*Z*)-dodecenoic acid and *Z*-3-hexanal from 13-hydroperoxylinolenic acid. The same enzyme is involved in the formation of 12-OXO-(9*Z*)-dodecenoic acid and *n*-hexanal from 13-hydroperoxylinoleic acid. This reaction, however, is inhibited by linolenic acid and 13-hydroperoxylinolenic acid.

All of the products of hydroperoxide lyase activity undergo further reactions. Alcohol dehydrogenase, for example, mediates the reduction of *Z*-3-hexenal and *n*-hexanal to *Z*-3-hexenol and *n*-hexanol, respectively, while 12-OXO-(9*Z*)- dodecenoic acid isomerizes to 12-OXO-(10*E*)-dodecenoic acid. *Z*-3-hexanal also isomerizes with formation of *E*-2-hexenal and *E*-3-hexenal. Further reactions, often involving pyrrolysis or auto-oxidation occur, especially during firing, resulting ultimately in formation of compounds, such as hexanoic acid, *E*-2-hexenoic acid and *E*-3-hexenoic acid (Figure 4.10).

Although linolenic and linoleic are the dominant unsaturated fatty acids of the tea leaf, palmitoleic and oleic acids are also present at relatively low concentrations. These acids are degraded to heptanal and heptanol and nonanal and nonanol, respectively.

Although formation aroma compounds continues throughout processing, a number play no part in the composition of the final aroma complex. This has been attributed to loss of lower boiling point compounds by volatilization during firing.

Although linalool is formed during carotenoid breakdown, an alternative route involves hydrolysis of β-D-terpene glycosides. Geraniol is also formed by this route and formation of alcohols, such as phenylethanol and benzyl alcohol, by hydrolysis of the corresponding glycosides may be important in aroma development during processing.

---

* Hydroperoxidation of both linolenic and linoleic acids occurs primarily at the C-13 carbon, resulting in the production of C_6 alcohols and acids. Cleavage of the two fatty acids at C-10 does occur, but only to a limited extent and yields small quantities of *E*-2,*Z*-6-nonadienal and *E*-2-nonenal from linolenic and linoleic acids, respectively.

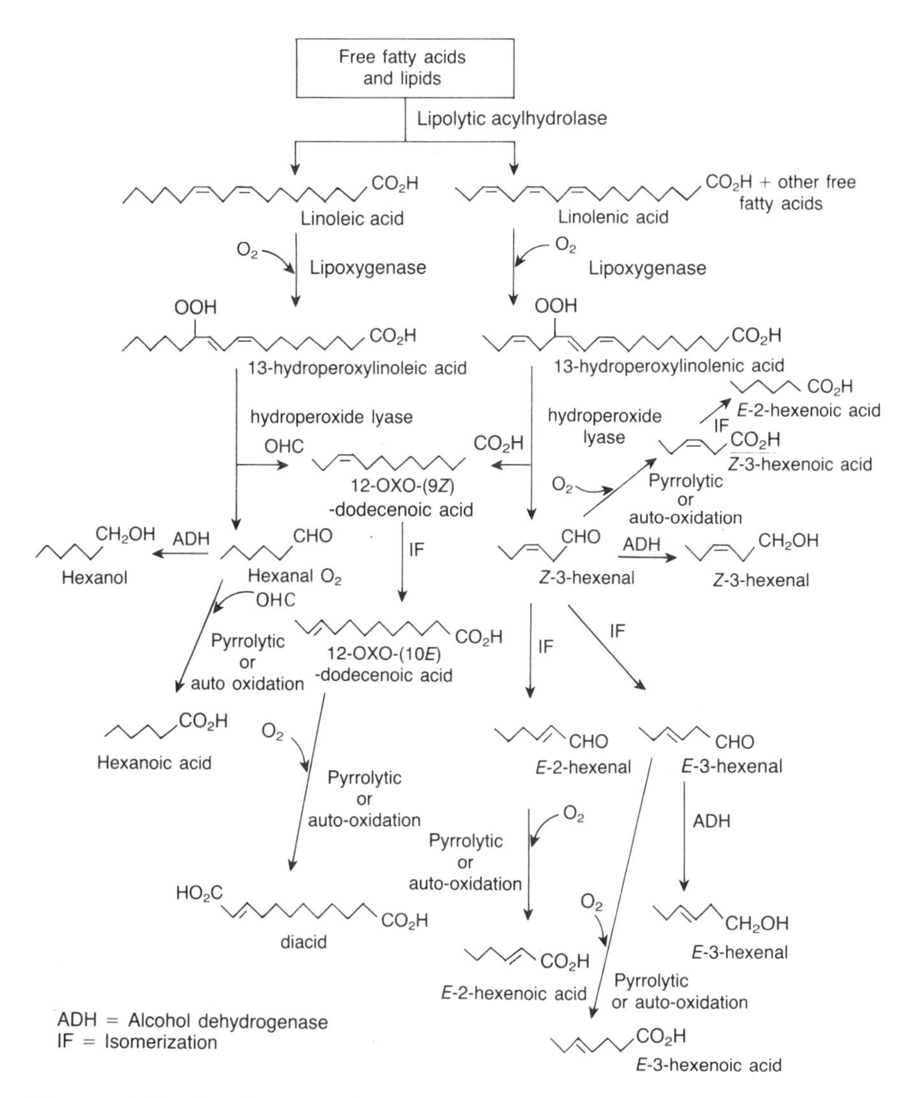

Figure 4.10 Production of aroma compounds from lipids. Reproduced with permission from Willson, K.C. and Clifford, M.N. (eds) 1992. *Tea. Cultivation to Consumption*. Copyright 1992, Chapman & Hall, London.

In addition to recognized precursors of aroma complex components, it is likely that other precursors exist, which may be of greater or lesser importance. Chlorophyll, for example, is present in relatively large quantities and degrades to phytol and other compounds which may play a role in the aroma complex. A positive correlation has been demonstrated between chlorophyll and tea quality. Chlorophyll is, however, of only limited significance as a determinant of tea quality, although this may be an artefact due to selection of light-coloured clones.

(c) Chemical determination of black tea quality

For many years the quality of black tea was determined either by expert tasters or by the price commanded at market. Expert tasters continue to play a very important role, in determining quality and suitability of individual lots of tea, in ensuring consistency of existing blends and in formulating new blends. Training of expert tasters is, however, a lengthy and expensive procedure and, inevitably, a greater or lesser degree of subjectivity becomes involved. The price commanded at market also continues to be regarded as a measure of quality, but is at best a comparative measure, since the major determinant of price is likely to be the balance between supply and demand.

A number of chemical parameters have been proposed for assessing the quality of both the fresh and processed leaf. A good correlation exists between total theaflavin content and quality with some, but by no means all, *Camellia* cultivars. A more satisfactory approach is to examine the relationship between individual theaflavins and quality, but this involves assay of a relatively large number of compounds. The contribution of theaflavins to astringency varies with the extent of galloylation and for this reason the concept of 'theaflavin digallate equivalent' (TDE) has been introduced as a quality parameter.

Although the TDE appears to be applicable to a wide range of clones and to correlate well with brightness and briskness, taste is generally recognized as being secondary to aroma. In recent years much use has been made of the semi-quantitative flavour index (see page 178). There appears to be a good correlation between high flavour index and good quality tea in African-grown plants, but some scepticism remains elsewhere. When assessing clones, it is necessary to be aware that climate also affects the value of the flavour index. The terpene index (see page 186) may also be used for assessing clones, but its value is limited to teas where strong floral notes are required.

4.5.2 Green tea

The inactivation of leaf enzymes as soon as possible after picking, means that flavour and aroma is almost entirely derived from compounds present in the green leaf, or compounds formed during firing. Further, the strength of both flavour and aroma is markedly less than that of black or semi-fermented tea.

(a) Flavour

Three major components of green tea flavour have been identified, astringency, bitterness and brothy flavour. Polyphenols, predominantly catechins (see pages 174–7) are responsible for *ca*. 75% of astringency and bitterness. Caffeine also contributes to bitterness by complex formation with catechins.

The brothy flavour of green tea is associated with leaf amino acids. As many as 18 amino acids may be involved in determining the brothy flavour, but L-theanine (N_5-ethyl glutamine), which accounts for *ca*. 40–70% of the total amino acid content of fresh tea leaves is dominant.

(b) Aroma

The aroma of green tea has been identified as consisting of fresh/flowery notes, derived from green leaf components, and roast notes derived from compounds produced during firing. Despite the lack of a fermentation, many of the compounds producing the fresh/flowery notes (Table 4.8) in green tea are also important in black tea. Linalool and, to a lesser extent, geraniol and *Z*-3-hexenyl hexanoate are largely responsible for flowery notes, while *Z*-3-hexenol and its esters and *E*-2-hexenal are important contributors to the fresh aroma. These compounds are formed by oxidative degradation of leaf lipids (see Figure 4.10, page 183).

The strength of the roast aroma is dependent on processing conditions and is greater in pan-fired tea than in Sen-cha tea. The characteristic aroma is largely produced by pyrroles and pyrazines, which are formed during pan-firing or drying by reaction between the Maillard intermediate products, α-dicarbonyls and free amino acids through Strecker degradation. In some types of tea (Gyokuro and Ceremony) dimethylsulphide, produced by degradation of *S*-methyl methionine is also an aroma-impact compound.

Table 4.8 Compounds associated with fresh/flowery notes in green tea

Group	*Examples*
Hydrocarbons	β-myrcene
Alcohols	1-penten-3-ol
	benylalcohol
	linalool
	geraniol
	E,*Z*-furanoid
	nerolidol
Aldehydes	*E*-2-hexenal
	benzaldehyde
	Z-jasmone
	β-ionone
Esters	*Z*-3-hexenyl formate
Miscellaneous	methylsalicylate

(c) Effect of clone and agronomic practice

The greater importance of leaf constituents in determining the character of green tea means that the effect of clone and agronomic practice are greater than with black tea. Catechin and amino acid content vary according to clone and there are also differences in the ratio of L-theanine to other amino acids, such as L-arginine. Variation in the content of glucosides (and thus monoterpene alcohols) are responsible for clonal differences in the flowery notes of green tea. The relative quantities of linalool and geraniol are of particular significance and the ratio is often defined using the terpene index:

$$\text{terpene index (TI)} = \text{linalool}/(\text{linalool} + \text{geraniol})$$

Cam. sinensis var. *assamica* contains a high proportion of linalool and some clones have a TI of *ca.* 1.0. In contrast, some *Cam. sinensis* var. *sinensis* cultivars contain very little linalool and have a TI as low as 0.1.

Highest quality Sen-cha tea is made from the first crop (spring harvested), which has a high content of L-theanine and other amino acids and a relatively low content of catechols. As the season progresses, the amino acid content falls and the catechol content rises and this is reflected in decreasing quality of the made tea.

Sen-cha tea is made from leaves grown unshaded, and has a relatively low amino acid content and a relatively high catechol content, when compared with shade-grown Gyokuro and Ceremony teas. Growth in shade also increases the *S*-methyl methionine content and consequently the contribution made to the aroma complex by dimethylsulphide.

4.5.3 Semi-fermented tea

Flavour and aroma of semi-fermented teas is dictated by the extent of fermentation and, for this reason, oolong has a considerably stronger flavour than pouchong, which undergoes a significantly shorter fermentation. The choice of clone is also important and oolong tea is made from clones which contain high levels of aroma precursors.

(a) Flavour

Neither oolong, nor pouchong, teas contain theaflavins. This is probably a consequence of the relatively slow fermentation. In oolong tea, however, the catechin content, especially (−)-epigallocatechin and (−)-epigallocatechingallate, falls during fermentation and there is some formation of thearubigins.

(b) Aroma

The slow fermentation during production of semi-fermented teas means that the rate of fatty acid oxidation is low and products, such as *E*-2-hexenal and *Z*-3-hexenol (see page 182), are present only in small quantities. Both oolong and pouchong tea, however, contain relatively large quantities of linalool and its oxides, benzylalcohol, 2-phenylethanol, geraniol, nerolidol, indole, jasmin lactone and methyl jasmonate. Differences in fermentation length is reflected in the relative quantities of aroma compounds. Oolong tea contains higher levels of linalool, geraniol, benzylalcohol and 2-phenylethanol than pouchong tea, but lower levels of nerolidol, indole, jasmine lactone and methyl jasmonate.

Firing appears to be relatively unimportant in the production of flavour and aroma compounds in semi-fermented tea. Significant quantities of 1-ethyl-pyrrol-2-aldehyde have, however, been detected in both oolong and pouchong tea.

4.5.4 Pickled tea

Relatively little attention has been paid to the flavour and aroma compounds present in different types of pickled tea. The relatively uncontrolled nature of the process also means that considerable variation is likely. In general flavour and aroma compounds are of two types, the first similar to those found in semi-fermented tea and derived by enzymic action, the second probably microbial in origin. Acetic acid and other volatile acids are important products of microbial metabolism in pickled tea and are largely responsible for the characteristic sour flavour. Secondary alcohols, such as 2-butanol, 2-heptanol, (*E*)-4-hepten-2-ol and 1-octen-3-ol, as well as acetoin are also present in significant quantities. Moulds are probably involved in production of 1-octen-3-ol, which has a musty odour. A component of the flavour spectrum of soy sauce, 4-ethyl phenol, is probably derived from coumaric acid, while phenols, such as 4-methylguaiacol, 4- ethylguaiacol and 4-propylguaiacol, are products of the microbial degradation of ferulic acid. Phenols impart a characteristic smoky flavour to pickled tea.

EXERCISE 4.1.

The feasibility of tissue culture of *Camellia* has been demonstrated on a laboratory scale, but there is currently little or no commercial impetus for further development. Consider carefully both the potential benefits of tissue culture and the problems associated with the technique on a large scale. What particular problems are likely to be encountered in scaling-up tissue cultivation of *Camellia* from laboratory to commercial scale? Consider the economic viability of tissue culture (compared with conventional cultivation) in the following situations.

1. A traditional tea-producing region, where a slight change in climatic conditions has led to major loss of output due to fungal disease.
2. A traditional tea-producing region, which wishes to significantly raise the quality of tea produced.
3. A new tea-producing area, where the climate is only marginally suitable (very low rainfall), but where a clone producing made tea of high and distinctive quality has been developed.

Further information on plant tissue culture may be obtained from Chen, Z. *et al.* 1989. ***Handbook of Plant Cell Culture***, vol. 6. ***Perennial Crops***. McGraw-Hill, New York.

EXERCISE 4.2.

A number of charitable organizations in the industrialized world are attempting to raise wages and improve conditions for tea pickers. You are employed by a charity, which has established a marketing organization for tea, profits of which are returned to the producing region. Develop a marketing strategy for the tea, identifying target socio-economic groups and defining an ideal product profile. Consider the relative merits of direct sales through existing charity shops and mail order systems, attempting to establish your brand in the independent grocery sector against existing national brands, or producing exclusively for a leading supermarket chain.

EXERCISE 4.3.

Production of teekvass is currently on a domestic or very small commercial basis. You are employed as microbiologist by a recently established company, which is developing a range of foods and beverages based on traditional fermented products. Teekvass has been identified as a possibility, but preliminary trials using an undefined inoculum have resulted in a product of highly variable quality and character. As a result you have been asked to develop pure culture starters from undefined inocula obtained from several types of teekvass. Outline the stages required for development of pure culture starters in this situation. Describe how you would determine the relationship between the different components of the microflora and the desirable properties of teekvass.

5

COFFEE

OBJECTIVES

When you have read this chapter you should understand

- The nature of the coffee tree, *Coffea* spp.
- Cultivation and harvesting of the coffee tree
- The processing of green coffee
- The technological objectives of grinding and roasting
- The manufacture of instant and decaffeinated coffee
- Coffee substitutes
- Quality assurance and control
- The physiological effect of coffee consumption
- The major constituents of green coffee
- Chemical changes during the processing of coffee

5.1 INTRODUCTION

Coffee, by repute, was discovered in the stone age and the stimulatory effects were noted early, possibly by observing the effects on animals grazing wild coffee berries. In historic times, coffee drinking became well established in Arabia, the name being derived from the Arabic *qahwah* (a poetic name for early wine). The coffee habit spread westward into Europe, becoming established in the UK in the 17th century. Coffee houses were established in Vienna and spread across Europe to London. Coffee houses became popular places of business and the insurance institution, Lloyd's developed from a meeting place of underwriters and ship owners. In the UK, coffee became replaced by tea as the favoured beverage during the 18th century, but coffee has remained dominant in continental Europe and North America. Coffee has always being recognized as being more difficult to prepare than tea and the US has, for many years, been at the forefront of developments in 'instant' coffee. The availability of this product was, to some extent, responsible for the increase in coffee consumption in the UK following the second world war.

The high caffeine content of coffee and concern over the effects of heavy consumption, has led to the development of decaffeinated coffee, while caffeine-free coffee substitutes are also available. These, however, seem unlikely to have the same impact on the market as herbal and fruit teas.

BOX 5.1 **A sense of value**

Coffee is not bought or consumed for nutrition. It may be purchased by bag, pound or cup, but weight (the quantity measure) has value only insofar as it has acceptable flavour (the intensity measure). Coffee has only one value: to give the consumer pleasure and satisfaction through flavour, aroma and desirable physiological and psychological effects. (Sivetz, M. and Foote, H.E. 1963. *Coffee Processing Technology*, vol. 1. AVI, Westport, CT).

5.2 AGRONOMY

5.2.1 The genus *Coffea*

Coffea is a member of the family *Rubiaceae* and comprises evergreen trees and shrubs. Funnel-shaped flowers are followed by a pulpy fruit, the 'cherry', which contains two seeds, the coffee beans. *Coffea* grows wild in Africa and Madagascar and the genus includes a large number of species. Only three, *C. arabica*, *C. canephora* (Robusta) and *C. liberica* have been successfully used in commercial cultivation. *Coffea liberica*, however, was devastated during the 1940s by epidemics of tracheomycosis, due to infection by *Fusarium xylaroides*, and commercial growth of this species has effectively ceased.

Both *C. arabica* and *C. canephora* are available in a large number of varieties and cultivars. A number of both intra- and interspecific hybrids have been developed, of which the Arabica–Robusta hybrid, Arabusta, is intended to produce a coffee of better quality than Robusta, while being more vigorous and disease resistant than Arabica. The beans are also of low caffeine content. Progress has been made, but Arabusta can be of variable performance, and also of poor bean size and coffee quality. Cultivation of this and other hybrids is currently restricted to small areas.

Although only *C. arabica* and *C. canephora* are grown commercially, the gene pool of *Coffea* consists of all species. Species such as *C. stenophylla* and *C. congenis* are thus of importance as sources of novel genetic material in breeding improved strains of *C. arabica* and *C. canephora*.

5.2.2 Cultivation of *Coffea*

(a) Coffea arabica

Coffea arabica may be cultivated over a wide geographical area delineated by the Tropics of Cancer and Capricorn. The plant is generally considered to be an upland species and the optimum temperature range is 15–24°C, photosynthesis being reduced at temperatures above 25°C. *Coffea arabica* is relatively disease-prone and the incidence of leaf rust also increases with increasing temperature. This is often the limiting factor to coffee production in lower altitude zones of growing areas. In common with all *Coffea* species, *C. arabica* is also adversely affected by low temperatures and is very frost-sensitive. Trees may also be damaged by high wind and low humidity, and it is necessary to use shade trees, wind breaks and frost protection measures.

The average rainfall in most areas of cultivation of *C. arabica* is 1500–2000 mm although, in East and Central Africa, cultivation is possible with a rainfall as low as 1000 mm providing that irrigation is available. A period of moisture stress before flowering is beneficial in concentrating flowering and providing a defined harvest season. The ideal pattern of rainfall is, therefore, even rainfall over a 9 month wet season and a 3 month dry season. Equitorial Kenya, northern Tanzania and Colombia, however, have two wet and two dry seasons resulting in double cropping.

Coffea arabica can be grown in a variety of soils of different geological origin. Volcanic soils of high base exchange capacity tend to be most suitable for all *Coffea* species, but soil must also be of the correct physical characteristics. Acid soils (pH 5.5–6.5) are preferred. Soils should be deep, friable, open-textured and permeable. Heavy or poorly drained soils are unsuitable due to the high oxygen requirements of the roots, while light, sandy soils lack sufficient water holding capacity.

Coffea arabica is predominantly self-pollinating and homozygous and is normally propagated by seed. Vegetative propagation by rooting green wood (*cf.* Robusta, pages 207–9), has been proposed for the introduction of disease-resistant hybrids in Kenya and Brazil and tissue culture has also been proposed as a means of rapidly bulking-up elite material. In Guatemala and El Salvador *C. arabica* scions are sometimes grafted onto *C. canephora* rootstock as a means of introducing resistance to nematode worms. Application of grafting is limited by the need for highly skilled labour.

Seeds are obtained from ripe coffee cherries taken from selected trees. The cherries are pulped and fermented (see pages 214–16) to remove mucilage, the seeds either being planted immediately or dried for use at a later date. Under normal storage conditions, a satisfactory level of viability is retained for *ca.* 6 months, but this period can be extended by controlling seed moisture content at 41% and the temperature at 15°*C.*

Coffee is grown as seedlings before transfer to the field. Polythene bags filled with potting mixture are now often used in place of traditional nursery beds. The use of bags is, however, expensive and some use is made of containers manufactured from local materials, such as banana leaves. The optimum size of seedlings for planting out remains a matter of debate. Smaller seedlings are cheaper to produce, require less nursery space and are planted out sooner. Large seedlings, however, survive better in dry conditions and come into production more rapidly.

Even in well established coffee growing areas it is necessary to exercise care when choosing new sites for field cultivation. In addition to ensuring the suitability of the soil, it is necessary to consider access to the site by mechanical equipment, if used, and

* Pregermination, a technique which involves keeping seed moist until the radicle appears and then planting, has been suggested as a means of ensuring equal development of seedlings. In Latin America, however, a simple and effective method has been developed in which seeds are sown thickly onto prepared beds and covered with *ca.* 1 cm depth of soil and a thin layer of mulch. The mulch is removed when the seedling emerges and the plant pricked out when the seedling is *ca.* 2 cm high with the cotyledons still enclosed by the parchment skin.

potential problems of erosion. The aspect of the slope is also important and plants facing the prevailing wind, or the afternoon sun, are likely to be adversely affected during the dry season. Frost pockets should be avoided and planting should be arranged so that there is no interference with the flow of cold air down the hill.

Depending on local practice, land may be cleared before planting with seedlings. After clearance it is usual to plant with annual crops or green manure for 1–2 years to eradicate perennial weeds and regenerating bush. Ground is not usually cleared in the high rainfall areas of Central America and Colombia. Native vegetation is slashed along the planned line of planting and weeds removed from a small circle around the site of each seedling. As the *Coffea* tree develops, the cover provided by native vegetation is gradually reduced.

Planting of seedlings must be timed to permit establishment before the onset of the dry season. In East Africa planting holes or trenches are used, the subsoil being weathered for 1–2 months and then refilled with a mixture of topsoil, phosphate fertilizer and cattle manure or compost. To minimize problems with *Fusarium* infections, the depth of planting must be the same as that in the nursery and mulch should be kept clear of the stems. Insecticides are applied at the time of planting where soil pests are a significant problem.

Protection from the sun is required, especially for bare-root seedlings. Short-lived legumes may be planted to give shade for up to 2 years. Properly managed, there is no competition with the coffee plants, while benefits are obtained from leaf-fall and enhanced soil nitrogen status. Spreading cover crops can be used to provide shelter in areas where full shading is unnecessary. Temporary shade shelters may also be contrived using grasses and leaves. Grasses are also used, where necessary, to construct frost shelters during the first winter after planting out. Frost protection may also involve earthing up of the stems or planting seedlings in groups. Wind can present a serious problem in exposed areas and in these situations shelter trees are planted at right angles to the prevailing wind.

The spacing at which seedlings are planted is primarily dictated by the need to maximize production by the adult trees. Under ideal conditions this implies a complete canopy to make maximum use of sunlight. Other factors must, however, be considered, including access for harvesting and other operations, the cultivar and the pruning system adopted. Two basic planting arrangements exist, conventional and high- density spacing.

Conventional spacing originally involved planting in straight rows, but this led to problems of soil erosion and contour planting is now more common. In this system trees are planted 2.0–2.75 m apart in the row, with rows separated by some 3.0 m. Contour planting is combined with other procedures, such as mulching, to minimize erosion. The major disadvantage of conventional spacing is the low yield in the early years after planting, when only a small proportion of the available soil and air space is utilized. Full production is not reached until 6–7 years after planting out, when the third or fourth crop is taken. This problem may be overcome by high-density spacing, in which trees are planted at much closer intervals. The high level of ground cover at an early stage also limits weed growth and reduces soil erosion, but high-density spacing can lead to excessive density. This requires either heavy pruning or thinning by removal of some of the trees.

A number of systems of high-density spacing have been investigated. Hedge planting, closer planting in rows separated by a wider avenue, is a popular system. Single-, double- and triple-rowed hedges have all been planted, but highest yield is obtained by single-row planting, which minimizes problems due to over-shading and competition. For normal sized cultivars, spacing of 1.5 m between the trees and 2.75 m between the rows gives high yield and permits access by tractor for spraying. Narrower spacings are used with dwarf cultivars.

Where coffee is planted at normal spacing, an early return on capital can be obtained by interplanting with annual crops. In large-scale coffee growing, yield is of greatest importance and interplanted crops should not interfere with the coffee tree. In contrast, the food crop is of greatest importance in smallholdings, where intercropping between mature trees is practised and a reduction in

* The adoption of high-density spacing has led to dramatic increases in yield in some countries of South America. In Kenya, however, yields were initially depressed because of a high incidence of coffee berry disease and leaf rust in trees planted at narrow spacings. Material resistant to coffee berry disease is now available and the use of high-density planting appears more promising.

coffee yield is accepted. Beans are suitable for intercropping in large-scale coffee cultivation, while many crops are grown between coffee trees on smallholdings, including yams, sweet potatoes and bananas. Interplanting of bananas is particularly common in Africa where the fruit is a dietary staple, but these plants are highly competitive with coffee trees, especially during droughts. Bananas are acceptable as shelter trees, but interplanting is not recommended.

Although the use of temporary shade plants is common in many countries, the use of shade for mature trees is controversial. Both advantages and disadvantages exist (Table 5.1), and decisions concerning shade should be largely based on climatic considerations. Ideally, shade trees should reduce light intensity, evenly, by *ca.* 25%. Deep rooted trees should be used to minimize competition for moisture and nutrients. Further, the use of deep rooted trees means that nutrients obtained from lower soil levels are transferred to the topsoil via dead leaves. The most important shade trees are *Inga* spp. in Central America and *Albizzia* spp. in Africa and Asia. More recently, ecological considerations have led to widespread use of *Leucaena leucocephylla*, which plays an important role in watershed conservation, and is a source of fire wood and livestock fodder. Shade can be provided cheaply by leaving some of the existing forest trees during ground clearance for coffee, but this method is often unsatisfactory due to competition with the coffee trees.

Table 5.1 Advantages and disadvantages of shade

Advantages
reduces die-back
can improve flowering and fruiting
increase leaf area
water loss reduced by lowering of wind speed and temperature
leaf fall a valuable addition to mulch
important source of fuel and timber, justifying loss of yield in small holdings
Disadvantages
compete for water, especially where rainfall low or marginal
compete for nutrients
can be foci for root diseases
may be prone to lightening strike

In many cases, shelter from wind is only required for the first 1–2 years after planting. Particularly exposed sites may, however, may require permanent shelter. *Grevillea robusta* is widely used for permanent windbreaks.

Growth of *C. arabica* in areas of low rainfall necessitates the adoption of special procedures, including mulching and irrigation. Mulching has both advantages and disadvantages (Table 5.2), but in low rainfall areas the advantages greatly outweigh the disadvantages and significant yield increases result. There may, however, be a reduction in coffee liquor quality. Little or no benefit to yield is obtained in areas where rainfall is high. In large-scale cultivation, mulch crops are specially grown. These include grasses such as elephant grass and Sudan grass. Alternatively natural grass may be harvested for mulch, or wheat or maize straw used. Mulching may be combined with a system of weed control by cultivation-free application of herbicides. In this case the additional mulch is provided by the dead weeds. This results in a long-lasting layer of gradually decomposing mulch, which encourages the development of feeder roots and increases nutrient uptake.

Irrigation is generally considered to be necessary where average annual rainfall is in the order of 1000 mm and is thus common practice in the coffee growing regions of Africa. Critical times when water stress must be minimized are flowering time, the period of berry expansion and the period of dry matter development. Irrigation at other times may increase yield, but the practice is expensive and can only be justified economically in defined areas of low rainfall. Irrigation does not negate the need for mulching,

Table 5.2 Advantages and disadvantages of mulch

Advantages
reduces moisture loss
provides nutrients and stimulates root growth
minimizes weed growth
reduces erosion and soil loss after heavy rain
Disadvantages
continuous mulching with same material leads to nutritional imbalance and increased fertilizer requirement
uneconomic if mulch crop grown on land suitable for coffee production
fire risk in dry season

and the highest economic return is obtained by use of irrigation in conjunction with mulch application, pest and disease control and fertilizer treatment.

In low rainfall areas, irrigation of seedlings is always necessary and proximity of an adequate water supply is an important factor in selecting a suitable site for a nursery. The scale of operations is relatively small and hand irrigation is usual, although sprinkler and other automatic systems are increasingly common. Irrigation may also be required, in the dry season, for the first 2 years after planting out. The extent of irrigation at this stage is critical and should be the minimum necessary to permit survival. This encourages the plants to develop a deep, extensive root system.

Various irrigation techniques are available. Trickle irrigation is most suitable, but the capital cost of installation is very high and either surface irrigation or sprinkler systems are most common. Surface irrigation can be of low capital cost and is economical in terms of water use. Irrigation of young coffee requires nothing more than buckets and a water tanker, but the mature crop requires a water distribution system using either furrows or pipes and hoses. Sprinkler systems have a relatively high capital cost, but a lower labour cost than surface irrigation systems. The absence of furrows means that mechanized equipment is unimpeded in use. Sprinkler systems are also suitable for use in hilly areas and, provided that the rate of application is not excessive, minimize soil erosion.

Application of fertilizer is required for succesful cultivation of *C. arabica*. Nitrogen is required for the vegetative development of the tree and affects the cropping level. Unshaded coffee is more prone to nitrogen deficiency and shows a greater response to application of fertilizer. Potash (K_2O) is of major importance, especially with respect to fruit development and the requirement is high. Soils of volcanic origin usually contain sufficient potash to support continuing coffee growing and fertilization is also not required where mulching is practised. In other soils, soil potash content often becomes limiting to yield after a few years cultivation. Phosphate is required for root growth and development of both flower and wood buds. The requirement for phosphate is low, but deficiency is relatively common in dry conditions. This is a consequence of unavailability due to the fixation of phosphate in the soil. In many cases, however, response to fertilization with phosphate has been poor and application is essentially a precautionary measure. Loss of added phosphate by fixation is a serious problem where soils are acidic. Under these conditions, phosphate should be applied in combination with mulching which minimizes fixation. Phosphate may also be applied as a foliar spray, which has a higher take up than soil application. Calcium and magnesium are both important nutrients. Calcium is required for the development of terminal buds and flowers. Many types of fertilizer contain calcium, which is also applied as lime to correct soil acidity. Application of special calcium fertilizer is not, therefore, required. Magnesium is required for efficient photosynthesis and deficiency is common. Deficiencies affect coffee quality rather than yield. The antagonistic effect between calcium, magnesium and potassium, means that fertilizer applications should achieve an ideal balance between these three elements. Rate of application of fertilizer is determined by soil analysis, the desired ratio K:Mg:Ca being 1:2–3:4–6. Deficiencies of minor elements are known, and boron, sulphur, iron, zinc and manganese deficiencies have all been described. Iron and manganese defiencies are associated with soils of high pH value and are best corrected by soil acidification rather than fertilizer application.

High concentrations of some minor elements lead to toxicity. Manganese toxicity is usually associated with aluminium toxicity and occurs in soils of extremely low pH value. Reduction of acidity by application of lime is an effective means of controlling this problem. The extensive use of copper-based fungicides can lead to toxicity due to build up in the soil following repeated applications.

Pruning is an essential part of the management of *C. arabica*. Pruning has a number of objectives, including the removal of dead and unproductive stems and branches and the encouragement of new crop-bearing wood. Pruning is also used to maintain the trees in manageable shape and to maintain a fairly open canopy of foliage. A dense canopy is undesirable, since pests and diseases are

* *Coffea arabica* has certain physiological characteristics which, to a considerable extent, dictate the requirements of the pruning regimen. Shoot growth is of two types, orthotropic shoots, which grow vertically to produce new shoots or plagiotrophic shoots, which produce new horizontal branches. Orthotropic stems can produce further vertical shoots or horizontal branches, but not fruit. Plagiotropic branches can produce fruit or horizontal branches, but not vertical stems. In either case apical dominance exists and suppression of secondary growth can be removed by cutting off the tips of stems and branches. Flower buds are normally produced only on wood in its second year; production is encouraged by light.

encouraged and flower end initiation is inhibited. A further important role of pruning is to regulate the biennial production cycle and to reduce dieback resulting from depletion of starch reserves in the heavy cropping year.

Several methods of pruning exist. Choice depends on many factors including spacing, growth of normal or dwarf cultivars, rate of growth, presence or absence of shade trees, incidence of disease and infestation and labour costs. Single stem pruning is the traditional method for shade grown trees but is labour intensive and yields are relatively low. Umbrella-shaped trees are produced with dense foliage in which pest and disease control is difficult. Single stem pruning involves restricting the height of the tree to 1.5–2.0 m and developing a permanent framework of primary branches. The crop is borne on primary and secondary branches.

Multiple stem pruning results in higher yields where coffee is grown unshaded and with high rates of fertilizer application. It is also a more economical process than single stem pruning. Various systems of multiple stem pruning are used, but the basic method is to bring up the young plant with between two and four vertical stems, which are replaced after 3–7 years, depending on the growth rate of the tree. Under conventional multiple stem pruning, stems are allowed to reach a height of 3–4 m, but the capping system was introduced in which the height of stems is limited to 2 m. Conventional multiple stem pruning involves yearly removal of lower primary branches after two crops have been borne. Heads of excessive height must also be removed and replaced by new heads. The length of the first pruning cycle varies from 5 to 7 years, and subsequent cycles from 5 to 7 years with dwarf cultivars and 3–5 years with normally sized cultivars.

Capped multiple stem pruning is initially similar to conventional, except that lower primary branches are not removed after the second crop. After the third crop trees are cut back from *ca*. 2.5 m in height to *ca*. 1.75 m. Pruning then follows the pattern for single stem pruning. Secondary growth is regulated and the centre of the tree opened up to prevent shading and to permit pest control. As with conventional single stem pruning, the tree tends to become umbrella-shaped and yield falls. These problems can be overcome by conversion to a new pruning cycle 3–5 years after capping. Problems occur, however, where coffee is planted at close intervals, since sucker initiation and development is inhibited by shade from existing heads. In this situation, a system known as rotational stumping is used. Stumping involves cutting back trees to a height of *ca.* 50 cm. Sucker development is stimulated by exposure to light and two or three vertical stems are allowed to develop. No pruning is required except for removal of unwanted suckers. After 3–5 years the tree is again stumped and the cycle repeated. Where stumping involves removal of all branches (clean stumping), the tree is entirely dependent on its starch reserves, until photosynthesis recommences with the development of new growth. Careful timing is required to ensure that starch reserves are high when stumping is carried out. An alternative procedure to clean stumping is to leave any primary branches below 50 cm to act as 'lungs' and to continue photosynthesis during the recovery period. In any circumstances, sucker growth tends to be more vigorous when 'lungs' are retained. Stumping means that trees carry no crop for up to 2 years and then crop heavily during the subsequent 2–3 years. It is common practice to maintain even cropping by stumping equal proportions of the total number of trees each year.

Weeds can be a serious problem in coffee plantations due to competition for water and nutrients. Perennial grasses and sedges are the most serious and also the most difficult to eradicate, sedges producing root exudates which are highly toxic to *C. arabica*. Weed control is required to avoid adverse effects on the coffee crop, but complete removal is often undesirable, since the risk of soil erosion may be increased.

Ideally the perennial grasses, couch, star grass and lalang, should be removed from the ground before coffee trees are planted. Once established control is difficult as the grasses spread by subsurface rhizomes. Removal of the rhizomes by digging is possible, but is labour intensive and can cause significant damage to the root system of the coffee tree. Application of systemic herbicides is expensive, but effective if correct procedures are followed. Dalapon has been widely used in the past, but glyophosphate is now popular and, although expensive, also controls sedges and broad leaved weeds. Sedges are a particular problem in East Africa and tend to colonize irrigated, unshaded coffee after eradication of perennial grasses and broad leaved weeds. Sedges spread by subsurface rhizomes and corms, which persist in the soil. Broad leaved weeds are the least serious, and where soil moisture and nutrient supply is adequate, total removal is not necessary. Growth is controlled during the rainy season by cutting back, which also prevents seeding. The weeds are then dug out at the beginning of the dry season and used as mulch. An alternative procedure, which has a number of advantages, is herbicide treatment without cultivation.

A great many types of pest may infest coffee trees, of which insects are the most serious. The greatest variety is found in Ethiopia, the origin of *C. arabica*. In the natural forest environment, however, an equilibrium exists between the coffee tree, its pests and parasites and predators of the pests. For this reason the epidemics encountered in countries where no similar equilibrium exists are rare in Ethiopia.

Infestation of roots by nematodes is widespread in all coffee growing areas. The significance of infestation of mature trees by the root knot nematode (*Meloidogyne africana*) in African countries is uncertain, but severe damage is caused in Latin America by root knot nematodes (*M. coffeicola*) and root lesion nematodes (*Pratylenchus coffeae*). Control of nematodes in nurseries is of particular importance and is primarily achieved by ensuring soil has not previously been used for coffee cultivation. Soil sterilizing agents such as dazomet may also be used. Some protection is also given to seedlings and young trees in the field by systemic insecticides, such as carbofuran, which are also effective against sucking and some leaf-eating insects.

A number of soil insects attack the base of the stem of nursey seedlings and young trees. These include the larvae of beetles and moths, snails and crickets. The persistent insecticide aldrin is used to provide protection in nursery beds. In the field, protection is obtained by painting the base of the stem with dieldrin solution. This procedure also indirectly controls scale and mealybugs by preventing attention by ants. This, in turn, leads to control by natural predators.

Stem and branch pests are usually the larvae of beetles (borers), which deposit eggs in the tree bark. Borers are of particular prevalence in East Africa, where the white borer (*Anthores leuconotus*) devastated low altitude coffee producing areas in the 1950s. This pest is now controlled by painting the base of the stem with dieldrin solution. The yellow-headed borer (*Dirphya negricornis*) is also a serious pest, which can seriously weaken the stems of coffee trees pruned by the multiple-stem system. Control lies primarily in shortening the pruning cycle to effect more frequent replacement of stems. Larvae in the stem may be treated directly with dieldrin solution.

Scales and mealybugs are the main pests attacking new shoots, but can usually be controlled by measures against soil insects (see above). Biological control was successfully used in control of the mealybug *Planococcus kenyae*, which caused catastrophic damage to Kenyan plantations during the 1930s. Biological measures are being implemented against another mealybug, *P. patersonii*, which has now become a serious pest in part of Kenya.

The most serious pests attacking the leaves of coffee trees are larvae of the *Lepidoptera*. The most serious damage, leading to severe defoilation, is caused by species of leaf miners (*L. meyricki* in Africa; *L. coffeella* in Central and South America). Infestation can be controlled by soil application, during the wet season, of a systemic insecticide such as carbofuran. Leaves are eaten by several species of caterpillar but most are of minor importance. Misuse of organophosphorous insecticides in East Africa, however, has led to serious outbreaks of infestation by the giant looper (*Ascotis selenaria reciprocaria*).

Antesia bugs (*Antestiopsis* spp.) are serious pests, causing shedding of flower buds and green berries and fan branching of stems and branches. Antesia bugs also introduce the mould, *Nematospora*, into the developing bean leading to rotting. A population of only two antesia bugs per tree can cause crop losses of 30–50% and regular testing programmes are in operation to enable control measures to be implemented when a critical level of infestation is reached.

Developing berries are also subject to attack by larvae of the Mediterranean fruit fly (*Ceratitis capitata*) and the coffee borer beetle

---

* Larvae of the Mediterranean fruit fly feed on mucilage within the developing berry. For many years, the presence of larvae in berries remaining on the tree was thought to be harmless. Subsequently it was found that larvae were responsible for the mysterious 'hidden stinkers' which caused so much harm to the Kenyan coffee industry. 'Hidden stinkers' were beans which developed a foul taste and faded appearance many months after processing. The underlying cause was introduction of micro-organisms with Mediterranean fruit fly larvae, leading to death of the embryo and tissue deterioration.

(*Hypothenemus hampei*). In each case direct loss occurs due to premature berry drop, while remaining infected berries are defective and adversely affect the quality of the coffee. The most effective control lies in removal of fallen berries from the vicinity of trees.

Coffea arabica is susceptible to a large number of diseases, most of which are caused by moulds. Control is usually possible using various fungicides, but such procedures are difficult, expensive and, in many cases, effective in the short-term only. Long-term solutions can only be offered by the breeding and selection of resistant cultivars and in some cases research programmes in this area have met with a considerable degree of success.

Roots of coffee trees planted on newly cleared ground may become infected by *Amillaria mellea*, or *Rosellinia* spp., which originate in the decaying roots of forest trees. The incidence may be minimized by procedures during clearing. In some parts of Africa, wilting and death of trees results from root infection by *Fusarium solani*. Trees grown in unfavourable soil are most susceptible and infection usually follows damage during hand weeding. A second species of *Fusarium*, *F. stilboides*, is responsible for three types of bark disease. *Fusarium* bark infection is economically significant in East and Central Africa, but there is some prospect of developing resistant strains. *Fusarium oxysporum* has been associated with a disease of *C. arabica* in Tanzania, Lyamungu dieback. Exhaustion of starch reserves is a major factor and a causative role for *F. oxysporum* is now doubted.

Ceratocystis fimbriata is responsible for fungal disease of coffee tree stems in Central America. The fungus enters through wounds in the bark and infected trees must be removed and destroyed. Fungal disease may also affect individual branches. Examples are thread blight (*Pellicularia koleroga*) and pink disease (*Corticium salmonicolor*). Although most stem and branch disease of coffee trees is fungal in origin, *Pseudomonas syringae* pv. *garcae* causes bacterial blight, a serious disease in parts of Africa and Central America. The bacterium is susceptible to copper-based fungicides, but not to most organic fungicides.

The foliage disease, leaf rust, is caused by the fungus *Hemileia vastatrix*. Leaf rust first appeared in Sri Lanka during the mid-19th century and has since spread to all coffee growing regions, finally

BOX 5.2 From the ashes

Sri Lanka (Ceylon) is renowned as a major tea growing country. The tea industry grew, in many cases, quite literally, from the ashes of the coffee industry. After its appearance, leaf rust spread rapidly and devastated coffee crops. The price of coffee on the world market rose steeply and this accelerated the trend in some countries toward tea drinking. In Sri Lanka, plantation owners faced bankrupcy, while workers, deprived of any source of income, faced near starvation.

appearing in Colombia in 1984. Leaf rust has been responsible for the abandonment of Arabica coffee cultivation in Sri Lanka, Indonesia and some other areas and its replacement with Robusta coffee or tea.

Leaf rust is spread by spores produced from lesions on the underside of the leaf. Infection occurs only during the wet season and the disease largely affects coffee grown at low altitudes, where wet season temperatures are high. Losses due to leaf rust can be minimized by spraying with fungicides, at 4–6 week intervals. Resistant cultivars are now available, but are unpopular due to their relatively low yield.

A number of other micro-organisms cause foliage disease of *C. arabica*. These are not generally of economic significance, although South American leaf spot, caused by *Mycena citricola*, can be a serious problem in shade-grown trees in South and Central America. The incidence of the disease is reduced by lowering the amount of shade and spraying with a copper-based fungicide is an effective control measure.

Coffee berry disease is an economically important disease, which was first recognized in Kenya during the 1920s, but which has since spread to most African countries growing *C. arabica*. The causative organism is a strain of *Colletotrichum coffeanum* of

* 'Tonic' spraying with copper-based fungicide is a routine procedure, in some parts of East and Central Africa, even though leaf rust is not present. Spraying is effective in preventing premature leaf fall due to unknown causes and, in some localities, increases yield by over 100%.

enhanced virulence, which invades the maturing berry. Control is by regular spraying with fungicides, of which captafol is most effective. Copper-based fungicides are also suitable, but systemic fungicides, such as benomyl, are not recommended due to development of resistance. Resistant cultivars are now available. Berries may also be affected with other fungal pathogens. Less virulent strains of *C. coffeanum* cause brown blight of berries as well as lesions on leaves and green wood. Infection with brown blight has an adverse effect on coffee quality, since affected berries are hard to pulp and beans are consequently prone to damage.

(b) Coffea canephora (Robusta coffee)

Coffea arabica and *C. canephora* are closely related and many aspects of husbandry are similar. Robusta coffee, however, is strictly cross-pollinating (allogamous) and propagation of high quality, 'elite', material must be vegetative to ensure that desired characteristics are reproduced. The ideal situation is the establishment of clonal fields of uniform and high quality plants. Various methods of propagation may be used, including cuttings, cuttings grown *in vitro*, grafting and layering. Of these, the use of cuttings is the normal commercial practice.

For commercial purposes, cuttings consist of a fragment of the upright (orthotrophic) stem which bears a piece of leaf and a dormant bud. Cuttings are removed from plants raised in clonal cutting gardens and rooted in propagators before planting out in plastic bags for hardening. A clonal cuttings garden reaches full production *ca.* 18 months after planting and provides up to 200 cuttings per tree. Orthotropic cuttings are taken from green, or non-woody, suckers. The terminal bud is suppressed by a preliminary capping some 12 days before taking the cutting and *ca.* 33% of the foliar lamina is cut off. The suckers are then cut into lengths immediately above each leaf insertion and *ca.* 5 cm below. Each length is then split longitudinally into two 'split' cuttings, which are ready for transfer to the propagator.

Propagators consist of simple brick bins covered by glass or plastic sheeting. Shade is arranged to reduce the light intensity by *ca.* 75%. The growing substrate must be well aerated to permit development of callusses and roots and yet must also retain sufficient moisture without any risk of waterlogging. A widely used arrangement is a basal layer of stones, over which is placed washed river sand, rice chaff, decomposed coffee husks, *etc*. Cuttings are usually planted in rows, *ca*. 3 cm apart, at intervals of 5–6 cm. The leaf petiole should be partially, but not wholly, buried. Propagators should be watered to maintain a relative humidity of 90%. Under favourable climatic conditions, calluses appear between 10 and 20 days after planting and roots are issued from the 30th day onward. The root system is well enough developed to permit planting out after 10–12 weeks, when the root length is between 5 and 10 cm. In dry areas, or where relative humidity is low, the process is extended by up to 6 weeks and the percentage of successful rootings may be less.

Suitable cuttings are planted out into plastic bags with a base perforated for drainage. The filling consists of a mixture of sifted humus and rich topsoil. Heavy shading is required immediately after planting out. Temporary shade isused to reduce the light intensity by *ca*. 75% for the first 3–4 weeks. Principal shade is constructed of locally available material, such as palm leaves or grasses. Shade is progressively removed and the plants exposed to full light for hardening at least one month before field planting.

Nurseries are usually weeded by hand, although herbicides can be incorporated in irrigation water. Irrigation water is also frequently used as a carrier for inorganic fertilizers. Pests, such as mealybugs and leaf miners, may cause problems, which can be controlled by pesticide application. Diseases are rarely significant, although inadequate shading can lead to infection with *C. coffeicola*.

Robusta coffee is now almost invariably planted out in poythene bags. Planting is carried out at the start of the rainy season. A small hole is dug for each plant, which is then topped up with surface soil. Digging in of manure, or compost, can be beneficial but fertilizer is usually applied to the soil surface. Herbicide application, to control weeds, is common practice before planting out.

---

* It is a common misconception that Robusta coffee cannot successfully be raised from seed. Selected seeds can be sown in the absence of facilities for use of cuttings. Generative selection can result in hybrid seeds which give a tree of good productivity, with special characteristics of quality or disease resistance. In general, however, trees grown from seeds are less vigorous and less productive than those grown from cuttings. Cultivation from seed is similar to that used for Arabica, but particular problems may be caused by *Rhizoctonia* rot of seedlings and damage by crickets.

Plants must be vigorous and well developed, bearing at least six pairs of leaves and two small branches. Larger plants are stumped to 20 cm some 10 days before planting. At planting, the plastic bag is removed, the twisted part of the taproot cut off and the neck region of the plant placed level with the soil. In most regions the plant is inclined at an angle of 30° to the vertical to encourage the development of suckers and produce multiple stems. A rather different procedure is adopted in high altitude planting, where the plant is initially placed in an upright position. At the beginning of the hot season, after recovery, the plant is bent over to an arch at the horizontal.

Immediately after planting, shade is provided, but is usually only required for for 2–3 weeks. In subtropical regions, premature fruiting can be a problem and shade hedges at a height of *ca*. 2.5 m are maintained for 3 years after planting. A number of shade plants can be used, but *Flemingia* is most common.

Bare root cuttings are occasionally used for convenience of transport in remote plantations with poor access. Nursery procedures should be modified to produce plants with maximum rootlet development. Care must be taken to minimize disturbance to the rootlets when lifting and protection during transport is provided by dipping in a clay slurry and wrapping in grass or leaves. After planting, a light application of fertilizer is made to the soil surface around the plant.

Spacing of Robusta coffee trees is generally similar to Arabica trees. Trees are grown in a square, or rectangular, formation rather than in paired rows. Under good cultural conditions, a spacing of *ca*. 1.75 m between trees in the row and *ca*. 3.0 m between rows is recommended. Narrower inter-row spacing is possible, but can lead to premature fall of berries due to shading of lower branches.

Multiple-stem pruning (see page 201) is common for Robusta, trees usually developing as three or four uncapped stems which are stumped every 5 years. Single- and double-stem pruning with capping is also used. A correlation exists between cultivar and the most suitable pruning method.

Although mulching and cover crops are used in Robusta coffee cultivation, extensive use is now made of black plastic sheeting as a ground cover. This is laid along the rows of coffee trees in strips *ca*. 1.25 m in width, trees being planted in the centre of the strip (see above). Use of plastic sheeting has a number of benefits and is especially useful as a means of water management. This results in a longer period of continued growth, and a lower death rate, of young plants during the first dry season after planting. Weed growth is also much reduced.

Cover crops are successfully used in Cameroon and the Central African Republic. Leguminous crops, especially *Mimosa invisa* and *Flemingia congesta*, are most widely used. In other areas, notably the Ivory Coast, cover crops are of no advantage, probably due to the marginal water supply. In windy regions, such as Madagascar, *Flemingia* is used as a shelter rather than for reasons of water economy.

The main pests of Robusta coffee trees are the berry borer (*Hypothenemus hampei*), the tailed caterpillar (*Epicapoptera* spp.) and the branch coffee borer (*Xylosandras compatus*). In Madagascar there are also serious problems with the root mealybug (*Formicococcus greeni*) and the coffee lacebug (*Dulinus unicolor*). In other areas, occasional infestations may occur during the dry season by berry moths (*Dichocrocis crocodera* and *Prophantis smaragdina*), leaf miners (*Leucoptera* spp.) and stem borers (primarily *Bixadus* spp.). The yellow and green locust (*Zonocerus variegatus*) causes serious damage in West Africa. Economic loss is considerable as a consequence of loss of production following severe damage to leaves, buds, green stems, and even small branches.

Robusta coffee trees are resistant to most diseases, including those which can devastate Arabica. There are no important root pathogens, with the exception of *Clitocybe tabescens*, which can cause major problems in new plantations, primarily in Madagascar. Diseases of leaf and fruit are rare and generally of no significance, with the exception of problems in central West Africa due to infection of specific clones by *Colletotrichum coffeanum*. Coffee rust disease does occur but the majority of clones are tolerant and problems are minimal.

5.2.3 Harvesting

To obtain the highest quality end-product, coffee should be harvested when the berries are fully red ripe. Under-ripe and over-ripe berries are difficult to process and result in a poor quality product. Coffee berries come to full ripeness over an extended period and it is usual to pick red berries individually and to repeat picking at intervals of 7-14 days.

Hand picking is highly labour intensive and of consequent high cost. Attempts have been made to improve productivity, but gains are usually only *ca.* 10%. A number of approaches have been taken to mechanization. Tree and branch shakers, which shake ripe berries off the tree and onto mesh nets, have been used in Hawaii and Puerto Rico. This method is relatively non-discriminatory, however, since a significant number of green berries and leaves are also removed. Attempts have been made in other regions to develop mechanized harvesting techniques based on removal of berries from stems cut from the tree. None have been fully successful.

An alternative approach to mechanized harvesting is the use of chemical ripeners to advance ripening. Costs can then be reduced by concentrating the harvest so that the same number of berries can be harvested in a smaller number of pickings. Alternatively treatment may be used to enable harvest of some trees to commence early, thus spreading the peak load. The most widely used chemical ripener is ethephon (Ethrel). This has been used successfully to advance ripening of hard green berries of Arabica coffee by 2-4 weeks, depending on application rate. When applied to Arabica or Robusta trees bearing soft green berries, the berries ripen while the bean is still immature. Ethephon application is, therefore, of value only where no soft green berries are present. This restricts use to areas, such as Central America, Hawaii and Puerto Rico, where cropping seasons do not overlap.

In most areas all remaining berries are stripped from trees and processed separately by sun drying. This is essential to interrupt the breeding cycle of the coffee berry borer. Harvest practice in Brazil differs from other areas in that harvesting is carried out in a single operation at the end of the season. All berries, including under-ripe and over-ripe are collected and sun dried (see below). This practice is more suitable for mechanization and machinery has been developed with vibrating fingers which remove berries while combing through the foliage.

5.3 TECHNOLOGY OF COFFEE PROCESSING

5.3.1 Green coffee processing

Green coffee is prepared from the ripe berries of the coffee tree. Processing involves a series of relatively complex operations, which are carried out entirely in the producing country. The fundamental purpose of green coffee processing is the recovery of the seed (bean), by removal of the various surrounding layers of the berry (Figure 5.1) and drying to produce a green bean with a moisture content below *ca*. 12%. Two basic methods of processing, dry (natural) and wet (washing) are used, together with ancillary processes including grading, cleaning and polishing. The processing of green coffee is summarized in Figure 5.2.

(a) Dry processing

Dry processing is the older of the two methods of green coffee processing and involves three basic stages: classification, drying of the whole berry and dehusking (hulling). Dry processing is used for virtually all Brazilian Arabica coffee and also for Robusta coffee in many areas. Dry processed coffee is generally considered to be of

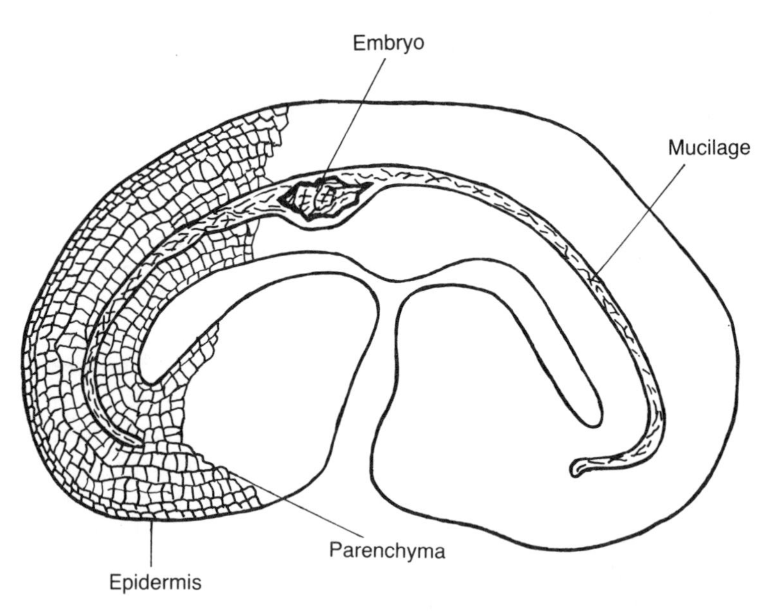

Figure 5.1 Structure of the coffee bean.

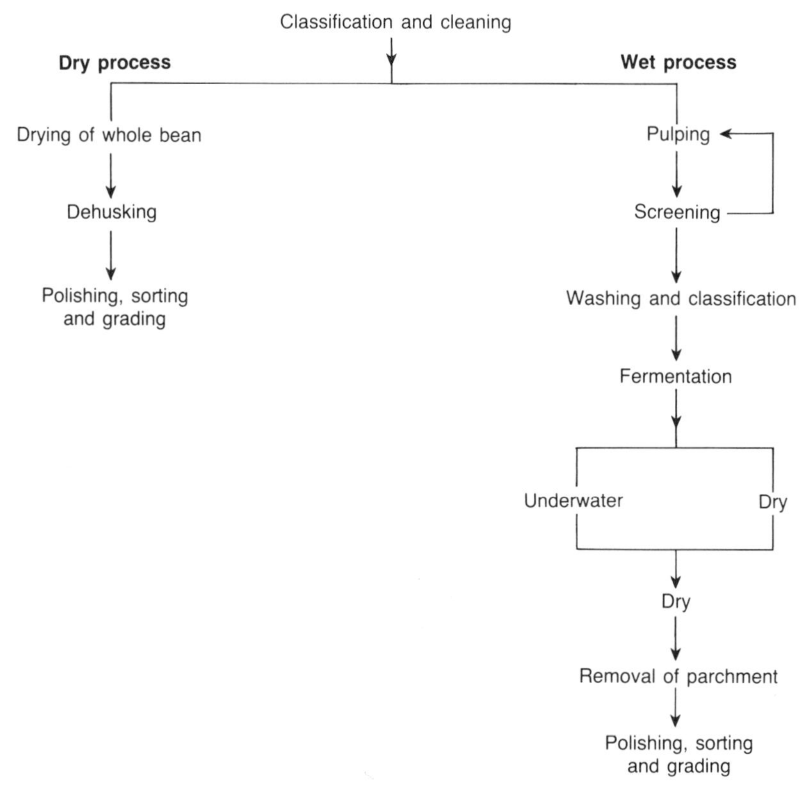

Figure 5.2 Processing of green coffee.

lower quality than wet processed and commands a lower price. The high quality Mocha coffee, however, is dry processed.

Classification involves the removal of unripe, overripe, or damaged beans. In some cases a preliminary classification, at least, is carried out at harvesting, but some further classification is required before drying. The use of water streams to convey berries to the drying areas means that procedures for classification by flotation and separation are readily incorporated. Such procedures have the additional benefit of washing the berries and it is also common practice to remove other extraneous material, such as stones and twigs, by sieving. In Brazil more attention is paid to classification either before drying or after dehusking.

Sun drying is universal in small-scale processing and is still common in large operations. Berries may be laid out onto matting, but concrete, or other impervious material, is normal in the large coffee 'patios' of Brazil. Economics dictate that the berries are piled onto the drying surfaces, but care must be taken to ensure that layers are not of excessive depth. Turning the coffee during the drying period, which lasts 8–10 days in favourable conditions, is essential. Under unfavourable conditions drying of the berries can be extended and, where relative humidity is high at night, the hygroscopic berry may re-adsorb a large proportion of the moisture lost during the day. Extended drying periods can result in the growth of micro-organisms with adverse effects on the flavour of the end product. Micro-organisms, however, probably play a role in degrading the pulp during drying and may contribute to the character of dry processed coffee.

Hot air drying is widely used in large-scale operations in Brazil and also in Africa. Hot air can be used for the entire drying process, which is reduced in length to *ca.* 3 days. Alternatively, especially where fuel is expensive, hot air may be used in combination with sun drying to reduce the final moisture content to below 12%. Air temperatures used in hot air drying should be below 30°C, to avoid overheating and a high incidence of defective beans.

The final stage in dry processing is dehusking; the removal of the dried husks to release the bean. This may be carried out manually or by machine, but in many cases dehusking is carried out at a central plant ('curing' works). Machinery used is similar to that used for removal of the dried parchment shell of wet processed coffee, but must be capable of handling the larger quantity of non-bean material present in dry processed coffee at this stage.

(b) Wet processed coffee

In general, wet processing leads to a higher quality product. With the exception of Brazil, virtually all Arabica coffee is wet processed and some use is also made of this method for Robusta coffee. The hybrid Arabusta is also wet processed.

Ripe berries only should be used for wet processing and classification carried out during harvesting should be supported by additional procedures at processing to remove underripe and overripe berries as well as hard, partially dried berries. Classification is also required to sort the cherries by size, to permit efficient operation of the pulping machines. Classification by flotation is most convenient and involves at least two stages. In the first stage, stones and dirt are removed, and in the second the berries classified. A single classification stage is not usually sufficient to fully distinguish between ripe and unripe cherries and additional stages, together with screening, are required.

Pulping involves mechanically 'tearing off' the skin and soft pulpy part of the berry. This stage is considered to be the most important stage in wet processing. It is essential that pulping commences as soon as possible to avoid the onset of fermentation and development of off-flavours. The process should be completed within 12–24 h after picking, but if this is not possible, the berries should be stored under water. A high standard of hygiene at the pulping stage is essential since dirty water or the presence of decomposing pulp leads to tainting of the coffee.

Screening is required after pulping to separate fully pulped beans from large fragments of pulp and partly pulped beans. Fully pulped beans pass to fermentation, while partially pulped pass to a second pulping machine for the process to be completed. Before fermentation, the beans pass through washing channels, where a further classification stage may be introduced if required.

Fermentation removes any residual adhering pulp and the mucilaginous layer, leaving the beans enclosed only by the parchment covering. The major changes involve pectin-degrading and various hydrolytic enzymes. These are present in the mucilage, although the ultimate origin may be the pulp or endosperm. Microbial growth occurs during fermentation, but while the presence of pectinolytic species, has been reported, the role of micro-organisms is limited. A microbial succession, involving members of the *Enterobacteriaceae*, species of *Enterococcus* and members of the 'lactic acid bacteria', plays at least some part in the lowering of the pH value from *ca.* 6.8 to *ca.* 4.3. Low pH values tend to inhibit the activity of pectinolytic enzymes, but prevent the growth of many micro-organisms with high spoilage potential. The role, if any, of micro-organisms in producing desirable flavour compounds during fermentation has not been proven, but extensive growth is likely to lead to development of undesirable flavours.

Fermentation is carried out on a batch basis in conventional practice, using a series of rectangular concrete tanks. The tank is filled with pulped coffee and water drained away (dry fermentation) to leave a sticky mass *ca.* 1 m thick. The length of fermentation is temperature dependent and, under extreme conditions, varies from 6 to 80 h. In most cases, however, the fermentation is complete within 24 h. After fermentation, the beans are washed in a rotating cylinder.

In a variant process, the water is not drained from the fermentation tank (under-water fermentation). This process is widely used in Kenya and is preferred for Robusta coffee, probably because some debittering occurs.

Continuing interest is shown in means of accelerating the fermentation process. Commercial pectinase preparations have been available, at relatively low cost, for a number of years and have been widely used in Central and South America. Pectinases were originally used in under-water fermentation, but many preparations function well in dry fermentations. Pectinase preparations are valuable in 'emergency' situations, where the fermentation process is proceeding too slowly but, on a routine basis, pectinase preparations are being replaced by use of sodium hydroxide.

Sodium hydroxide (0.1 M) is an effective means of rapidly removing the mucilaginous layer of coffee beans. The process requires careful control and supervision and the beans must be thoroughly washed as soon as removal is complete. Beans processed using NaOH are of poor appearance and a two-stage process has been proposed to overcome this problem. This involves mucilage removal by NaOH followed by washing and immersion in water for 24 h to enhance the final appearance of the beans.

After the fermentation stage, the coffee consists of the bean within the wet parchment covering (parchment coffee). Drying is necessary before the parchment can be removed. This may involve sun drying, hot air drying or a combination of the two methods. Parchment coffee has a relatively constant moisture content of *ca*. 57% and sun drying usually takes 8–10 days before the required content of 12%, or less, is reached. Hot air drying is in wide use in large-scale operations. Most dryers consist of rotating drums through which hot air is blown. For quality purposes it is necessary to use a temperature of less than 30°C in the initial stages. A recommended protocol is the use of ambient temperature air until the moisture content is *ca*. 43%. At this stage, the air temperature can be raised to 60°C and the coffee dried to a moisture content of 10%. This enables the drying process to be completed in 8–9 h. It is commonly thought that, while hot air drying is more efficient than sun drying, a period of exposure to the ultraviolet component of the sun's irradiation improves the colour of the final bean. For this reason a combination of air drying and hot air drying is often recommended, and also has economic benefits through reduced fuel costs. It has been claimed that sun drying to *ca.* 44% moisture, or less, followed by hot air drying is a highly effective combination. The chemical rationale of the effect of sunlight on pigmentation tends to be confused, although it is accepted that under some conditions excess exposure to sunlight bleaches pigments.

Much interest has been shown in fluidized bed drying (see Chapter 4, page 154), but adoption on a commercial scale has been slow. Fluidized beds much reduce the exposure time to high temperatures in the second stage of drying and thus produce a high quality product. Sun drying to 15.5% moisture content, then completing drying in a fluidized bed dryer, is considered to produce an exceptional coffee.

Removal of the dried parchment (hulling, milling, shelling) is required to release the bean. Like dehusking of dry processed coffee, it is usually undertaken at a central 'curing works'. Two types of machine are used; the most common consists of a long rotating screw, the helical pitch of which increases towards the discharge end. The coffee is transported through the hulling machine by the screw, back pressure being controlled by a gate at the discharge end. During transit the parchment is broken by friction and is either removed by a vacuum applied to a perforated section in the base of the machine, and/or by directed air jets as the coffee leaves the hulling machine. The second, less common, type of machine is based on impact milling in which the parchment is physically broken before removal. The contact surfaces of hulling machines are traditionally made of phosphor bronze, which is believed to improve the colour of the finished beans.

(c) Ancillary processes

Green coffee beans undergo additional processing, including polishing, sorting and grading. Polishing involves removal of the silverskin, except that retained in the centre cut of the beans. The purpose is purely cosmetic, to improve the appearance of the beans. A separate polishing stage may not be required for wet processed Arabica. Dry processed coffee, especially Robusta, has a very tough and tenacious silverskin, which requires a wet polishing operation to remove.

Sorting is required to remove any defective beans remaining after processing. A certain amount of extraneous material, including whole berries, twigs, stones and fragments of husk and parchment, may also be present and requires removal. Sorting may be carried out physically by blasting air upward through the beans. This process, air lifting, will remove defective beans and extraneous material which has a significantly lower density than sound beans. The process is of relatively low discrimination and a proportion of the sound beans is usually removed with the defective. In some cases beans removed by air lifting are exported as low-grade *triage* coffee.

Hand sorting of wet processed Arabica coffee is a traditional process. Hand sorting is highly labour intensive, but the cost can be justified by the high quality of hand sorted coffee. Expert sorters are employed and the highest quality is obtained by sorting in stages, removing the easily recognized black beans and discoloured sour beans at separate sorting points. Discoloured sour beans are difficult to recognize, but complete removal is essential for high quality, since only a small number adversely affect the flavour of the brew.

Electronic sorters have been available for a number of years, and developments both in optical systems and in electronics have resulted in sophisticated machines capable of high throughput and a high level of discrimination. Electronic sorters sort by both colour and, through image analysis, by shape, unsatisfactory material being removed by air jets. It is now common practice to operate electronic sorters in tandem with inspection under ultraviolet excitation to detect 'hidden stinkers'. Beans fluorescing in the correct wavelengths are removed automatically. The use of a combination of electronic and ultraviolet sorting has enabled hand sorting to be totally replaced in some large-scale operations.

The size of green coffee beans varies and forms the basis of grading systems. Coffee is marketed according to grade specifications which are assessed by laboratory test screen analysis. Most green coffee is graded into different sized portions by screening, although some is sold unscreened. Reel graders are most commonly used for screening. These consist of rotating drums fitted with perforated screens of varying dimensions. Coffee beans enter at one end, where the screen dimensions are finest, and pass over increasingly coarse screens, falling through the screen corresponding to the bean size.

(d) Storage and transportation of green coffee

Green coffee is stored in silos for a short period before packing for transport to the consuming country. Jute sacks, containing 60 kg of coffee, are still most widely used, although transport in bulk containers is increasingly popular.

Green coffee is susceptible to adverse storage conditions, undergoing deleterious changes in flavour and appearance. Problems are likely to be particularly severe where ambient temperatures are in excess of 25°C or where moisture uptake in conditions of high relative humidity leads to the water content of the bean rising above 12–13%. Mould growth occurs under these conditions causing rots and development of off-flavours.

Infestation and damage to beans by insects may occur during storage. A large number of insects are potential pests of stored coffee, but the beetle *Araecerus fasciculatus*, is the most important. Fumigation is applied where necessary. Rodents and birds may also cause problems, which require routine application of control measures.

5.3.2 The conversion of green coffee into beverage

Conversion of green coffee into beverage involves a number of stages (Figure 5.3). Of these, only the initial stage, roasting, is invariably carried out on a centralized basis. Other stages may be carried out either centrally or at point of consumption, including the home.

---

* German consumers are highly conscious of the possibility that gastric disturbances can result from coffee drinking. A number of German coffee manufacturers use a limited treatment by solvent washing or steaming to remove the outer coffee 'wax'. This consists primarily of 5-hydroxytryptamide which, possibly with other substances, is considered responsible for gastric disturbance. A small quantity of caffeine is also removed.

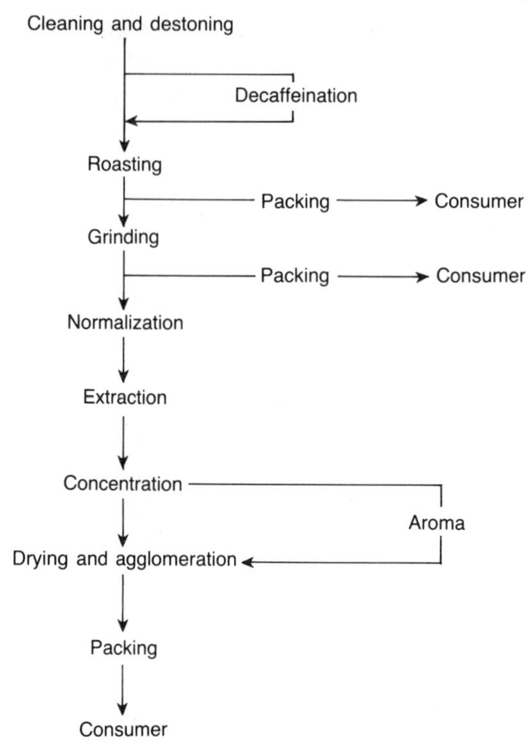

Figure 5.3 Conversion of beans into beverage.

(a) Roasting

Coffee roasters offer various blends and degrees of roast. Many also offer individual coffees. Coffee for roasting normally requires little pre-preparation except for choice of beans and simple cleaning and destoning. Decaffeination, where applicable, usually takes place before roasting (see page 230).

Roasting is a time-temperature dependent process, which initiates significant chemical changes. A loss of dry matter occurs, primarily as gaseous CO_2 and other volatile pyrolysis products. Many of the pyrolysis products are important in determining the flavour of coffee. A high proportion, however, is retained within the coffee together with *ca*. 50% of the CO_2. Dry matter losses can be broadly correlated with degree of roasting. A light roast loses 3–5% (in addition to moisture lost), a medium roast 5–8% and a heavy roast 8–14%.

Degree of roasting determines many of the flavour characteristics of the brewed coffee. The colour of the beans also varies with degree of roast and may be used as the basis of a simple classification system. Physical changes occur, including loss of density due to 'popping'. This is a function of rate of roasting as well as extent and is reflected in a lower bulk density in the ground coffee.

Roasters are ovens, which operate either on a batch or a continuous basis. Heating is normally carried out at atmospheric pressure, the usual medium being hot air and combustion gases. Heat may also be applied by contact with heated surfaces and in some designs of roaster, this is the primary, or sole, means of heating. Older roasters used air at a temperature as high as 540°C, but temperatures of 375°C, or lower, are now common, but require higher air velocities. The most common design, which can be adapted for either batch or continuous roasting, consists of a horizontal rotating drum. In most cases, beans are tumbled in a current of hot air flowing through the drum, although in some roasters a cross-flow of hot air is also possible. Air is heated directly by gas or oil fuelled burners and in modern designs air recirculation systems are used. Air recirculation systems incorporate secondary burners (afterburners), the prime purpose of which is to reduce atmospheric pollution. Operating costs are also reduced.

The economics of large-scale coffee roasting dictate that the process should be as short as possible. Shortening the roasting time has been achieved largely by increasing the velocity of the hot air flow, although mechanical means such as the use of rotating paddles to increase the rate of heat transfer to the beans are used in some roasters. To some extent, the scope for further improvement of conventionally designed roasters is limited and a number of novel approaches have been taken. A batch roaster based on the centrifugal principle is available, for example, in which beans and hot air are intermingled in a rotating bowl. The most effective means of utilizing high velocity air flows currently available, however, is fluidized bed roasting. A number of designs are now commercially available and it seems likely, for large-scale operations at least, that fluidized bed roasting will replace other types.

Although most roasters operate at atmospheric pressure, interest has been shown in presssure roasting. Pressure roasting is considered to be potentially more efficient and is generally considered to increase the acidity of the brew. Nitrogen is used as pressurizing agent.

The first stage of roasting involves the removal of moisture. Roasting proper begins when the temperature of the green beans is *ca*. 200°C. Subsequent chemical reactions are exothermic and the temperature rises rapidly. The length of roasting varies from *ca*. 5 to *ca*. 30 min. The exothermic nature of chemical reactions means that control of the process must be carefully applied. The roasting process is usually verified by instrumental assessment of the colour. Samples are taken immediately after roasting and ground using a rigidly standardized method before reading the colour. Provision of a rapid and efficient cooling stage is also necessary. A small proportion of water (quench water) is added to the beans at the cooling stage. This assists rapid cooling and also leads to a high degree of particle size uniformity in subsequent grinding. In batch roasters, quench water is added to beans in the roasting drum at the end of the process. The beans are then discharged into a perforated cooling tray through which air is blown. Continuous roasters of the drum type incorporate a cooling section, separated by a heat lock, into which the beans pass after roasting.

Roast coffee beans can be retailed directly without any further processing, except packaging. Packaging may consist of nothing more than paper bags, when coffee is retailed from specialist outlets and use by the consumer is expected to be more or less immediate. A longer shelf life is required for coffee retailed through non-specialist outlets and vacuum packing is common to prevent oxidative deterioration. A major problem is that roast coffee contains up to 15 volumes CO_2. This is released during storage and can cause swelling of non-rigid packs. Current practice is to vacuum pack in cans of sufficient strength to withstand the pressure developed, or to vacuum pack in pouches of very poor barrier properties, which permit release of CO_2, but admit oxygen.

* Alternative packaging systems have been proposed to overcome the problems of CO_2 release. These include incorporation of a one-way valve into the pouch, which permits the CO_2 to escape without lowering the vacuum. Alternatively, active packaging can be used to scavenge CO_2. This involves inclusion of a sachet of calcium hydroxide, which combines with CO_2 to form calcium carbonate.

(c) Grinding

Coffee beans may be ground in the home, at point of sale or in industrial-scale operations. The principles of efficient grinding are, obviously, the same in each case. In the home, process efficiency is generally of little importance and in many cases an attractive grinder design takes precedence over effective grinding. In industrial-scale coffee grinding, however, efficient performance is of key importance. The subsequent discussion is concerned only with industrial-scale operations.

> BOX 5.3 **Swap shop**
>
> Although coffee grinders often feature as part of an ideal kitchen, their domestic use is limited. In a number of surveys, coffee grinders have been named as the least used pieces of domestic equipment. Coffee grinders feature prominently, alongside rowing machines and exercise bicycles, in swaps organized by local radio stations, etc. - the fate of unwanted gadgetry.

Coffee grinding is necessary to permit rapid extraction of the soluble material by hot water. The particle size of ground coffee can be varied and different types of extraction, especially on a domestic scale, require different particle sizes. Percolators, for example, require relatively coarsely ground coffee to minimize the quantity of fine particles entering the coffee cup.

Large-scale coffee grinding almost invariably involves the use of roller grinders. A cutting action is required rather than crushing and this is achieved by paired rollers with special serrated surfaces. Single-stage grinding is only rarely used since the process produces a wide range of particle sizes and a high proportion of very small particles. Multiple roll grinders are most widely used and consist of two to four sets of paired rollers. Coffee is progressively reduced in size during passage through each set of rollers, the degree of grinding being determined by the number of sets used. It is not possible to attain the ideal situation of all ground particles being of the same size, but variation in size is much less where multiple roll equipment is used. An alternative is closed circuit grinding. This system is based on single stage equipment, but the coffee passes over a sieve after grinding and oversized particles are returned to the feed for regrinding.

For special purposes it is necessary to grind coffee to very fine particles (50 μm or smaller). This is possible using multiple roll equipment, but a large number of sets of rollers is required and the process is basically unsatisfactory. Cryogenic grinding, in which the beans are cooled to low temperatures by solid CO_2 or liquid nitrogen, is in general use for this purpose.

A significant quantity of chaff (remaining silverskin) is released when coffee beans, especially Robusta, are ground. Chaff may be removed by air lifting and other separation methods, but this results in loss of soluble solids. In many cases, chaff is dispersed among the ground coffee by mixing with rotating blades mounted in a trough (normalization). To some extent the bulk density of the ground coffee can be controlled by varying the rotational speed of the blades at this stage. Incorporation of chaff has the further advantage, especially with dark roasted coffee, of improving the flow properties by absorbing exuded oil.

Grinding releases a proportion of CO_2 from coffee. A high proportion is released during and immediately after grinding. A significant quantity may be retained, however, especially with coarse ground coffee. Vacuum packing in tins or flexible pouches is used where an extended storage life is required. With fine ground coffee, vacuum packaging immediately after grinding is possible without any special precautions against development of pressure due to CO_2 release. In the case of coarse ground, however, it is common to delay packaging for some hours to permit continuing release of CO_2. This practice can lead to rapid deterioration of the packed coffee due to the absorption of oxygen, which may not be removed when the vacuum is applied at packaging. Flushing with an inert gas is reduces the oxygen level, but is relatively expensive. A further solution is to introduce a vacuum degassing stage to rapidly remove CO_2 without O_2 absorption.

(c) Extraction

Of the various processes involved in the conversion of green coffee to beverage, extraction is that most commonly applied in domestic or catering environments and a wide variety of methods are available.

BOX 5.4 A bonfire of vanities

'Real' coffee is seen as reflecting a particular lifestyle and leads, inevitably, to coffee snobbery. The 1960s cult film, the *Ipcress File* opened with Michael Caine, as the laid back spy, lovingly making a cup of 'real' coffee. A generation of students followed suit, happily consuming home-percolated coffee which, all too often, had a taste and texture evocative of a suspension of iron filings.

Very large-scale extraction is an essential stage in the manufacture of instant coffee. In this situation operating costs and yield of soluble solids are important considerations as well as taste and quality of the end product.

Extraction of coffee is primarily a physical process which involves the mass transfer of diffusible components from a solid to a liquid phase. In model systems it can be demonstrated that extraction yield increases with increasing solvent temperature and decreasing particle size. Yield decreases with higher degrees of roasting and with increasing grounds: solvent ratio. The effect of agitation is minimal. Extraction is not a simple diffusion phenomenon and has two phases.

1. *Washing phase*. This phase extracts the 'free solubles'. Surface soluble substances instantaneously dissolve and are transferred to the extract by a convection mechanism.
2. *Diffusion phase*. This phase involves soluble substances within the cellular particles.

More than 90% of the yield of soluble substances is obtained during the washing phase. For this reason it is only this phase which is of practical importance.

The composition of the extract is affected by yield and thus by factors influencing yield, primarily solvent temperature and particle size. Some chemical reactions may occur at the high temperatures commercially used, including cleavage of high molecular weight polysaccharides. Formation of monosaccharides by hydrolysis would seem, however, to be minimal. Under any commercial procesing conditions, only a very small proportion of coffee oil is recovered in the extract. This effectively means the loss of volatile components and separate isolation procedures, applied prior to water extraction, have been devised. These usually involve mechanical pressing, followed by recovery of volatiles by inert gases or steam, but have not been adopted on any significant commercial scale.

Two main methods have been used for large-scale coffee extraction. The first, slurry extraction, involves free contact between ground coffee and hot water in a series of pressurized tanks. Coffee particles are separated from the extract by centrifugation. Finely ground coffee may be used, but the capital cost is very high. Where used, slurry extraction is normally a secondary operation of low throughput and restricted to processing the proportion of very fine particles resulting from grinding. The second method involves percolation batteries. These consist of vertical vessels, or columns, containing a bed of ground coffee through which hot water flows. A battery typically consists of five to eight percolation columns, which are operated on a batch-continuous basis, with one column being out of use for recharging during each extraction cycle. Flow of water through the columns is usually counter-current. Continuous extraction is also possible. Many designs of continuous extractor have been proposed, but a widely used type consists of a horizontal vessel through which a 'bed' of ground coffee is moved by a screw conveyer against a flow of hot water.

Degree of grinding of coffee for percolation columns and continuous extractors is determined by the conflicting requirements of high yield (fine powder) and good solvent flow (coarse powder). A medium grind is applied to coffee for extraction in percolation columns and a rather coarser grind for extraction in continuous extractors.

Extraction of solubles for instant coffee manufacture originally involved use of water at 100°C. This resulted in a yield of *ca.* 30% and, after drying, a powder of poor flavour. Many manufacturers attempted to improve quality by adding up to 50% carbohydrate, usually derived from corn syrup, which also improved the flow properties of the powder. It was subsequently realized that, providing the efficiency of extraction could be increased, sufficient carbohydrate can be obtained from the coffee itself. A markedly increased yield can be obtained either by extracting with dilute acids at 100°C or by extracting with water at temperatures of up to 175°C. The latter method has been adopted, the first large-scale plants being built in the early 1950s. High-temperature extraction plants must be operated under pressure, to prevent the water boiling. It is now general practice to establish a temperature gradient across an entire percolation battery. The feed water is used at the highest temperature to extract the most difficult solutes from the almost spent grounds in the final percolation column, while liquor from previous columns is used at *ca.* 100°C to recover the most easily extractable solutes from the fresh 'coffee'. Similar principles have been adopted in continuous extraction equipment. Temperature controllers can be used to establish a suitable gradient in a single extractor, or extraction can be carried out in two separate units. In this case one unit operates at *ca.* 175°C and the other at *ca.* 100°C, partially extracted grounds must be transferred externally from one unit to another.

Following extraction, residual insoluble material is removed from the liquor by centrifugation. The liquor is then cooled by plate or tubular heat exchangers and passed to refrigerated storage tanks.

(d) Concentration and drying

Although it is possible to dry coffee extracts directly, the resulting powder is of poor quality and the process uneconomical. In common with other drying operations, it is customary to remove as much water as possible during preliminary concentration. Concentration by thermal evaporation is most commonly used, although there is increasing interest in freeze-concentration.

Modern evaporator installations are of high thermal efficiency, and correspondingly low operating costs, although capital costs are relatively high. The degree of concentration possible is limited by the high viscosity of concentrated coffee liquor and it is usual to install evaporators capable of handling viscous material. Wiped-film evaporators are capable of concentrating to as high as 70% solids. Such equipment is, however, expensive and many manufacturers prefer more conventional designs. Plate, rising and falling film evaporators (see Chapter 2, page 40) have all been used for concentration of coffee extract. Multiple-effect plant fitted with thermal, or mechanical, vapour recompression is now common. A major problem common to all types of evaporator is the loss of volatile substances, with adverse effects on flavour and aroma. This problem can be overcome by stripping and recovery by condensation of the volatiles. This may be achieved by placing a preliminary aroma recovery unit ahead of the main evaporator, but most modern evaporator plant incorporates an integral aroma recovery section. Aroma concentrate is then added back to the product flow ahead of the drying stage or added to the dried powder.

Freeze-concentration involves cooling the coffee extract to below its freezing point and removing the ice crystals formed. There is no loss of volatile substances and no deterioration of flavour due to heating. Freeze-concentration plants, however, are of higher capital and operating costs than thermal evaporators. Degree of concentration possible is limited by the viscosity of the concentrated extract.

Reverse osmosis has found no significant application in concentration of coffee extracts. The process does, however, have a secondary function for preliminary concentration of very dilute extracts.

In modern commercial practice, spray drying or freeze drying is used. Spray drying is used in production of 'standard' quality instant coffee powder. The bulk density of the powder is of considerable importance, since for convenience in use it is required that a teaspoon should dispense the correct quantity for an average strength drink (usually *ca.* 2 g). This requirement can be met by a powder of average particle size 300 μm, production of which requires a tall narrow drying chamber. Pre-filtered extract is introduced into the drying chamber through a pressure nozzle designed to produce a centrifugal spray pattern. Such nozzles contain an internal cyclone-shaped chamber (spin chamber) into which extract is forced at *ca.* 690 kPa by reciprocating feed pumps. The extract enters tangentially and takes up a circular path before emerging from the bottom of the nozzle as a cone-shaped sheet or curtain. This breaks up a short distance from the exit orifice to produce a hollow spray. Dryers are of the co-current type in which hot air enters the top of the dryer where the extract is most dilute, the air usually being heated by direct combustion of gas or refined fuel oils. Heavy powder particles are removed by gravity from the base of the drying chamber, while fine particles leave in the air stream and are recovered by a system of cyclones and filters.

Coffee powders are of relatively poor wettability and thus can be difficult to dissolve. In recent years, instant coffee granules have become available and are now generally preferred. Coffee granules are agglomerations of powder particles produced by the so-called instantizing process. Instantization processes originally involved moistening powders with steam or water and re-drying in a fluidized bed dryer. It is now, however, possible to produce instantized powder in a single operation by re-introducing fine powder particles into the dryer with the atomized extract.

Spray drying inevitably involves loss of flavour volatiles and freeze drying is preferred for 'premium' quality coffee. In some cases spray dried and freeze dried powders are blended. The removal of water by freeze drying is an expensive process and a concentrated extract is required both to minimize cost and to ensure a high quality product.

Freeze drying, at its simplest, involves heating the frozen extract under high vacuum conditions. Conditions must be set to ensure that water loss is by sublimation and a vacuum below 1 mm Hg is required. Water vapour is removed by condensation. Semi-continuous and continuous freeze dryers are now in common use and radiant heating is sometimes preferred to conduction. Coffee extract is usually frozen in thin slabs which are ground in frozen form to produce the granules preferred by most customers. The majority of the granules produced will be of the required size, but some form of recycling system is required for those over- or under-sized. A variant process, normally used with highly concentrated extracts, involves foaming the extract directly before freezing. Loss of flavour volatiles is less than during spray drying, but occurs to some extent. The extent of loss depends on a number of factors including rate of freezing and conditions during sublimation.

(e) Packaging

Dried coffee extracts present relatively few problems at the packaging stage. Glass jars have largely replaced tins in Europe and the US, the mouth of the jar being sealed with an impermeable paper diaphragm to prevent pick up of moisture from the atmosphere. It is also necessary to minimize the oxygen content to reduce oxidative deterioration. In some cases, especially 'premium' quality freeze dried coffee, a vacuum is applied before sealing and the pack back-flushed with nitrogen, CO_2 or a mixture of the two gasses. Coffee extract is also packed as individual measures in foil-laminated pouches. A major use is in the provision of coffee-making facilities in hotel bedrooms, but pouches are now available for domestic purposes.

5.3.3 Decaffeination of coffee

Concern over the possible adverse effects of caffeine has led to growing demand for decaffeinated coffee, although the process was used as early as 1905. Decaffeinated beans, ground coffee and 'instant' soluble extracts are all available. Proprietary technology is normally used in decaffeination and details of processes are rarely available. Decaffeination, however, is almost always carried out on the green beans, although processes exist for decaffeination of roast beans and of soluble extract. The process involves solvent extraction of caffeine from the green beans followed by recovery of the caffeine and re-use of solvent. Older processes involve moistening the beans with water followed by extraction with organic solvents, typically ethyl acetate or methylene chloride. Residual solvent can be almost entirely removed by steaming and drying the extracted beans. The use of organic solvents, however, is widely considered inappropriate in foods and alternative processes have been developed. The first of these used an aqueous extraction, designed and controlled to minimize the quantity of flavour compounds co-extracted with caffeine. Solvents are used to recover caffeine from aqueous solution, but have no contact with the coffee. The second, and generally preferred, method involves the use of supercritical CO_2.

Decaffeination processes are typically able to reduce the caffeine content of coffee to below 0.1% on a dry weight basis. The decaffeination process is not, however, totally selective and a proportion of flavour compounds are inevitably removed. The difference in the final brew is usually small, however, providing that good process control is applied.

5.3.4 Other coffee products

(a) Speciality extracts

The majority of 'instant' coffee extracts are based on blends, which usually include significant quantities of Robusta coffee. In the past

---

* Removal of caffeine and other bitter xanthines by salting has been proposed. Dicalcium-disalicylate salts of xanthines are tasteless, bitterness being totally removed. A potential exists for the removal of bitterness in cocoa and tea as well as coffee. Salicylic acid may, however, be toxic. (Roy, G.M. 1990. *CRC Critical Reviews in Food Science and Nutrition*, **29**, 59–72).

product diversity has been largely restricted to degree of roasting, method of drying and whether or not decaffeinated. Further diversification initially involved coffee extracts manufactured from beans, usually Arabica, obtained from a given country, or geographical area. More recently, instant 'cappucchino' and instant 'espresso' coffee granules have become available. Instant 'cappucchino' granules consist of a blend of dried coffee extract with whitener (whey powder, low-fat skim milk and vegetable oil), while instant 'espresso' granules are a blend of dried coffee extract with dark roast ground coffee beans.

> BOX 5.5 **Grounds in my coffee**
>
> Instant cappuccino and instant espresso coffees are good example of added-value products, in which the unit cost of a basic ingredient is significantly increased by the addition of small quantities of additional ingredients. The additional ingredients may themselves be of low inherent cost, but the new product is able to command a significant market premium. A significant part of the success of many UK multiple retailers stems from their brilliant exploitation of the added-value concept.

(b) Extracts with extenders

Coffee extracts can be extended with materials such as chicory and figs, which may also be used as direct substitutes (see below). Extenders are relatively easy to handle and have a high soluble solids content. The use of extenders is not entirely for economic reasons since there may be consumer preference for blends of coffee with chicory, etc., over coffee alone.

(c) Coffee essence

Coffee essence is a well established product. A very high sugar concentration is required to permit storage at ambient temperatures without spoilage and traditional types of coffee essence are intermediate moisture foods. Occasional problems of spoilage due to osmotolerant yeasts occur, usually due to contamination after opening the container. The flavour of traditional essences is distinctive and may be preferred to that of the more 'authentic' dried extracts by older persons. Such essences are now effectively niche products, which find a use not only as a beverage, but as flavouring for home made cakes, etc. In addition to traditional sugar-preserved coffee essence, attempts have been made to market deep frozen essences. These have generally been unsuccessful.

(d) Chilled, bottled brewed coffee

In parts of northern Europe it is common practice to keep brewed coffee at high temperatures for several hours. This is not possible with Italian espresso-type coffee, which is consumed cold in summer, and it has become common domestic practice to store brewed coffee at low temperatures before consumption. Attempts have been made to produce a bottled ready-to-drink chilled coffee. The technology is relatively straightforward and involves preparing an infusion of the correct strength, pasteurizing at 70°C and filtering. The brew is then filled into pre-sterilized bottles under clean conditions. Products of this nature have been marketed on a small scale. Shelf stability is highly dependent on the initial pH value and the fall in pH value during storage.

5.3.5 Coffee substitutes

Over the years a wide variety of plants have been used to prepare coffee substitutes. In many cases, substitutes were developed in response to high prices or non-availability of coffee itself. There is also, however, a desire by some consumers to avoid caffeine, while some coffee substitutes, such as acorn, may actually pre-date coffee drinking in Europe.

Chicory (*Cichorium intybus*) root is the most common coffee substitute. The plant is perennial and, in Europe, the roots are lifted in the late Autumn. Preliminary processing involves washing, slicing and drying in brick kilns or rotary dryers to *ca*. 13% moisture content. The chicory roots are then roasted in rotating drums, 1% vegetable oil usually being added before heating commences. After

---

* Ultrafiltration has been proposed as a means of obtaining both a pre-concentrate for subsequent freeze drying and a permeate. It has been suggested that the permate, which is a highly shelf-stable weak brew, could be marketed as a convenience beverage to be consumed as an alternative to coffee. Organoleptically, the permeate has a full, clean coffee aroma, but with less body than coffee and little bitterness. (Zanari, B. and Pagliani, E. 1992. *Food Science and Technology*, **25**, 271–4).

roasting the roots are ground to powder and either packaged directly, or extracted and spray dried.

Roast cereals, especially barley, but also malt, rye, wheat and maize, have achieved a considerable measure of popularity in the health food market where the prime concern is avoidance of caffeine. Grains may be roast whole or ground to flour, mixed with water to form a paste, cooked and coarse milled before roasting. Coffee substitutes prepared from roast dandelion roots are also available in the health food market.

At various times a wide variety of other plants have been used as coffee substitutes. These include fig, acorns, seeds of various legumes, carob and beetroot. With the exception of fig, which is used as an extender in Viennese coffee, such substitutes are not now in general use.

5.4 BIOLOGICAL ACTIVITY OF COFFEE

Coffee consumption has been associated with an extremely large number of chronic conditions, ranging from arrythmias to various forms of cancer (Table 5.3). In many cases, however, supportive evidence is lacking and the general conclusion may be drawn that the risks associated with moderate consumption (up to five cups daily) are very low. A heavy coffee drinker may, however, develop temporary headaches and lethargy if consumption ceases.

5.4.1 Pharmacologically active substances in coffee

The major pharmacologically active compound in coffee is the methyl xanthine, caffeine (structure, page 237), which is known to have effects on a number of functions including stimulation of the central nervous system. A number of related alkaloids, such as chlorogenic acid are also present and are also pharmacologically active.

* An unusual outbreak of food poisoning occurred in Romania following consumption of a coffee substitute. The causative organism was an enteropathogenic strain of *Escherichia coli*. The source of infection was not known. (Doyle, M.P. and Padyhe, V.V. 1989. In *Foodborne Bacterial Pathogens* (ed. Doyle, M.P.). Marcel Dekker, New York).

Table 5.3 Chronic conditions associated with coffee consumption

Anxiety
Arrythmias
Birth defects
Cancers
Fibrocystic breast disease
Gout
Hyperlipidaemia
Hypertension
Osteoporosis

Coffee contains a number of other compounds, which have been shown to have pharmacological activity. The significance in coffee, however, is doubtful since levels are very low. Glycosides of atractyligenin and related compounds, for example, inhibit oxidative phosphorylation by blocking the transport of ATP generated in the mitochondrial membrane in exchange for ADP generated outside, but the effect cannot be demonstrated in coffee. In addition, the effect of unidentified opiate metabolites, previously thought to be of major significance in coffee, is now known to be limited.

5.4.2 Physiological effects of caffeine

Caffeine has been shown to stimulate the central nervous system, to stimulate cardiac muscle and to relax smooth muscle, especially bronchial muscle, and to act on the kidney to produce diuresis. Caffeine also produces a slight increase in the basal metabolic rate and increases the capacity for muscular work.

Definitive statements concerning stimulation of the central nervous system are difficult to make, partly because doses of caffeine ingested by most coffee drinkers are very low and partly because of marked variation in the response of individuals. There have also been surprisingly few systematic studies. Despite this, it is well established that caffeine delays the onset of sleep and produces a disturbed sleep pattern during the first 3–4 h. Caffeine has been associated with hyperactivity in children, but this normally results from consumption of caffeine-containing soft drinks.

Caffeine can have a restorative effect on functions such as physical endurance, writing and monitoring of equipment when these are

Table 5.4 Relationship of coffee consumption with various cancers: Current status of knowledge

Organ	*Relationship*
Bladder	disproved
Breast	disproved
Colon	no association
Kidney	disproved
Ovaries	no scientific evidence
Pancreas	possible weak association; further work required
Prostate	no scientific evidence
Stomach	possible protective effect; further work required

diminished by fatigue. It has also been shown to improve sporting performance when determined by parameters such as the power of a boxer's punch or the aerobic and anaerobic capacity of an athlete.

5.4.3 Epidemiological effects of coffee drinking

Epidemiological studies have shown that consumption of coffee at an acute level produces a number of short-term responses, including increase in blood pressure, plasma renin activity, urine production and gastric acid secretion. There are also increases in levels of catecholamines and free fatty acids in the serum. None of these responses, however, is produced by consumption of coffee at a chronic level. In particular, older users show no increase in blood pressure or heart rate and continuous heavy ingestion does not lead to hypertension. There also appears to be no epidemiological evidence to support most of the other allegations concerning coffee consumption (Table 5.3), including the association with some types of cancer. It does appear possible, however, that coffee consumption is associated with arrythmia is susceptible persons.

* In many cases the importance of arrythmia is probably overstated. 'What is not appreciated is that ventricular premature beats are innocuous in the overwhelming majority of persons. They no more augur against sudden death than a sneeze portends pneumonia'. (Graboys, T.B. and Lown, B. 1983. *New England Journal of Medicine*, **308**, 835–6).

Although reports have tended to concentrate on the adverse effects of coffee consumption, epidemiological studies have demonstrated a number of beneficial effects of coffee consumption. Coffee is recommended as a booster of pain-free walking time in persons with chronic pain-free angina and also prevents post-prandial hypertension in the elderly. The beverage is also an effective bronchodilator in young persons with asthma and is a good dietary source of potassium.

5.4.4 Carcinogenic properties of coffee

Coffee consumption has been linked to cancers in many organs, but there currently appears to be little, or no, supportive evidence for these contentions (Table 5.4). A number of components of coffee have been shown to have mutagenic activity during *in vitro* testing. Mutagens are believed to be formed during roasting from amino acid and carbohydrate precursors. Mutagenic activity has been determined in both home-brewed filter coffee and commercial spray dried extract. Aliphatic dicarbonyl compounds, such as methylglyoxal, are believed to be responsible for activity. Coffee, roasted at high temperature, has also been thought to contain 2-amino-3,4-dimethylimidazo [4,5-*f*] quinoline and possibly other heterocyclic amine-like mutagens. Detection has not, however, been possible either in commercial coffee extracts or in over-roast home-brewed coffee.

Results of *in vitro* testing for mutagens should be treated with caution, since there is not necessarily a correlation with cancer in humans or animals. In the case of coffee, it appears likely that metabolic inactivation of potential mutagens offers a significant degree of protection.

5.4.5 Anti-microbial properties of coffee

Coffee extract has been shown to have bactericidal activity against a number of pathogenic micro-organisms, including *Staphylococcus aureus*, enteropathogenic *Vibrio* spp. and *Aeromonas* spp. Coffee extract also exhibited activity against the thermostable direct haemolysin of *Vibrio parahaemolyticus*. The anti-microbial effect of coffee was markedly less than that of tea (see Chapter 4, page 171).

5.5 CHEMISTRY

5.5.1 Constituents of green coffee

(a) Nitrogenous compounds

Caffeine (Figure 5.4) is of major importance with respect to the physiological properties of coffee and is also an important factor in determining the bitter character. The caffeine content of green beans varies according to species. Robusta coffee contains *ca.* 2.2%, on a dry matter basis (dm), and Arabica *ca.* 1.20% dm. The hybrid Arabusta contains 1.72% dm. Environmental and agricultural factors appear to be of minimal importance in determining caffeine content.

A number of other nitrogenous bases have been reported to be present. These may be placed in two groups. Compounds inherently stable at roasting temperatures, primarily ammonia, betaine and choline, which are present in trace quantities (less than 0.1% dm) in green beans and compounds which decompose during roasting. Major compounds in the second group are the 5-hydroxytryptamides (serotonin amides), which are mainly present in the wax (see pages 241-2) and trigonelline (Figure 5.5).

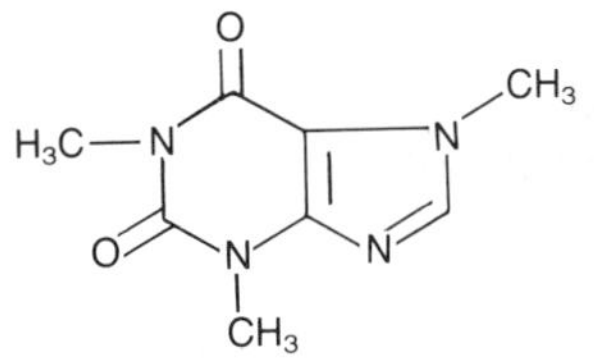

Figure 5.4 Structure of caffeine.

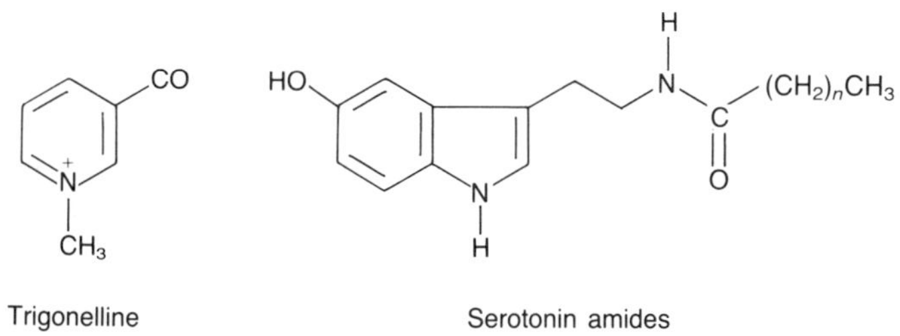

Figure 5.5 Structure of 5-hydroxytryptamides and trigonelline.

Trigonelline degradation products are important with respect to both flavour and nutrition (see page 244). The compound is present in Arabica at levels of *ca.* 1.0% dm and in Robusta at levels of *ca.* 0.7% dm. Trigonelline has physiological activity affecting the central nervous system, bile secretion and intestinal mobility, but this is unlikely to be significant at the levels present in brewed coffee. There is a weak bitter taste, *ca.* 25% that of caffeine, which has only a slight effect on the brew.

Proteins and free amino acids of coffee beans have received relatively little attention. There is only minor difference in protein content between species, Arabusta containing *ca.* 9.2% dm and Robusta *ca.* 9.5% dm. Proteins predominantly exist in unbound form in the cytoplasm, or bound to cell wall polysaccharides. Some of the minor proteins exhibit enzyme activity (e.g. amylase, catalase, lipase, peroxidase, protease, etc.) and attempts have been made to associate activity of some enzymes with sensory quality of the brew. In general, these attempts have been unsuccessful, although high polyphenol oxidase content in Robusta does appear to be a predictor of poor quality brews.

The free amino acids of green coffee are probably the single most important group of compounds in relation to the organoleptic quality of the final brew. Despite this levels in beans are low, varying between *ca.* 0.15 and 0.25% dm. Levels tend to be higher in Robusta, although Arabica contains 50% more glutamic acid. Dipecolic acid is found only in Arabica and may be used to differentiate between the two species of bean.

(b) Carbohydrates

Carbohydrates are present in green coffee both as low molecular weight free sugars and as polysaccharides. Sucrose is the major free sugar, the quantity present varying according to cultivar, state of maturity, processing applied and storage conditions. Arabica contains *ca.* 6–8.3% dm and Robusta *ca.* 3.3–4.1% dm.

Other simple sugars are present in green beans, including reducing sugars. Quantities are small, total reducing sugar content being *ca.* 0.1% dm in Arabica and *ca.* 0.5% dm in Robusta. Sugars such as arabinose, galactose, raffinose, rhamnose and ribose are present only in trace amounts, while slightly higher concentrations of fructose and glucose are present.

The content and nature of sugars in green beans is important in the development of flavour and pigmentation during roasting (see page 246). The sugar profile of green beans is likely to change during decaffeination. This normally involves a slightly decreased sucrose content, but a marked increase in reducing sugar content.

Polysaccharides are important constituents of green beans and comprise 40-50% of dry matter. The naturc of coffee polysaccharides has not been fully elucidated, but there appear to be some unusual features. Mannose has been shown, after hydrolysis, to be the predominant monosaccharide, although arabinose, galactose and glucose are also present in significant quantities. Some doubt remains over the nature of the polysaccharides present and there may be considerable variation. Cellulose (glucan), mannan and arabinogalactan are most probably the constituent polysaccharides.

Cellulose is primarily a β-D-(1-4) glucan and is present at a level of *ca*. 5% dm. Mannan is also present at a level of *ca*. 5% dm and contains 94% D-mannose and 2% galactose. The molecule is thought to be a linear chain of β-D-(1-4) mannopyranose units. Mannan appears to important in conferring hardness to the bean. Arabinogalactan contains L-arabinose and D-galactose in the approximate ratio of 2:5. The backbone of the molecule appears to consist of β-(1-3) linked D- galactose units. Arabinogalactan is present at *ca*. 8.5% dm.

(c) Chlorogenic acid

Green coffee beans contain small quantities of free quinic acid, but most occurs as a series of esters. These are collectively known as chlorogenic acids. A number of appellations have been applied to chlorogenic acids and the trivial names are often confusing. Quinic acid and chlorogenic acids should be treated as cyclitols, with the general structure shown in Figure 5.6. Under this system the naturally occurring isomer is 1L-1(OH),3,4,5-tetrahydroxycyclohexane carboxylic acid. The most important chlorogenic acids in the coffee bean are caffeoylquinic acids, dicaffeoylquinic acids, feruloylquinic acids, *p*-coumaroylquinic acids and caffeoylferaloylquinic acids.

The content of the chlorogenic acid complex (CGA) of the coffee bean is much higher than that of other plant organs. The CGA is

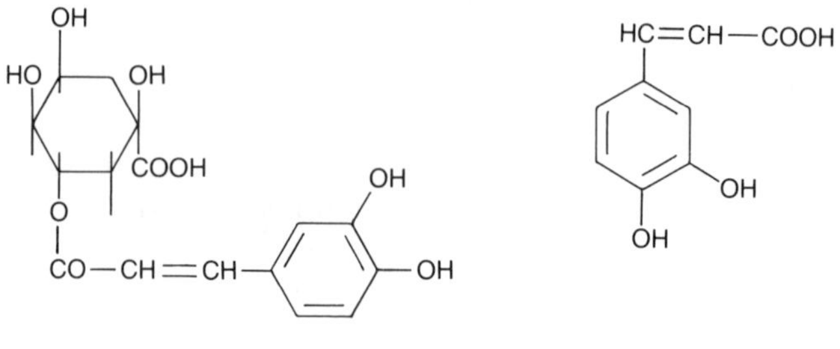

Chlorogenic acid

Caffeic acid

Figure 5.6 General structure of cyclitols.

known to be involved in control of indole acetic acid levels, but the high content in the bean suggests that there may be further functions. Green beans and coffee pulp has been shown to contain an indole acetic acid oxidase- inhibiting CGA–protein complex, which may be of physiological significance during formation and germination of the bean. It has also been postulated that CGA is a precursor in the biosynthesis of lignin and in the synthesis of the initial protective layer formed at the site of physical damage. The high CGA content may also protect against predation by birds, insects and mammals and have an anti-microbial function.

Analysis of the CGA content is difficult, but, it is generally recognized that Robusta coffee has a higher content than Arabica coffee. Reported levels for Robusta are 7.0–10.5% dm and for Arabica are 5–7.5% dm. Levels of CGA appear to be dependent on species and are unaffected by differences in agronomic practice or method of processing.

The organoleptic properties of chlorogenic acids have received relatively little attention, although it is recognized that there is little, if any, contribution to acidity. Various sensory descriptors have been applied to chlorogenic acids, but 'lingering' and 'metallic' appear to be the key terms. The superior quality of

* The white fluorescence, exhibited by 'stinker' beans under ultra-violet excitation, has been attributed to caffeic acid. This is released from the caffeoylquinic acid component of CGA by mould-derived hydrolytic enzymes. Glycosides, such as scopolin, are also subject to mould hydrolysis and may be an alternative source of caffeic acid.

Arabica coffee has been attributed to the lower CGA content, but this may be an over-simplification.

Phenols other than chlorogenic acids may be present, including scopolin. Traces of free cinnamic acids, especially caffeic acid, are found in all green beans. Levels in excess of 0.05% dm, however, would indicate CGA hydrolysis.

(d) Carboxylic acids

Aliphatic carboxylic acids are present in green beans, together with smaller quantities of alicyclic and heterocyclic acids. The carboxylic acid profile of Arabica coffee includes 0.5% dm citric acid, 0.5% dm malic acid, 0.2% dm oxalic acid and 0.4% dm tartaric acid. There appears to be no equivalent data for Robusta coffee.

Acidity increases during storage of green beans. This has been attributed to the release of free fatty acids by enzymic hydrolysis of lipids. Evidence for this is not entirely convincing and further work is required.

(e) Phosphoric acid

Phosphoric acid is the major determinant of acidity in green coffee.

(f) Lipids

The lipid component of green coffee beans comprises coffee oil, which is primarily present in the endosperm, and 'coffee wax', which is present on the outer layer. Arabica coffee contains *ca.* 15% dm oil and Robusta *ca.* 10% dm. The oil contains triacylglycerols and considerable proportions of other lipid components (Table 5.5).

There is no significant difference between Arabica and Robusta coffees, with respect to fatty acid composition. Saturated fatty acids tend to esterify preferentially with diterpene alcohols. During ageing the free fatty acid content increases from *ca.* 0.5 to 1.89% dm.

On the basis of solubility in petroleum ether, coffee wax can be separated into two components. The soluble component, which comprises 37% of the total, is generally similar to coffee oil. Differ-

Table 5.5 Composition of the lipids of green coffee

	%
Triacylglycerols	75.2
Esters of terpene alcohols and fatty acids	18.5
Diterpene alcohols	0.4
Esters of sterols and fatty acids	3.2
Sterols	2.2
Tocopherols	0.04 - 0.06
Phosphatides	0.1 - 0.5
?	0.6 - 1.0

Note: Based on extraction at 7% dm. Composition may differ if higher extraction rates are used.

ences exist in fatty acid composition (Table 5.6), while the wax component also contains a lower level of saturated hydrocarbons.

The petroleum ether-insoluble component of coffee wax comprises caffeine and phenolic compounds. The phenolic compounds are derived from a combination of the primary amino group of 5-hydroxytryptamine with either arachidic acid, behenic acid or lignoceric acid in the ratio 12:2:1.

Table 5.6 Composition of fatty acids of coffee oil and the petroleum ether-soluble component of coffee wax

	Fatty acids %	
	Oil	Wax
C14:0	1.5	trace
C16:0	24.9	31.1
C18:0	6.5	9.6
C18:1	4.8	9.6
C18:2	23.8	43.1
C18:3	trace	1.8
C20:0	14.1	4.1
C22:0	21.0	0.9
C24:0	3.7	trace
Saturated C18:0 - C24:0	45.3	14.6

(g) Volatile compounds

The volatile compounds of coffee are largely responsible for the aroma. Green beans are often thought to have no agreeable flavour or aroma, but a large number of volatiles are present. Many of these increase in concentration during roasting, while the concentration of other volatiles falls due to degradation.

Aliphatic hydrocarbons are derived from the oxidation of green bean lipids during storage and transport prior to roasting. These mainly comprise lower alkanes and alkenes. Pyridines, quinoline, pyrroles, arylamines and polyamines have been detected in all green beans examined, while the characteristic note of green beans is provided by methoxypyrazine. Three polyamines are present: putrescine, spermine and spermidine. The presence of some volatiles, including alcohols, carbonyls, esters, furans, phenols and thiols is dependent on the species or variety.

Comparisons of Arabica and Robusta coffees have shown significant differences. Arabica contains lower concentrations of furans, pyrazines, benzene and naphthalene derivatives, 2-butanone, 2-heptanone, 3-heptanone, 2-methylpropanol and 2- and 3-methylbutanal. Conversely, levels of terpenes, 3-methylbutan-1-ol and 2- and 3-octanone are higher in Arabica than Robusta coffee. More specific relationships have been found between volatile components and coffee type. The 'solai' flavour of Kenyan coffee, for example, has been correlated with high concentrations of ethanol and dimethyl sulphide.

5.5.2 Changes during roasting

(a) Nitrogenous compounds

Theoretically, sublimation of caffeine should lead to a considerable loss during roasting, but in practice loss is small, unless roasting is very severe. On a dry matter basis the caffeine content increases by up to 10% as a consequence of loss of water and degradation of other components of dry matter. Loss of caffeine is probably limited by an elevation in sublimation temperature due to increase of pressure within the bean and a low rate of diffusion of caffeine vapour through the bean.

Trigonelline is rapidly degraded during roasting and losses in the order of 50–85% may be anticipated. The extent of degradation is strongly dependent on the temperature and length of roasting. Rapid degradation occurs at 230°C, with loss of *ca.* 85% trigonelline. At 180°C, little loss occurs during the first 15 min of roasting but the rate of degradation increases between 15 and 45 min, and a total loss of *ca.* 60% of occurs.

Trigonelline is degraded to nicotinic acid, *N*-methylnicotinamide and methylnicotinamide, together with 29 volatile compounds, of which nine have been identified in coffee aroma. The volatile compounds comprise 12 pyridines (46%), four pyrroles (32%) and nine bicyclics (21%).

Nicotinic acid in roast coffee is derived primarily from trigonelline, the vitamin being present at levels of only 1.6–4.4 mg/100 g in green coffee. Nicotinic acid is formed during roasting by progressive demethylation of trigonelline. The reaction is temperature dependent and formation of nicotinic acid only becomes significant above 160°C. The vitamin itself, however, is degraded at temperatures in excess of 220°C. Demethylation accounts for only 1.5% of the trigonelline degraded during roasting, the nicotinic acid content of roast coffee being *ca.* 14.9 mg/100 g. Higher quantities may be present in dry- than in wet-processed coffee, while decaffeinated coffee contains less nicotinic acid, probably as a result of loss of trigonelline during decaffeination.

Proteins are denatured during roasting and degraded to lower molecular weight fragments. Some react with carbohydrates in the Maillard reaction and there may also be reactions with phenolic compounds. In some cases protein-containing complexes formed during the early stages of roasting are broken down to amino acids in later stages. Loss of protein during roasting is usually 20–40%, depending on roast severity, but can exceed 50%.

Free amino acids are severely degraded during roasting and only traces remain. Mechanisms of degradation vary according to the nature of the amino acid and include simple pyrolysis, interaction with carbohydrates, or with α-diketones (Maillard intermediates) during Strecker degradation.

The sulphur-containing amino acids cysteine, cystine and methionine are degraded by pyrolysis as well as reacting with Maillard

Figure 5.7 Structure of thiophenes and thiazoles .

intermediates. Pyrolysis of cysteine leads to formation of hydrogen sulphide, ammonia and acetaldehyde, while the products of reaction with Maillard intermediates include thiophenes and thiazoles (Figure 5.7). Hydrogen sulphide may react with the double bond of 3-methyl-2-buten-1-ol (prenyl alcohol) to form 3-mercapto-3-methylbutanol. Methionine is involved in Strecker degradation reactions with α-dicarbonyl compounds, which results in decarboxylation and transamination to an aldehyde containing one less carbon atom. This reacts to form a pyrazine or oxazole. Methional is also formed during Strecker degradation and is further degraded to methanethiol (methyl mercaptan) and dimethyldisulphide.

The hydroxy amino acids serine and threonine yield more than 200 heterocyclic compounds. Alkylpyrazines are particularly abundant. These are probably formed from hydroxy amino acids which have built up to larger molecules in the early stages of roasting. Unsaturated dihydrocyclopentapyrazines are formed at low levels and hydroxy amino acids also act as precursors for alkyl pyridines (*cf.* trigonelline degradation.

Pyrroles are formed by reactions between proline and hydroxyproline and Maillard intermediates, including α-dicarbonyls, unsaturated carbonyls, cyclic enolenes, acetylfurans and formylfurans. Proline and hydroxyproline may also play a direct role in the Maillard reaction. The structures of pyrazines, pyridines and pyrroles are illustrated in Figure 5.8.

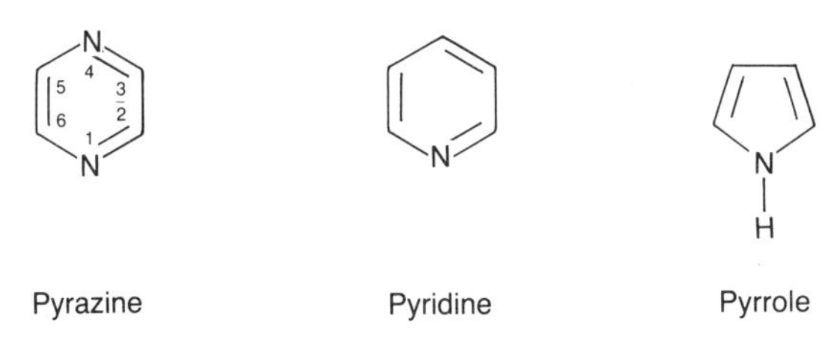

Figure 5.8 Structure of pyrazines, pyridines and pyrroles.

(b) Carbohydrates

The carbohydrates of coffee undergo major changes during roasting. The extent of change progressively increases with degree of roast from light to dark. Of the low molecular weight sugars, sucrose is rapidly lost. Only 3.4% of the sucrose in the green bean remains after light roasting. As little as 1% sucrose remains in medium roasted coffee, while the sugar may be undetectable in very dark roasts. Other simple sugars, especially glucose, fructose and arabinose are progressively destroyed as the degree of roasting increases. Relatively little data is available which directly compares quantities in the green coffee with those in the roast. It has been shown, however, that glucose levels decreased from 0.185 to 0.07% dm and arabinose levels from 0.16 to 0.08% dm during medium roasting. In the very early stages of roasting an initial increase in glucose and fructose concentration can be demonstrated. This results from inversion of sucrose and is accompanied by generation of small quantities of arabinose and rhamnose by hydrolysis.

The fate of sucrose during roasting is still not entirely clear. The sugar undergoes caramelization on heating beyond the melting point of 130°C, the transformation being accompanied by formation of CO_2 and water. Observed losses of weight are primarily CO_2, water and generated volatile compounds. Furan derivatives are the principal products of decomposition of monosaccharides and higher sugars, typical degradation products at the start of pyrolysis being vinylfurans, furfural and 2,4-pentadienal. Furans undergo complex reactions with other compounds, which may be catalysed by phosphates, acids or alkalis. Sulphur-containing amino acids react strongly with furans forming sulphur derivatives, such as the organoleptically significant furfurylthiol (see page 248). Water-soluble, brown pigments of unknown origin are also products of pyrolysis. Reducing sugars also undergo pyrolysis, but this reaction is likely to be of secondary importance to Maillard-type reactions with amino acids and protein degradation products.

The early stage of the Maillard reaction primarily involves formation of glycosylamines by condensation of aldose sugars with amino groups of amino acids and other compounds. Ketose sugars are also involved via parallel formation of ketosylamines, but the reaction proceeds to a much lesser extent. Subsequent stages involve formation of Amadori rearrangement products, which may degrade to hydroxymethylfurfural via C1,2 degradation and to C-methyl reductones via C2,3 degradation. Aroma volatiles, such as pyruvaldehyde and 2,3-butanediene, are formed by fission.

Roasting causes significant loss of polysaccharides. Loss correlates with degree of roasting and only *ca.* 75% of long chain polysaccharides remain after medium roasting. Structural changes are indicated by an increase in polysaccharides of shorter chain length. Glucose units appear to be least susceptible to loss on roasting and arabinose units most susceptible.

There is only limited formation of monosaccharides from polysaccharides during roasting. Anhydride formation, such as production of glucosans from glucans occurs, and there is also formation of polymeric 'condensation products' and complexes. These are frequently linked by covalent bonds to proteins, protein fragments and chlorogenic acids and their degradation products. Condensation complexes are commonly referred to as 'melanoidins', 'humic acids' and 'Maillard products'.

(c) Chlorogenic acids

The quantity of detectable chlorogenic acids falls considerably during roasting. A number of possible mechanisms exist, including acyl migration, hydrolysis, oxidation, fragmentation and accompanying loss of fragments with flue gases, polymerization and association with denatured, or degraded, proteins. It is likely that different chlorogenic acids differ in susceptibility to the various mechanisms.

In medium roast coffee, the fate of *ca.* 50% of chlorogenic acid is as ill defined pigmented material, formed by reaction of CGA and CGA degradation products with protein, free quinic acid and free, low molecular weight phenolic compounds. Quinic acid is formed during roasting, quinic acid lactone has been detected and epimerization may be predicted. Quinic acid itself is degraded, yielding primarily catechol, quinol, pyrogallol and 1,2,4-trihydroxybenzene. Catechol is also formed as a consequence of the degradation of caffeic acid, but major degradation products are 4-methylcatechol, 4-ethylcatechol, 4-vinylcatechol and 3,4-dihydroxycinnamaldehyde. The quantity of phenols in roast coffee tend to directly reflect the quantity of their immediate precursors in green coffee.

(d) Carboxylic acids

Content of total aliphatic acids is at a maximum after *ca.* 15% weight loss during roasting. This corresponds to a medium roast. Acetic and formic acids increase in the early stages of roasting to a maximum at a roasting loss of *ca.* 14–15%, but after this levels fall rapidly. The maximum quantity formed is unpredictable and highly dependent on the conditions during roasting. Citric and malic acids increase in concentration in the early stages of roasting, but are subsequently degraded. In general, pressure roasting leads to greater acidity in the roast bean. This is considered desirable for Robusta coffee, but not for high quality, wet processed Arabica.

(e) Lipids

Relatively little degradation of lipids occurs during roasting. Terpenoids degrade to monoterpenoids, such as linalool and myrcene, while higher terpenoids, such as squalene, may be oxidized and yield furans. Autooxidation of lipids results in formation of volatile aldehydes and compounds such as 2,3-butanedione, hydroxyacetone and glyoxal.

(g) Volatile compounds

Roasted coffee contains more than 600 volatile aroma compounds (Table 5.7), the largest number identified in any food and drink. In general, however, little is known of the contribution of individual compounds to the overall flavour and only a relatively small number (*ca.* 60) may make a significant contribution to flavour. In a number of cases, volatile compounds are considered desirable at low, but undesirable at higher concentrations and the situation is complicated further by synergism and antagonism between different volatiles. There can also be significant differences in the volatile composition depending on coffee species/variety, conditions during cultivation and harvesting, and method of processing. Sulphur-containing compounds are present at relatively low, but have very low detection thresholds. Furfurylthiol has an aroma of freshly roast coffee at concentrations of 0.01–0.05 ppb, but has an undesirable sulphurous and stale aroma at higher concentrations. Kahweofuran also has an aroma associated with roasted coffee, which becomes sulphurous at higher concentrations, while dimethyl sulphide is an essential part of a high quality aroma in all circumstances.

Table 5.7 The volatile components of roast coffee

Furans		84
Pyrazines		70
Hydrocarbons		
aliphatic	35	67
aromatic	32	
Pyrroles		66
Ketones		
aliphatic	57	62
aromatic	5	
Phenols		39
Esters		29
Oxazoles		28
Thiazoles		27
Thiophenes		27
Aldehydes		
aliphatic	16	21
aromatic	5	
Acids		19
Alcohols		
aliphatic	16	18
aromatic	2	
Sulphides		15
Amines		11
Pyridines		11
Quinoxalines		11
Disulphides		10
Lactones		9
Thiols		6
Indolines		4
Quinolines		4
Anhydrides		3
Pyrines		3
Trisulphides		3
Dithiolines		2
Nitriles		2
Thioesters		2
Acetals		1
Imides		1
Oxines		1

Furfurylmethyl disulphide has a bread-like aroma, which may be considered either desirable or undesirable. Thiophenes tend to be associated with undesirable sulphurous, mustard-like and onion-like aromas, although their esters and aldehydes may produce desirable caramel or nutty flavour notes. Thiazoles containing both nitrogen and sulphur in a heterocyclic ring can be associated with a wide range of aromas. These are often green and vegetable-like, but with increasing substitution nutty, roasted and meaty notes become dominant. Alkyl derivatives of thiazoles have green, vegetable-like aromas, alkoyl derivatives are sweet to nutty and acetyl derivatives nutty to cereal-like.

Three sulphur-containing volatile compounds, 3-methyl-2-butene-1-thiol (prenylmercaptan), 3-mercapto-3-methylbutenal and 3-mercapto-3-methylbutyl formate, have 3-methyl-2-buten-1-ol (prenyl alcohol) as common precursor. The aromas of these volatiles are objectionable in large quantities, but may play a positive role at low concentrations.

Pyrazines are abundant in roast coffee and are associated with bitter, bittersweet and corn-like aromas. Alkyl substitution results in volatiles with nutty, burnt, grassy and pungent aromas. The presence of a thiol group in a pyrazine imparts nutty, biscuit-like notes, while the additional presence of a furfural group results in an aroma which may be perceived as either roast meat- or coffee-like. Low molecular weight methyl and dimethyl pyrazines are present in larger quantities than higher molecular weight alkyl pyrazines. The aroma imparted by the pyrazine complex as a whole varies with concentration and there are both antagonistic and synergistic effects with other aroma components.

Pyridines are present in greatest quantities in dark roast coffee, but can be associated with unpleasant flavours in less severe roasts. Pyridine itself has a sharp taste and a characteristic 'dirty' odour. Some pyridines have a fruity and, possibly, a coffee-like note. In most cases, however, the predominant aroma is bitter-green and astringent. Roasted overtones may also be present. Methyl, acetyl and vinyl derivatives are all present and of particular importance are 2-methylpyridine, which has an astringent, nutty aroma and 3-ethylpyridine, which provides green, buttery and caramel notes. Despite the structural similarity of pyrroles and pyridines, the aroma qualities are distinct. In total, dark roasted coffee contains the greatest quantity of pyrroles, but the types formed differ with temperature and time of roasting and the pH value of the coffee. The aroma qualities of pyrroles can differ markedly according to concentration. Acylpyrroles have a bready, cereal aroma, which becomes sweet, smokey and slightly medicinal as the concentration increases. Similarly alkylpyrroles have a basic sweet, slightly burnt

Figure 5.9 Structure of furans and oxazoles.

aroma, which becomes petrol-like at high concentrations. Furfurylpyrroles, however, provide green, mushroom-like notes at any concentration above detection threshold.

Numerous furans (Figure 5.9) are present, including acids, alcohols, aldehydes, ethers, esters, ketones, sulphides and thiols. Furans are also present in combination with heterocyclic compounds, such as pyrazines and pyrroles. The wide variety of structures of furans is reflected in the numerous and varied aromas. Many furans have a predominant burnt sugar, caramel aroma and furfuryl alcohol is of particular importance in providing the burnt notes of dark roasted coffee. Furan itself and 2-methylfuran have an ether-like aroma and furfural is hay-like. A characteristic of furans is that perception of the aromas varies markedly on a person-to-person basis.

Aldehydes are present in relatively high quantities in freshly roast coffee, but are volatile and prone to oxidation. Methanal, ethanal and pyruvaldehyde have pungent, stinging aromas, which are undesirable in large quantities, but longer chain aldehydes provide floral notes.

Ketones are also lost during storage of roast coffee. The aroma of those present in coffee varies considerably. Cyclic ketones (Figure 5.10.), such as 3-hydroxy-2-methyl-4-pyrone (maltol) and 2-hydroxy-3-methylcyclopent-2-enone (cyclotene), are derived from caramelization of sugars and have sweet, fruity, burnt aromas. Propanone

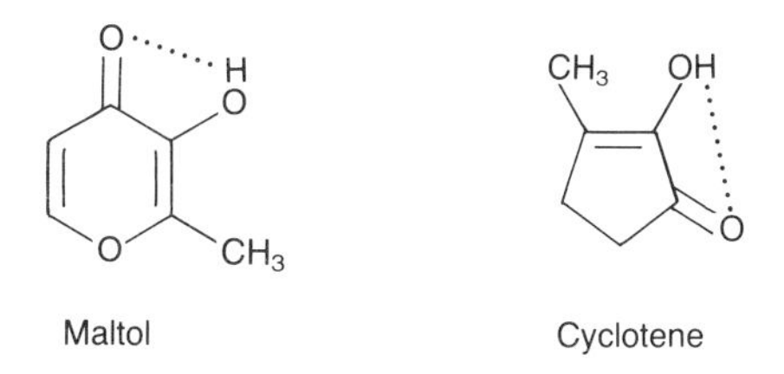

Figure 5.10 Structure of cyclic ketones in roast coffee.

and some other ketones have a sweet, pungent aroma, while 2,3-butanedione has buttery notes.

Phenolic compounds are generally characteristic of dark roast and Robusta coffees. Astringent and smoky aromas dominate and are present in combination with burnt, spicy and clove-like notes.

5.5.3 Composition of fresh brewed coffee and instant coffee

The quantity and nature of soluble solids in the final consumer product are of obvious importance to the coffee industry. Total solids content varies according to a number of factors. Home brewing is less efficient than industrial extraction, but with home brewing there is a wide range of efficiency depending on method used, water: coffee ratios and coarseness of grind. Total solid yields vary from *ca.* 15 to 28% dm, high yields being obtained with some types of automatic coffee makers using very finely ground coffee. Robusta coffee usually yields higher solids contents than Arabica.

(a) Nitrogenous compounds

The solubility of caffeine increases rapidly with increasing temperature. Under industrial conditions, extraction rates are very high and virtually all of the caffeine present in the roasted beans passes into instant powder. Levels in instant powder vary from *ca.* 2.8 to 4.6% dm. Nicotinic acid is also highly soluble and is almost entirely extracted during domestic coffee brewing and during manufacture of instant coffee.

Little information is available on the protein content of coffee brews. Proteins, however, are effectively denatured during roasting and are of very low solubility. A number of surveys of the protein content of instant powders have been made and show variation from *ca.* 0.5 to 5.5% dm. Different analytical techniques were used, which may account for some of the variation. Hydrolysis of denatured proteins takes place during infusion, resulting in production of free amino acids. The free amino acid content of brews is, however, very low.

(b) Carbohydrates

The total quantity of carbohydrates present is highly dependent on the method of extraction used. Industrial-scale extraction, at temperatures in excess of 100°C, yields a markedly higher level of solids than domestic extraction. Polysaccharides account for a considerable part of the increase, especially when extraction temperatures exceed 130°C. There is also an increase in the level of monosaccharides during extraction at higher temperatures.

Analysis of instant coffee has demonstrated the presence of small quantities of arabinose, galactose and mannose, and trace quantities of several other sugars, including sucrose, xylose and ribose. Glucose and fructose have also been detected, but may have been derived from chicory. The levels of arabinose, galactose and mannose in instant coffee are significantly higher than levels in equivalent roast beans. This is probably due to hydrolysis of araban during extraction at temperatures above 100°C. Some loss of monosaccharides may occur during drying if excessively high temperatures and long residence times in the dryer are permitted.

Although polysaccharides are present in significant concentrations most, if not all, are probably bound to Maillard products. Carbohydrate conversion products will also be extracted and sucrose caramelized compounds are particularly water-soluble.

(c) Chlorogenic acids

There have been no known reports of degradation during commercial extraction or brewing in the home. A level of hydrolysis, acyl migration and, possibly, oxidation may, however, be expected. Levels of chlorogenic acid in instant coffee vary from *ca.* 3.5 to 12% dm. The acid is highly ionized in brews and may be expected to contribute to the organoleptic character.

(d) Carboxylic acids

Levels of carboxylic acids in the brew, or in instant coffee, reflect the type of bean and the degree of roasting. Levels in the brew are usually slightly higher than in the roast bean, especially with acetic and citric acid. Loss of volatile acids occurs during drying and up to 75% of the acetic and 50% of the citric acid content may be lost in this way.

(f) Volatiles

A considerable loss of volatiles occurs during manufacture of instant coffee. This loss accounts for differences between freshly brewed and instant coffee and has been the subject of much research within the coffee industry. Volatiles are often added back to instant coffee, but concentration and drying at high temperatures can lead to the presence of degradation compounds with resultant effect on flavour. The relatively poor permeability of volatiles at the surface layers of drying droplets means that the concentration stage is likely to be of greatest importance with respect to loss of volatiles during manufacture of instant coffee. It is probable that the greatest losses occur during evaporation (or drying) from dilute solutions. In the case of freeze-dried instant coffee, slow drying is usually thought to minimize loss of volatiles

Although preparation of instant coffee probably involves a large number of changes in the composition of volatiles, a relatively small number of compounds may be of key importance. Instant coffee is often considered to have a characteristic 'tang', which is absent from the fresh brew. This has been attributed to the presence of relatively high levels of furfural and pyridine. At the same time freshly brewed coffee has a distinctive and eponymous characteristic which is absent from instant coffee. This characteristic has been attributed to *E*-2-nonenal, which also reduces less desirable caramel and astringent notes.

Changes in the composition of volatiles continues during storage of instant coffee in air. Relatively little information is available, but it appears that the relative proportions of propanal, 2-methylpropanal, pentonal and 3-pentanone increase, while methyl acetate, ethyl acetate, butanal/butanone and 3-methylbutanal/3-methylbutan-1-ol decrease. The overall organoleptic effect is a reduction in the brewed coffee characteristic and an increase in 'metallic' and 'unnatural' notes. Changes under an atmosphere of CO_2 are negligible.

EXERCISE 5.1.

You are manager of a coffee plantation in a South American country. A species of stem borer has been introduced into your country and, although controllable by dieldrin, is causing significant loss of yield. Your employers have commissioned two separate reports concerning long-term control of this problem. The first recommends replacement of existing trees with a more resistant cultivar, while the second recommends the introduction of a predator and the setting up of a biological control programme. Although you are not an expert in pest control, it is your responsibility to decide whether to adopt one, or both, of these recommendations, or to persist with chemical control. Discuss the options available and write a short report for your management outlining, what you consider to be, the most suitable action.

EXERCISE 5.2.

On a number of occasions, it has been suggested that coffee packaging should carry a warning of the dangers of consumption, similar to that printed on cigarette packets. Do you consider that such a warning would be justified in the current state of knowledge or that warnings would be effective? Consumption of a number of other foods and beverages have also been associated with ill health. What criteria would you use for determining the need for health warnings, both for specific groups and the population as a whole.

6

COCOA, DRINKING CHOCOLATE AND RELATED BEVERAGES

OBJECTIVES

After reading this chapter you should understand

- The nature of the cocoa plant, *Theobroma cacao*
- The processing of the raw bean and the key role of fermentation
- The roasting procedure
- Processsing of the roast bean
- The nature and manufacture of the different consumer beverages
- The physiological effects of cocoa consumption
- Chemical changes occurring during the various stages of processing and the relationship with quality
- The microbiology of cocoa processing and the role of micro-organisms in fermentation
- The role of cocoa as a potential source of *Salmonella*

6.1 INTRODUCTION

Theobroma cacao, the cocoa tree, was first cultivated in Central America by the Mayas of Yucatan and the Aztecs of Mexico. The Aztecs believed the tree to be of divine origin, a belief reflected in the appellation *Theobroma* – 'Food of the Gods' – given to the genus by Linnaeus. A preparation, *chocolatl*, made by mixing roasted ground cocoa nibs with water, maize and spice, was consumed as a luxury in the court of the Aztec emperor Montezuma and was widely thought to have aphrodisiac properties.

Cocoa beans were brought to Europe by Columbus, but were initially thought to be of novelty value only. The first recognition of the potential commercial importance was made by another Spaniard,

Don Cortes. The Spanish were able to produce a beverage from cocoa which was acceptable to European tastes, and during the mid-17th century the chocolate drink spread to Italy, Holland and France and, subsequently, to England. The drink, however, was expensive and could only be afforded by the very rich. In England, prices fell following the introduction of factory scale processing by Fry of Bristol in 1728, but consumption was limited by the fatty and rather crude nature of chocolate drinks then available.

From a technological viewpoint, the greatest breakthrough occurred with the invention of the cocoa press in 1828, by the Dutchman van Houten. This removed a high proportion of fat from the cocoa powder resulting in a beverage which was both more palatable and easier to prepare. To the 19th century consumer cocoa, as made using van Houten's press, was effectively a new product and consumption increased dramatically. Consumption of cocoa was, however, always small in comparison with tea and after the 1890s, chocolate confectionery became of overwhelming importance in the commercial exploitation of the cocoa tree.

In the years between the first and second world wars, cocoa beverages became regarded as a bedtime drink, especially of less affluent social groups. This image has largely remained until the present day. The high fat and, in the case of drinking chocolate, high sugar content has also limited consumption in recent years.

BOX 6.1 **Cocoa is a vulgar beast**

The downmarket image of cocoa means that, while some brands have always been favoured, there is no consumer association between the origin of the cocoa, the method of processing, or the means of preparing the beverage and quality. This is unlike the situation with coffee and tea and does mean, at least, that the world is spared the 'cocoa snob' to accompany the 'coffee snob', 'tea snob' and 'wine snob'. From the manufacturer's viewpoint, however, it does limit the possibilities for product diversification, since there is no market for cocoa from a particular region, or which has undergone a particular roasting process.

Despite these difficulties, new cocoa beverages have been developed. These include reduced-calorie products made from defatted cocoa powder and artificial sweeteners, such as Aspartame™. Added value, flavoured products have been introduced alongside drinks of perceived higher quality based on chocolate crumb rather than cocoa powder.

6.2 AGRONOMY

6.2.1 *Theobroma cacao*

The cocoa tree, *T. cacao*, is a member of the family Sterculiaceae. The genus *Theobroma* consists of over 20 species, but only *T. cacao* is of commercial value. Other species may, however, have a role as sources of genetic material in breeding improved strains. Two distinct subspecies were recognized in the late 19th century, *criollo* and *forastero*. Very little *criollo* cocoa is in world commerce, which is dominated by *forastero*. A third subspecies, *trinitario* is a hybrid of *criollo* and *forastero*. Unless planted in selected clones, *trinitario* develops as hybrid swarms exhibiting a wide range of combinations of *criollo* and *forastero* characteristics. *Trinitario* cocoa is traded on a global basis, but to a lesser extent than *forastero*.

Theobroma cacao bears flowers in small groups on the trunks and lower main branches of the trees, which may grow as high as 9 m. Only a small proportion of the flowers are pollinated, the main agent of pollination being a midge, *Ceratopogonidae*. Pollinated flowers develop into berries ('pods'), which mature over a 5-6 month period. During this time a significant number of berries drop prematurely. The berry is a drupe 2.5-4.0 × 1.25-1.75 cm in size. When ripe the berry contains between 20 and 40 seeds, surrounded by a mucilaginous pulp (Figure 6.1).

6.2.2 Cultivation of *Theobroma cacao*

Theobroma cacao is a native of the Amazon rainforests, but it is now grown, on a commercial basis, in South America and the West Indies, West Africa, Malaysia and New Guinea. Most of the cocoa in

---

* The tree, its berries and its seeds are all referred to as *cacao*. The term 'cocoa' is used to describe the bulk, fermented and dried beans, as well as the powder produced from the beans. (Minifie, B.W. 1989. *Chocolate, Cocoa and Confectionery: Science and Technology*, 3rd edn. Van Nostrand Reinhold, New York.

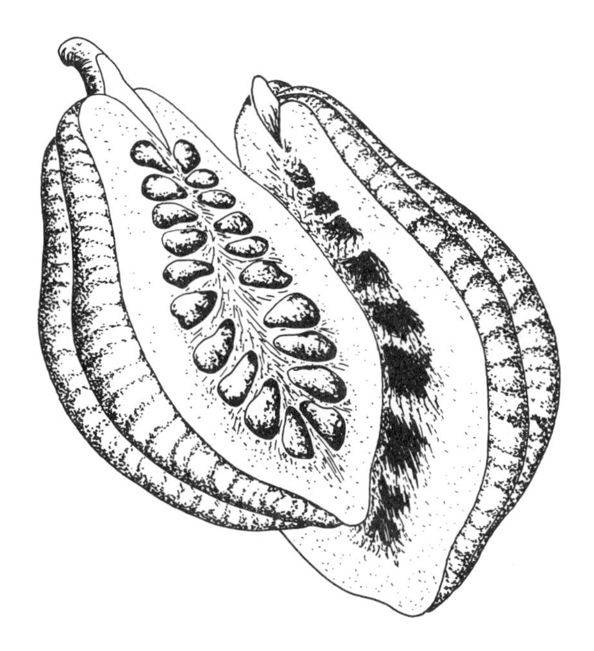

Figure 6.1 Section of cacao pod.

international trade is grown in West Africa and Ghanaian cocoa is considered to be of definitive quality. Malaysia has developed rapidly in recent years, but the cocoa is often of poor quality and commands low prices. Although grown over a wide geographical area, the tree has relatively specific requirements with respect to altitude, latitude and humidity and some 75% of commercial cultivation is within 8° either side of the equator. The tree is favoured by a hot, moist climate and has an optimal growing temperature of 18–32°C and an optimal rainfall of 1500–2000 mm per annum. Under these conditions, the relative humidity will be 70–80% during the day and rise to saturation point at night. *Theobroma cacao* is essentially a lowland tree and cultivation above 3000 feet is unlikely to be successful.

Seeds are the most common means of propagation, but vegetative methods are increasingly used for propagating high-yielding clones or clones with particular quality attributes. In West Africa seeds may be planted directly by simply pressing into the soil, but this practice tends to be unreliable and a better system is to use seedlings raised in a nursery. When the nursery system is used, seedlings are planted out at between 4 and 6 months old. A spacing of

7–10 feet is required and young trees require special protection against pests. The area around the young trees should be kept free from weeds and temporary shade is required. Cash crops, such as cassava, maize and plantain are suitable. In many areas shade is also provided for the adult tree, although the value of this practice is doubted. Light shading only is required providing soils are of reasonable fertility, while in rich, highly fertile soil shading is not usually necessary. Shade-grown trees tend to produce berries with a higher pulp sugar concentration and this is reflected by a higher acidity in the resulting cocoa powder.

Theobroma cacao is subject to attack by more than 1500 different insects. The most serious infestations involve the capsids, *Distantiella theobroma* and *Sahlbergia singularis*, which attack the young shoots and berries. In severe cases the tree is killed, but effective control is possible using pesticide sprays. Other insects pose a secondary, although still significant, threat to the cocoa tree. Various species of thrips, leaf miners, stem borers and ants are most commonly involved.

BOX 6.2 **The noisome pestilence**

The discovery of the organochlorine pesticide Lindane in chocolate confectionery bars has led to scrutiny of pest control practice in the cocoa growing areas of Africa and the Pacific basin. It has been alleged that Lindane and no less than 29 other pesticides are used and that pesticide usage has increased in an attempt to increase yield and offset falling prices. A confectionery trade association, however, claims that low cocoa prices mean that pesticides are too expensive for use by most African farmers. (*The Guardian*, April 19, 1993).

Disease has caused widespread economic loss in cocoa plantations and surveys have suggested tree mortality rates of 30% or more in some areas. The viral disease, swollen shoot disease, caused widespread devastation in Ghana and, to a lesser extent, Nigeria during the late 1940s and early 1950s. The causative virus affects roots, leaves and berries as well as shoots and has a high mortality rate, death occurring *ca*. 3 years after infection. A pool of infection exists amongst some species of native tree, viruses being transmitted to *T. cacao* by sap-sucking mealy bugs. The disease can also be spread over a wider geographical area by movement of seedlings and cacao seeds and strict quarantine must be observed. Attempts to develop resistant strains of *T. cacao* or to eliminate mealy bugs by insecticides have not been entirely successful and the most effective means of control lies with destruction of infected trees. It is essential that destruction includes *all* infected trees, including those in the early stages of the disease which are still yielding acceptable quality beans. To be acceptable to farmers, this requires an education programme together with realistic financial compensation.

Swollen shoot disease is potentially devastating, but black pod (*Phytophthora palmivora*) is of more common occurrence. The fungus invades leaves and young shoots as well as berries, but losses stem from infection and consequent loss of the berry. Total loss of crop can occur in areas of high humidity, but copper fungicides offer an effective means of control. A number of other fungal diseases of *T. cacao* are of economic significance. These include witches' broom disease caused by *Crinipellis perniciosa*. This disease results in production of distorted shoots (the brooms) and in loss of yield. Witches' broom disease was originally localized in the Amazon areas of South America, but is now also prevalent in the West Indies. Short term control is possible by cutting out and burning the broom, but resistant clones have now been developed.

6.2.3 Harvesting

Theobroma cacao normally begins to bear berries after 3 years and yield reaches a maximum after 8–9 years. Trees simultaneously bear flowers, developing berries and mature fruit and harvesting is spread over several months. Harvesting is manual in nature, mature fruit being cut from the tree by a long knife or cutlass.

6.3 TECHNOLOGY

Processes involved in the production of cocoa and other chocolate based beverages are summarized in Figure 6.2. It should be noted that the processing of cocoa beans involves risk of contamination with *Salmonella* and that prevention of contamination from the raw to the heat processed (roast) material is an essential part of production management.

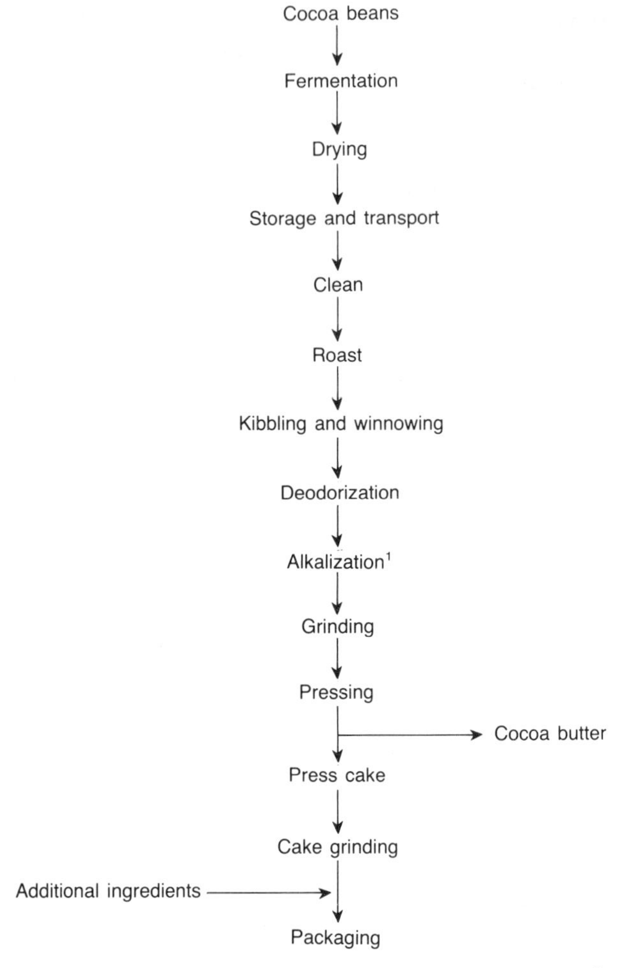

Figure 6.2 Flow diagram: processing of cocoa powder. [1] Alkalization may be carried out at several stages.

6.3.1 Fermentation

After harvesting, the mature berries are transported to a convenient point, cut open and seeds together with adhering pulp removed. In most countries the seeds are removed as soon as convenient after harvest, but it is traditional practice in Ghana to hold the berries for up to 7 days before processing. During this time respiration of pulp sugars results in a decrease in the pulp: cotyledon ratio and reduced acid production during the fermentation. Such a pulp pre-conditioning stage is now recommended as a means of controlling the problem of cocoa acidity common with Malaysian beans (see page 266). Storage of berries for seven days, or more, markedly increases the incidence of mould in the dried beans.

Fermentation proper has three purposes: liquefaction and removal of the mucilaginous pulp, killing of the bean and initiation of the development of aroma, flavour and colour. The fermentation stage is of major importance in determining the quality of cocoa powder and chocolate confectionery.

Following harvest, beans with adhering pulp are transferred to heaps, baskets or boxes for fermentation to take place. In some cases, beans are transferred to a central site for fermenting and delays during transport can result in initiation of an uncontrolled fermentation. Actual practice adopted varies; heaps tend to be used by small producers in Africa, while boxes are general in large plantations and are also used by small producers in South America and the West Indies. In Nigeria, baskets lined with leaves are commonly used. Heap fermentation is a cheap process, but boxes permit improved control.

Heap fermentation involves forming *ca*. 300 kg of beans into a flat cone. Banana or plantain leaves are used as a base and to cover the cone. Fermentation boxes are of various shapes and sizes, although all must be sufficiently large to prevent excessive heat loss and boxes typically contain *ca*. 1 tonne on large plantations. Bases are perforated to permit drainage of the liquefied pulp. In all methods of fermentation, beans are turned to permit adequate aeration. Fermentation lasts five to seven days, the process being quicker with *criollo* beans than with *forastero*. Pulp is liquefied and runs away as 'sweatings' early in the fermentation, but chemical changes affecting flavour and colour development require a longer period. Extending the fermentation beyond 7 days leads to the growth of spoilage micro-organisms and marked loss of quality.

In contrast to 'fermentation' of tea and coffee, micro-organisms play an important role in the fermentation of cocoa beans processing. A wide range of types of micro-organisms are present during fermentation, but only a small number have a known role. The initial stage of fermentation involves an alcoholic fermentation by yeasts. Several species can be involved, including *Candida*, *Debaryomyces*, *Hansenula*, *Kloeckera*, *Pichia*, *Rhodotorula*, *Saccharomyces* and *Torulopsis*. The alcoholic fermentation is followed by a significant increase in numbers of *Lactobacillus*, primarily *L. plantarum*, *L. collinoides* and *L. fermentum*, with accompanying lactic acid production. The third stage is intense acetification by the acetic acid bacteria *Acetobacter* and *Gluconobacter*. The heterogeneity of cocoa undergoing fermentation is such that the three stages progress concurrently in different parts of the heap or fermentation box. The pH value of the pulp, which is initially 3.5–4.0 may fall initially, but rises to between *ca*. 5.1 and 5.9. This is due to oxidation of strongly ionized citric acid and its replacement by less ionized acetic and lactic acids. Towards the end of fermentation there is also some oxidation of acetic acid.

Oxidation of ethanol to acetic acid is accompanied by evolution of heat and the temperature rises to 50°C or higher. Incipient germination may occur and has been considered important for development of flavour precursors. There is, however, little evidence to either support or refute this contention. Germination proper is undesirable, due to reduction in the content of cocoa fat and an increased susceptibility to microbial infection following exposure of the micropyle.

For a number of years it was thought that the bean was killed by a combination of rising temperature and increasing acidity. More recent opinion, however, suggests that acetic acid and ethanol are responsible for death of the bean. At the same time the cotyledons gain moisture and undergo textural changes to form a fissured structure and amino acid and peptide flavour precursors are formed through proteolytic enzyme activity (see page 286). The colour of the beans changes slowly during fermentation and subsequent drying from purple to the characteristic chocolate brown. The temperature must not be allowed to fall during colour development. Towards the end of fermentation species of *Bacillus* increase in importance, although numbers present can vary considerably. *Bacillus stearothermophilus*, which can grow at 65°C may be of particular importance in high temperature fermentations. Species of

* *Zymomonas mobilis* may be of importance during fermentation in conditions of restricted oxygen supply. The bacterium ferments sugars by an unusual pathway involving the Entner–Doudoroff mechanism, followed by pyruvate decarboxylation. The major products of fermentation are ethanol and lactate, together with small quantities of acetaldehyde, acetylmethylcarbinol and glycerol. In this case it is assumed that ethanol and acetaldehyde would be responsible for killing the bean rather than acetic acid.

Leuconostoc are often present and are assumed to be involved in lactic acid production. *Micrococcus* spp. are also present in significant numbers and may reduce acidity through oxidation of acetic and lactic acids. Mould growth on the surface of fermenting beans is undesirable and involves a risk of mycotoxin formation.

Of the various reactions occurring during fermentation acidification has the greatest effect on quality. Acetic acid is absorbed by the cotyledons (nibs) which, if excessive results in the end product having an undesirable acid character. Further, activity of proteolytic enzymes is much reduced when the pH value of the bean falls below 5.0. This results in failure to develop the characteristic cocoa flavour and aroma. Equally, however, acetic and lactic acids, at moderate concentrations, play a positive role in determining quality and conditions which favour reduction in acidity lead to a low quality product.

Continuing problems of acid nibs and poor flavour development with Malaysian cocoa beans have led to extensive investigation of fermentation parameters. Cultivar, method of fermentation and duration of pulp preconditioning (post-harvest storage) all affect the chemical and physical profiles of fermentation. Results of different investigations, however, often appear contradictory and in many cases chemical differences are not reflected in perceived quality. Despite this, generalizations are possible and it is apparent that the extent of aeration in the early stages of fermentation is of considerable importance in determining development of acidity. Adequate aeration suppresses the initial anaerobic phase and increases the ratio of respiration to fermentation. This results in a reduction in alcohol and, consequently, acetic acid production. The quantity of lactic acid produced is also reduced. For this reason cocoa fermented in boxes, which offer better control of aeration, tends to be less acid. The effect of aeration, however, is limited since the pulp layer restricts diffusion of oxygen and creates local environments in which alcohol production continues. Cocoa made from cultivars such as Upper Amazon hybrids, which have a high pulp:cotyledon ratio, tends to be more acidic than that made from cultivars such as Amelonado, which have a low pulp:cotyledon ratio. Problems are exacerbated by the higher sugar content of the pulp of Upper Amazon hybrids and, potentially, greater lactic acid production. A number of solutions have been proposed to problems of over acidity including methods for degrading acids by oxidation after production. These include pressing the beans to remove residual pulp and subjecting the beans to an air blast in the late stages of fermentation. The concentration of acetic, but not lactic acid, is reduced by these means, but the end product tends to be of poor quality and pulp preconditioning (see page 262) appears to offer a better solution.

BOX 6.3 **Occam's razor**

Cocoa produced in Malaysia has suffered major problems of over-acidity, which were thought to be specific to that country. A prominent English microbiologist, the late Dr J.G. Carr, was, however, able to demonstrate that acid cocoa was produced in Ghana when Upper Amazon hybrids were the source of the raw material. He concluded that cultivar was of overwhelming importance, but noted that this simple explanation 'appeared too facile' for the chocolate industry, who persisted with unsuccessful modifications to fermentation procedures. (Carr, J.G. 1983. *Journal of Applied Bacteriology*, **55**, 383–401).

Fermentation is often carried out under unhygienic conditions and serious contamination by micro-organisms can occur. In theory a number of pathogens can contaminate cocoa beans at this stage, but in practice *Salmonella* is the cause of any problems. Vegetative micro-organisms are destroyed during roasting, but the possibility of cross-contamination from raw material exists. Cocoa beans may also contain significant numbers of *Bacillus*, including thermophilic species. Endospores of these bacteria survive roasting and can cause problems when cocoa powder is used as an ingredient in sterilized products, such as chocolate milk.

6.3.2 Drying

Drying is necessary to prevent deterioration of the beans during storage and transport. It is also of importance in determining quality. In many countries sun drying is used, but this is not possible where rainfall and humidity are high during the main harvest season. Sun drying is generally considered to produce higher quality beans of lower acetic acid content and higher pH value. Where artificial drying is necessary the quality of the beans can be improved by reducing the drying rate to balance the rate of evaporation of liquid from the surface of the bean with the rate of diffusion of moisture from the cotyledons.

Sun drying involves spreading the beans onto trays in layers *ca.* 5 cm in depth. These must be raked frequently to ensure even drying. Canopies must be available to protect the beans during rain and at night and these, or the trays, are often mounted on rails for ease of movement.

A large number of types of artificial dryer are used. These vary widely in sophistication and, in some cases, are locally constructed from scrap materials. An essential design requirement of all types of dryer, however, is to ensure that contamination of the beans with combustion products is totally avoided. Platform dryers are simple in design and construction. Industrially made platform dryers consist of a rectangular platform of perforated steel or other material onto which the beans are placed. The space beneath the platform is constructed as a closed chamber through which warm air is blown upwards through the bed of beans. Heating of the air can be by direct combustion, but a heat exchanger is more common and avoids any possibility of contamination of the beans. Platform dryers are also constructed locally for use by small farmers and consist of a series of drying platforms arranged around an inclined flue. In many cases the flue is constructed from oil drums.

Large diameter fans powered by diesel engines provide an economical means of drying. Drying is largely achieved by the high air flow, heat usually being supplied only by the diesel engine. Drying efficiency is limited in areas of high humidity and spoilage by moulds can occur during the extended period required for drying high moisture content beans. Problems of spoilage can be avoided by a preliminary drying in shallow layers, followed by finishing in a deeper bin.

Most dryers used in cocoa processing operate on a batch basis, but continuous dryers have been proposed. Of these only the Buttner dryer has been widely adopted. This consists of a vertical column containing a chain elevator which carries a series of perforated trays. Air, heated by a heat exchanger, is blown upward through the vertical chamber, while loaded trays pass slowly downward through the air stream. Trays are consecutively loaded and unloaded at the base of the column.

Spoilage of dried cocoa beans can result from growth of moulds during drying, or from failure to dry to a sufficiently low moisture content (*ca.* 6.5%). Mould growth is accompanied by marked organoleptic deterioration. The possibility of mycotoxin production also exists, despite some claims that raw cocoa beans are a poor substrate for mycotoxin production.

6.3.3 Storage and transport

Hessian sacks, the traditional means by which cocoa beans are stored and transported, remain widely used. Storage and transport facilities in many producing areas remain relatively primitive and moisture pick up is a significant problem where humidity is high. In such areas the use of polythene liners in the sacks is recommended. Use is also made of bulk containers but, while effective, these are unsuited to use where producing units are small.

Moisture pick up during shipping is also a significant problem and results from the condensation which occurs when the ship enters cold waters. Direct contact between bags of cocoa beans and the ships sides should be avoided, but the most effective control is by forced air circulation. Condensation is also a problem with bulk containers and facilities for ventilation must be fitted.

Cocoa beans are subject to damage by insects and rodents during storage and strict precautions should be taken in warehouses and on ships. In the past infestation with *Ephestia* larvae was a major problem, but this pest is now largely controlled. Fumigation of incoming cocoa beans with methyl bromide is used on a precautionary basis by many factories, especially in the US. Indiscriminate fumigation may, however, lead to high levels of pesticide residues if the beans have already been fumigated at the country of origin.

6.3.4 Cleaning of raw beans

Raw beans contain surface grit, occasional small stones and hessian fibres. Contaminants are removed before further processing by a variety of mechanisms including rotating brushes, air blasts and screening. Immature beans and clusters of beans are removed at this stage and a detector should be fitted to remove any metal fragments.

6.3.5 Roasting

Roasting is a critical stage in the manufacture of cocoa and chocolate products. From a technological viewpoint, the function of roasting is drying of the nib, removal of undesirable flavour compounds, development of the final flavour and colour (see pages 288–9), and loosening of the shell. Traditional roasting involved the whole bean, but some manufacturers prefer to roast the nib separately. Roasting is also a critical stage in determining the safety of cocoa and chocolate products, since it is usually the only stage at which heating is sufficient to inactivate *Salmonella* and other vegetative micro-organisms.

(a) Conventional roasting

Roasting conditions vary according to the nature of the end-product and the type of equipment used. Single stage processes can involve temperatures as high as 150°C, although lower temperatures are used for those cocoa powders in which a red colouration is desirable. Two-stage processes have also been introduced. These typically involve a preliminary heating at less than 100°C, which loosens the shells but has virtually no roasting effect. However, the moisture distribution resulting from the initial heating is ideal for the removal of undesirable volatile compounds. The second stage involves heating at temperatures of 125–130°C during which the required chemical and physical changes associated with roasting take place.

Equipment used for roasting cocoa beans evolved over a long period. The first modern design, which remains in wide use, was the Sirocco-type batch roaster. In this type of roaster, a batch of beans is placed in a bowl, or rotating drum, through which hot air is blown. Air flow through the beans is turbulent and individual beans are heated to a consistent degree. In most designs cooling takes place outside the roaster, the beans being discharged onto a perforated plate through which air is blown.

* Dust resulting from the cleaning process contains large numbers of micro-organisms and is a potential source of contamination of roast beans. The bean cleaning operation must be situated away from the handling of the roast material and there must be entirely separate air flows.

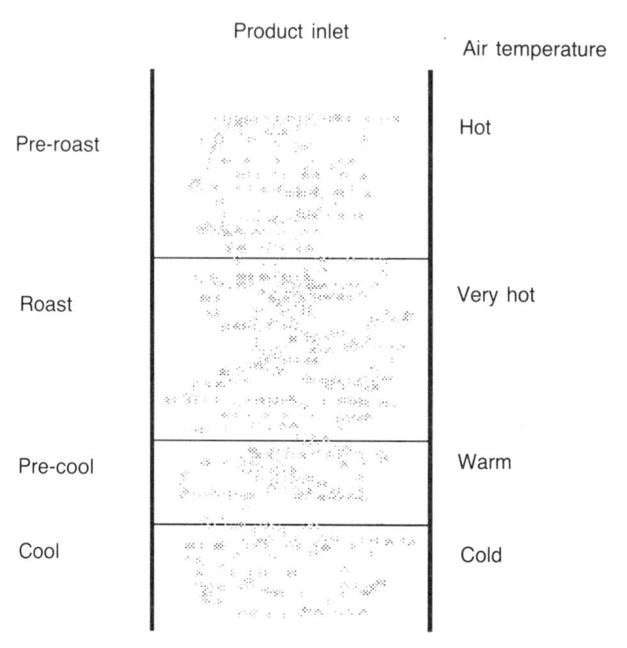

Figure 6.3 Schematic diagram: continuous roaster.

Various types of continuous roaster are now available. These include vibrating fluidized bed designs and various tower designs in which beans fall through currents of heating and cooling air. Heat transfer is entirely by convection and air flow is directed to ensure even heating of individual beans. Roasters of this type usually consist of three stages; pre-heating, roasting and cooling (Figure 6.3).

(b) Nibs, alkalizing, roasting and sterilizing process

The nibs, alkalizing, roasting and sterilizing (NARS) process has been developed relatively recently as an alternative technology to conventional roasting, winnowing and alkalization. The NARS process is summarized in Figure 6.4. A key feature of the process is roasting of the nibs rather than of the whole bean.

In the first stage of the NARS process, the beans are heated by infrared and radiant heat in a Micronizer™. This is a continuous process in which beans pass through a vibrating steel trough under gas or electric radiant heaters. Exposure time varies from *ca.* 60 to 120 s during which the shells are rapidly heated, dry, crack and

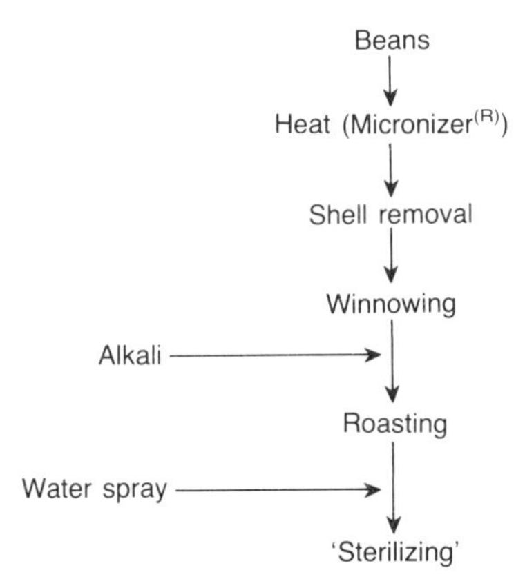

Figure 6.4 Nibs, alkalizing, roasting and sterilizing process.

become detached from the beans. The nibs receive much less heating, although there is slight expansion which assists the detachment of the shell. The moisture content is reduced to 5-6%. Transfer of fat from nib to shell is less than in many conventional processes and the tendency of the nib to break up is also considerably reduced. This reduces the quantity of nib dust and minimizes loss during winnowing.

Following micronization the cocoa beans pass to winnowing machines (see pages 272-3) where removal of the shell is completed. The nibs are then roasted in a two-stage process designed to optimize product quality. A Tornado-type roaster is most commonly used in the NARS process. This consists of an externally heated rotating drum in which nibs are heated by a combination of conduction and convection. In operation, the drum is charged with beans. Where alkalization is used (see pages 275-6) a pre-heated solution of alkali is sprayed onto the nibs before heating commences.

The first stage of heating involves temperatures slightly below 100°C and lasts for *ca.* 10 min. During this period the moisture content is reduced to 2-3%. The temperature is then raised for the second stage which lasts 15-20 min. The temperature used in the second stage can be as high as 130°C depending on the nature of the end product. The indirect heating used in Tornado-type dryers avoids case hardening of the nib surface, a problem with circulating air dryers, and permits even drying and diffusion of moisture and volatile compounds from the nib interior.

Table 6.1 Precautions against contamination of roast cocoa beans

Physical separation of areas handling raw and roast beans
Direct air flows from roast to raw bean handling areas
Prevent movement of personnel between areas handling raw and roast beans

The final stage of heat treatment by the NARS process involves inactivation of bacterial endospores by exposure to moist heat. This is achieved by injecting a fine spray of water into the drum shortly before the end of the second stage of the roasting process. True 'sterility' is not achieved, but numbers of mesophilic endospores are reduced to less than 100/g.

There is no heating stage after roasting which is sufficient to inactivate *Salmonella*, or other microbial pathogens, and strict precautions must be taken to ensure that contamination of roast material does not occur. This effectively means complete separation of the plant handling unheated material from that handling heated. Precautions are summarized in Table 6.1.

6.3.6 Kibbling and winnowing

Kibbling and winnowing is a combined process designed to separate the shell as completely as possible from the nib. Kibbling (cracking) involves lightly crushing beans which have been roasted, or partially dried, to loosen the shell. It is important at this stage to retain both nib and shell as large pieces (85–90% of particles should be over 3 mm in size) and to avoid creating fine particles and dust. This is best achieved by use of impact rollers, which are pairs of hexagonal rollers, rotating in the same direction, which throw the beans against metal plates. Some older machinery, which used toothed rollers to break the beans is, however, still in use.

Winnowing is a separation process which depends on differences in density between the nib and the shell. Two basic fractions are

* Moist heat is considerably more effective than dry heat in inactivation both of vegetative cells of micro-organisms and endospores. This is attributed to the greater efficiency of heat conduction in moist air. A 12 log cycle reduction in a population of mesophilic endospores, for example, which could be achieved by heating at 120°C for *ca*. 15 min by steam would, require 2 h or more in dry air.

produced: nib, which should contain the smallest possible quantity of shell and germ, and shell, which should contain the smallest possible quantity of nib. Modern equipment consists of a vertical stack of sieves, mesh size decreasing from top to bottom. Beans pass direct from the impact rollers to the top sieve. At each stage shell pieces are removed by suction at the overflow from the sieve, while nibs pass to exit chutes. Blockage of the sieves is avoided by imparting a vibratory motion to the stack and by fitting mechanical rakes and sweeps. Germ separators are available but are generally ineffective and not widely used.

Nibs can contain as much as 1.5% shell and this causes particular problems in superfine grades of cocoa for beverage use due to rapid settling out in the made drink. This problem can be minimized by diverting the largest nibs, which have the lowest content of shell and germ, from the main product stream and reserving these for manufacture of superfine cocoa.

6.3.7 Grinding

Nibs contain *ca.* 55% cocoa butter, which exists in solid form within the cells. Grinding ruptures the cell walls releasing the cocoa butter, while at the same time frictional and/or applied heat raises the temperature above 34°C, the melting point of cocoa butter. The particle size of the non-fat portion is progressively reduced and a fluid mass, the cocoa liquor is produced. This consists of solid particles in suspension in a continuous fatty phase. Grinding may be carried out by several different types of equipment. Until fairly recently, all were evolutionary developments of the original stone mills, which used sets of horizontal stone disks. Nib or partly ground liquor was fed to the centre of the stones and ground during movement towards the periphery. Efficient grinding necessitated the cutting of grooves in the surface of the stone disks, which required frequent refacing. The process also generated a considerable amount of heat, which resulted in strong flavours in the cocoa liquor. Stone disks were gradually replaced by grooved metal disks. Early liquor mills of this type had vertically mounted grinding discs, but more modern designs have readopted a horizontal configuration. The most common type employs three sets of grinding discs, the nibs being ground in three stages. Mills are constructed to provide very accurate setting and a very fine liquor can be produced. The mills are cooled to prevent liquor overheating.

Pin mills offer an alternative means of grinding. These consist of two contra-rotating discs each of which is fitted with hardened steel pins. Nibs are subject to a pulverising action, which produces a mobile paste of particle size between 90 and 120 μm. This requires further reduction and the paste passes through a set of three refining rollers. Roll pressures are hydraulically controlled and the rollers cooled by circulating water.

The most recent development in cocoa nib grinding is the vertical ball mill. Grinding involves frictional and impact forces generated by balls of hardened steel moving freely within a defined space. Vertical ball mills consist of a steel column containing a motor driven spindle fitted with a series of horizontal agitator arms which carry the grinding balls. Cocoa beans, direct from winnowing, cannot be handled, and the feed consists of a mobile paste produced by a single stage disk or pin mill. The paste is pumped up through the base of the mill and ground by the steel balls, set in motion by the spindle and agitator arms. The agitator arms also serve to mix the paste and ensure even grinding.

6.3.8 Deodorization

Deodorization of cocoa liquor is most widely applied where the end use is in milk chocolate confectionery. The process is, however, sometimes applied during beverage production, especially where high quality drinks based on chocolate crumb are involved. The simplest system involves heating a thin film of liquor in a steam-heated trough or drum. In some cases *ca.* 2% water is added to the liquor before heating. Temperatures of 80–110°C are used, undesirable volatile compounds being removed with the steam.

Treatment of thin films of liquor by scrubbing with hot, humidified air is the basis for a number of systems. Scrubbing usually takes place as the liquor flows down a heated column against an upward flow of hot air. Agitators are fitted to ensure turbulent flow. Single-

---

* Deodorization of cocoa liquor for confectionery manufacture may be combined with other parts of the chocolate manufacturing process. Equipment is available, for example, which permits combined deodorization and alkalization. Sugar syrups also may be introduced to initiate the Maillard reaction. In some cases the heat treatment is referred to as 'pasteurization'. If used to control *Salmonella* and other micro-organisms, however, the process must be validated and operations fully controlled and verified.

stage equipment has been used in the past, but is inefficient and modern equipment normally uses three sequential stages.

Removal of volatiles is achieved more efficiently under vacuum and evaporators, usually of the climbing film type, have been adapted to this purpose. In one design flushing with nitrogen or CO_2 is incorporated to reduce the oxygen level. Vacuum deodorization systems are often considered likely to remove a significant proportion of desirable flavours as well as the undesirable.

6.3.9 Alkalization

The term 'solubilization' may be used synonymously with alkalization, but this usage is incorrect since there is no significant change in solubility. Alkalization does, to some extent, improve dispersibility but the major purpose is to change the colour of the cocoa. There is also an effect on flavour, but disagreement as to whether this is desirable or undesirable.

BOX 6.4 **The fault of the Dutch**

Alkalization was introduced in Holland by Van Houten in 1828 and the process is sometimes still known as 'Dutching'. More recently, 'Dutching' has been used to describe the practice of shipping shellfish, such as prawns, from the UK to Holland for decontamination by irradiation. This practice was clearly illegal under UK Food Law, but has still been employed by at least one major company.

Potassium and sodium carbonates are most widely used in alkalization. Other alkalis have been used but generally offer no advantage. In many countries the chemical nature of alkalis and the quantities used are controlled by law. Within legal restrictions, quantities added vary according to the desired colour of the cocoa, larger quantities being required where red colours are favoured. The pH value rises in proportion to the amount of alkali added, although in some countries food-grade acids may be added to control pH and to marginally enhance colour formation.

Alkalization can take place at any stage from whole beans to cocoa cake. Alkalization of nibs is widely used, especially where a red cocoa powder is required and equipment is available which combines alkalization with drying and roasting. A nib can take up as much as half its weight of alkali solution, but this requires a considerable time. The process can be significantly shortened by mixing the nibs with more concentrated alkali solutions in pressurized vessels.

Alkalization of cocoa liquor has been widely practised in the past. It is necessary to use concentrated solutions of alkali to avoid adding excessive quantities of water and it is not possible to obtain red coloured powders. After alkalization, water is removed by heating thin films of liquor.

Alkalization may also be carried out on the pressed cake. Best results are obtained using cake prepared from well fermented beans, roast at low temperatures. A heated, rotating drum is used, which is maintained under vacuum to preserve red colours. Such equipment provides good mixing and low temperature drying. Ammonium compounds are sometimes used as alkalizing agents, excess ammonia being driven off by heat.

Alkalization of whole beans involves applying an alkali to beans in a drum roaster. Roasting is at a low temperature where a red coloured cocoa is required. Alkalization of whole beans is only rarely used in commerce.

Techniques for alkalizing cocoa have been considerably refined in recent years. Precise control is possible, but variations still occur. For this reason, where colour is critical, consistency is attained by blending powders.

6.3.10 Pressing

(a) Hydraulic pressing

Pressing of the liquor is necessary to separate the cocoa butter from the particulate non-fat material. Hydraulic presses are used consisting of a horizontally mounted series of up to 14 press pots, each of which contains a metal filter screen. At the start of each cycle, liquor at 95–105°C is forced into the press. The pressure applied during filling is sufficient to release some of the cocoa butter, but the pressure is then increased to as high as 12 000 lb/in^2. Yield of cocoa butter (and thus fat content of the press cake) is controlled by the pressure/time cycle and the distance the hydraulic ram travels. After the required time period the direction of the hydraulic ram reverses, the press pots open and press cake falls into containers or onto a conveyer belt. The minimum fat content of cocoa produced by presses of this type is *ca.* 10% under practical conditions. Large presses have an output of 600-800 kg/h at 10% fat; higher output being possible where the cocoa contains more fat.

(b) Expeller pressing

Expeller (screw) presses, similar to those used in the edible oil industry, are also used for production of cocoa butter from whole beans and from winnowing products, such as dust and small nibs. Steaming is necessary before pressing to soften the material. Although the primary concern is the recovery of cocoa butter, cocoa press cake is also produced.

Expeller presses operate by using a rotating screw to force the material being extracted into a tapering tube. The tube has slits along its length and terminates in a cone-shaped device which can be adjusted to vary the size of the outlet. As material passes along the tube it is subject both to continuously increasing pressure and to the shearing action of the screw. Cocoa butter is expressed through the slots in the sides of the tubes, while flakes of press cake emerge from the outlet at the end of the tube. The butter contains a variable quantity of cocoa powder, which is removed by centrifugation. Recovery of cocoa butter is generally more efficient than with hydraulic presses and the fat content of the press cake is 8 to 9%.

Cocoa powder prepared using an expeller press is suitable for beverage use, provided it is made from winnowed nibs of low shell content. The shearing action, however, leads to powder of a much greater absorbency and different reconstitution properties.

6.3.11 Cocoa press cake grinding

Press cake must be ground to powder for use either as a beverage, or for ingredient use in other products. Grinding would appear to be a simple physical process, but is complicated by the presence of cocoa butter. Friction causes a significant temperature rise during grinding which melts the fat at temperatures above 34°C causing adhesion of particles and clogging of machinery. At lower temperatures some of the acylglycerol fractions melt and while clogging does not occur, the powder is 'unstable' with an unattractive greyish tinge and a tendency to cake. Problems are greatest with cocoa containing 20% or higher levels of fat and where powder is intended for beverage use and appearance is of major importance. Optimal conditions for obtaining a powder of good colour and general appearance are those in which cocoa butter present in the cake is cooled from a liquid to a solid form during the grinding process.

Press cakes produced by hydraulic presses are very dense, especially when of low fat content. Before grinding the cakes are broken by passage between sets of paired rollers with intermeshing teeth. This reduces the cake to small lumps of less than 5 mm diameter, which pass to the grinding stage. This usually involves use of pin mills.

Cake enters the pin mill at *ca.* 45°C. Flow rate is important and is automatically controlled in modern equipment. The mill is cooled during operation by circulation of air through ducts in the walls. Powder leaves the mill and is conveyed pneumatically to a cooler. This consists of a series of jacketed horizontal pipes cooled by a circulating refrigerated liquid. Cocoa powder passes through the cooler in the air stream and then through a tempering tube, where the colour is fixed, to a cyclone, which separates the finished product from the air. In recent years it has become common practice to employ a further cooler and stabilizer directly before filling. The most common design consists of a cylindrical fluidized bed cooler, which not only assists in production of cocoa powder of good quality, but also minimized compaction of the powder during storage.

Large quantities of air are required for cooling and transporting cocoa powder. In many cases control of humidity is required since air with a relative humidity of more than 50-60% will lead to moisture pick up and the possibility of mould growth. A very low

---

* Grinding at this stage involves reduction in the size of compacted aggregates and not reduction of the mean particle size of the powder. The particle size of powder leaving the cake grinder is determined by the extent of grinding during cocoa liquor production. In general, it is not possible to produce a fine powder at this stage if the initial particle size was coarse.

relative humidity, however, leads to powder particles acquiring an electrostatic charge and difficulties during packaging.

6.3.12 Packaging

Packaging of cocoa beverage products is straightforward. Significant use is still made of moistureproof paper bags within cardboard outer packets and of cardboard drums with metal, or plastic, ends and lids. The inner surface of the cardboard is laminated with a metallic film to minimize entry of moisture. Screen-printed tin plate containers have been re-introduced, apparently to exploit the 'nostalgia' factor. Individual-portion foil laminate sachets are used for packing some added-value products.

6.4 COCOA-BASED BEVERAGES

6.4.1 Cocoa

Traditional cocoa powder contains no other ingredients whatsoever. In other cases flavouring is added, although in the UK and some other countries, it is not permitted to add flavouring which imparts the flavour of chocolate or milk fat. Vanilla, cinnamon and cassia are widely used flavourings. Flavourings are usually added as powders with a particle size equal to, or smaller than, that of the cocoa. This minimizes problems of flavour deterioration due to oxidation. Some cocoa powder contains a small quantity (0.5%) NaCl, which is commonly added during alkalization. Sweetened cocoa with added sucrose or other permitted sugar is also produced (*cf.* drinking chocolate) and, in the UK, has a minimum cocoa content of 32%.

Fat reduced cocoa powder is now available and has a minimum cocoa butter content of *ca.* 8% depending on national legislation. Defatting involves solvent extraction of press cake.

6.4.2 Instant (cold dispersing) cocoa powder

Traditional cocoa powder is of poor dispersibility and requires use of boiling milk or water. This is a significant disadvantage in catering use or where the increasingly popular cold chocolate beverages are required. The simplest approach is the addition of a food grade wetting agent, lecithin being most widely used.

As much as 3% lecithin may be required to obtain full wetting of cocoa with a fat content in excess of 22% and it is essential both that the lecithin used is effectively tasteless initially and that no off-flavours develop during storage. Vegetable lecithin, for example, has been used but tends to develop strong flavours during storage and specially modified lecithins are preferred. Lecithin solution can be sprayed onto press cake at the time of grinding. Alternatively lecithinization can be combined with instantization (agglomeration), or added in powder form to instantized powder.

6.4.3 Drinking chocolate

Drinking chocolate has to some extent replaced cocoa powder for beverage use. Drinking chocolate is generally easier to prepare than cocoa and may be made as a cold beverage, or used in vending and catering, without lecithinization. In its simplest form, drinking chocolate contains *ca*. 70% sucrose, or other permitted sugar, and *ca*. 30% cocoa powder. The minimum permitted cocoa content varies according to national legislation. Basic formulations consist of no more than a mixture of the two components, but it is more common practice for the mix to undergo co-agglomeration. Details of processes are usually proprietary, but general principles are known. One recipe involves boiling a sucrose syrup to a supersaturated state, mixing with cocoa powder and drying. It is generally considered, however, that a better quality product is obtained by using only a portion of the sucrose to prepare a syrup. The remainder is graded to remove fine particles, mixed with the syrup and cocoa powder and the mixture instantized.

Milk powder can be incorporated in drinking chocolate to produce a complete, truly instant product. Full cream powder is not normally used due to limitation of the shelf life by oxidative deterioration. Skim milk powder combined with vegetable oil has been used in the past, but it is now common to use a coffee whitener-type product containing whey powder and caseinates in addition to skim milk powder and vegetable oil. These have superior functional properties in terms of providing mouthfeel and foaming. The use of milk powder or a coffee whitener-type product also permits the sugar content to be reduced. Low calorie ('light') products are made using fat reduced cocoa powder and the artificial sweeteners Aspartame™ and Acesulfame K™. Additional 'body' is required where artificial sweeteners are used and maltodextrins are often added for this purpose. Separation of the cocoa and other compo-

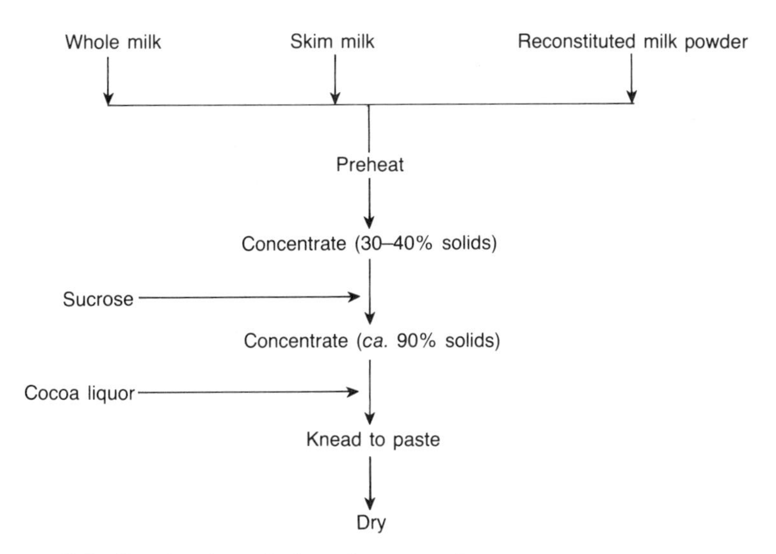

Figure 6.5 Production of chocolate crumb.

nents is a problem when beverages of this type are prepared and xanthan gum is often incorporated as stabilizer.

Although most 'milk' chocolate beverages are made with cocoa powder, some utilize chocolate crumb. Chocolate crumb is widely used in milk chocolate confectionery manufacture and is available as a commodity. Manufacture is summarized in Figure 6.5, the key stage being the Maillard reaction between milk protein and sugars which produces a characteristic flavour. The starting material of chocolate crumb can be whole milk, skim milk, or reconstituted dried milk, which is pre-heated to 75–80°C and concentrated in an evaporator to *ca.* 30–40% solids. Multiple effect falling film evaporators (see Chapter 2, page 40) with vapour recompression are now normally used. The second stage involves dissolving sucrose in the concentrated milk and further concentrating to *ca.* 90% solids, at which level sucrose crystallization commences. Older technology involves the use of batch vacuum pans for dissolving and crystallizing the sucrose during boiling at 75°C. The milk/sugar concentrate is then combined with cocoa liquor in a heavy-duty kneader to form a very stiff paste in which sucrose crystallization continues during kneading. The paste is then dried in either batch or continuous vacuum ovens at temperatures between 75°C and 105°C for 4–8 h. The desired flavour is attained by varying the time/temperature combination, precise control being required.

The use of batch pans is now obsolescent, although still in use in some factories, and sometimes considered to produce the highest quality crumb. The basic process is the same in most modern variants, differences lying in the substitution of continuous processes for batch pans. A typical process of this type involves use of a continuous vacuum concentrator in which concentrated milk from the evaporators is mixed with sugar and cocoa liquor in metered quantities. More water is removed and the resulting paste passes to a continuous mixer/kneader where sucrose crystallizes. The paste is then dried, under vacuum, in a rotating drum dryer. In some cases final drying and additional caramelization take place as the crumb falls down a column against an counter-flow of hot air.

Specialist equipment is also available, which manufactures crumb without the need for vacuum. Sweetened condensed milk or an equivalent product made from reconstituted powder is the starting point. This is mixed with cocoa liquor and additional water and sucrose if required. The mixture passes through a steam-heated, tubular heat exchanger where the total solids content increases from *ca*. 70 to 95%. Caramelization also occurs at this stage. The paste leaving the heat exchanger passes to a water-jacketed crystallization unit, where the sucrose crystallizes and the crumb structure is formed. The crumb then passes to a fluidized bed dryer for final drying.

When used in manufacture of chocolate beverages, crumb is mixed with cocoa powder and additional non-cocoa solids such as skim milk powder, whey and caseinates. Xanthan gum is usually present as a stabilizer. Chocolate crumb is currently used only in premium products which, for the UK market, are intended to resemble the chocolate beverages of continental Europe.

6.4.4 Added-value products

Although flavouring, such as vanilla, has traditionally been added to some types of drinking chocolate, added-value products have been introduced in which a significant part of the product character is derived from flavouring. Flavours chosen are generally those considered complementary to chocolate, such as mint and orange. In some products the flavouring is added as discrete particles, which have a strong impact in the mouth. A chocolate/coffee combination is also available, in which coffee flavour is provided by grains of dark roast coffee.

6.4.5 Ready-to-drink cocoa beverages

Cocoa powder is widely used in production of chocolate milk, but a ready-to-drink cocoa beverage has also been produced. The product, which does not appear to have been widely distributed, was based on cocoa powder, dried whey powder, skim milk powder and vegetable oil. Processing was by ultra-heat-treatment (UHT) and the beverage was shelf stable. Despite the presence of xanthan gum and carrageenan as stabilizers considerable sedimentation occurred during storage.

6.4.6 Cocoa substitutes

Historically the high cost of cocoa powder has meant a demand for low cost substitutes. In this context, substitutes were normally used as extenders at a level of up to 30%. More recent interest in cocoa substitutes has arisen from the wish to avoid caffeine and theobromine, rather than because of cost. Use of substitutes in chocolate confectionery has been of particular interest, because of the alleged connection between chocolate consumption and hyperactivity in children but substitute beverages have also been produced.

Although a number of substitutes have been proposed, only one, carob powder, has been used on a significant commercial scale. Carob powder is produced from the pods of the leguminous tree, *Ceratonia siliqua*, which is grown in eastern Mediterranean countries. The tree is also the source of carob gum (locust bean gum), which is used in foods as a stabilizer and gelling agent. Carob undergoes a similar processing to cocoa, which it resembles in colour, appearance, taste and flow properties. Beverages are available in which cocoa has been entirely replaced by carob, although in some cases up to 20% cocoa is present.

6.5 BIOLOGICAL ACTIVITY OF COCOA

6.5.1 Physiological effects of cocoa

Cocoa contains both caffeine and the related theobromine (Figure 6.6). Theobromine has similar physiological effects to caffeine (see pages 234–5) and has been associated with hyperactivity in children.

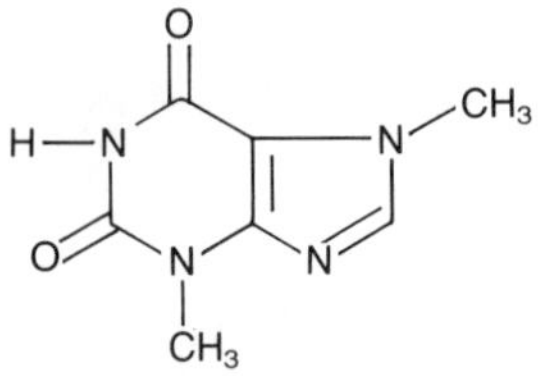

Figure 6.6 Structure of theobromine.

> **BOX 6.5 The sport of kings**
>
> Horse racing has a long and dishonourable history of 'doping'; the illegal administration of drugs to make the horse run faster or slower. Doping is still practised, despite the use of drug testing and sophisticated security measures at stables and theobromine is one of the prescribed substances. In one case, at Ascot races during 1989, traces of theobromine in the blood sample taken from a horse was attributed not to illegal and sinister activities, but to the horse having stolen a chocolate confectionery bar from a handler.

Consumption of cocoa-containing products has also been associated with histamine-like reactions, although confectionery has usually been implicated rather than beverages. Cocoa and chocolate do not contain detectable quantities of histamine, but phenylethamine, formed by decarboxylation of phenylalanine, can be present at levels of up to 10 mg/l. Reaction to phenylethamine most commonly involves flushing and a rash of face and neck, a headache and, in some cases, heart palpitations. Less common reactions include intestinal upset, visual disturbances and dizziness. Phenylethamine is usually removed from blood by amine oxidase. This enzyme is inhibited by many common antidepressant drugs and persons receiving such medication are at particular risk of reaction.

6.6 CHEMISTRY

6.6.1 Chemical changes during processing of cocoa

(a) Fermentation

Removal of the pulp surrounding the beans provides visual evidence of the course of fermentation. The pulp is largely

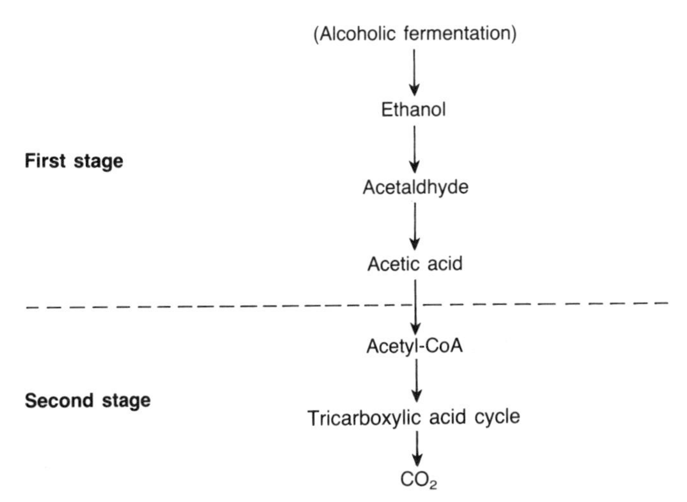

Figure 6.7 Oxidation of ethanol to acetic acid. *Note*: *Acetobacter* grows rapidly on ethanol and acetic acid accumulates. The second stage ('over-oxidation') of oxidation of acetic acid to CO_2, occurs much more slowly. *Gluconobacter*, the second major member of the acetic acid bacteria, lacks a functional tricarboxylic acid cycle and is unable to oxidize acetic acid to CO_2.

composed of sugars and is subject to both alcoholic and lactic fermentation. Sucrose is also hydrolysed to reducing sugars. Lactic acid is not usually metabolized further during cocoa fermentation, but ethanol is oxidized to acetic acid by acetic acid bacteria (Figure 6.7). Excessive acidity is a fault in cocoa beans (see page 266). Production of acetic acid in the early stages of fermentation restricts the final quantity formed due to inhibition of yeasts and limitation of the quantity of ethanol available as a substrate. Acetic acid is itself oxidized, either by acetic acid bacteria or other microorganisms and the concentration tends to fall towards the end of fermentation, especially if the beans are highly aerated. A substantial proportion of acetic acid is lost during drying and roasting, but lactic acid is less volatile and tends to be responsible for acidity in cocoa powder.

The death of the bean is a progressive process, which occurs cell by cell as acetic acid accumulates. The intra-cellular pH value falls from *ca*. 6.5 to 4.2, which obviously has profound effects on cellular components. Not all changes, however, can be explained in terms of reduced pH value. When sufficient acetic acid has penetrated the membrane, the integrity of intracellular compartmentalization is lost. The underlying mechanism is not known, but the result is the mixing of water soluble substances, including enzymes, substrates, inhibitors and reactive molecules. The collapse of the polyphenol storage vacuole tonoplasts and the plasmalemmas of all cells results in the mixing of enzymes and substrates which are normally never in contact. In the intact cell, enzymes for a given pathway are often bound to membranes in close proximity to each other. Disruption of the cell and loss of integrity within organelles probably reduces the activity of some pathways. The fall in pH value has a differential effect on enzyme activity, activating some and inhibiting others. The rise in temperature may also be expected to have differential effects on enzyme activity.

During early phases of bean death, the tissue is anaerobic and polyphenols are reduced. Reversible protein–polyphenol reactions occur, which reduce enzyme activity. During the later stages of fermentation and during drying, the tissue becomes aerobic as oxygen penetrates the seed coat. Quinines are formed through reactions mediated by polyphenol oxidases. Quinines irreversibly complex with proteins and the catalytic properties of most enzymes are lost.

During fermentation water-soluble compounds, including acetic acid, amino acids, ethanol, caffeine and polyphenols, migrate from the bean and are deposited on the shell. The mechanism of migration cannot be explained in terms of aqueous diffusion, since following cell death inversion of the hydrophobic and hydrophilic phases within the cell occurs. This results in fat vacuoles, which are dispersed in the aqueous phase of live cells, fusing to form a continuous hydrophobic phase containing dispersed aqueous inclusions. Migration continues, however, through the hydrophilic cell walls. These are composed mainly of cellulose and hemicellulose and remain relatively intact after bean death.

Intrinsic enzyme activity during fermentation involves an initial anaerobic-hydrolytic phase, which is followed by an aerobic oxidative phase. Proteolytic enzymes, which are of major importance in production of flavour precursors, are active during the anaerobic phase. Activity is indicated by a large decrease in ethanol insoluble nitrogen and a corresponding increase in free amino acid nitrogen. The extent of proteolysis is less in beans fermented during the dry season, probably due to evaporation reducing protease activity. The consequences for flavour, however, appear to be limited.

Glycosidases are also active during the anaerobic phase and play the initial role in the development of the characteristic colour of the fully fermented bean. Unfermented, or slightly fermented, beans are slate, or blue-grey, in colour due to the anthocyanins 3-β-D-galactosidyl cyanidin and 3-α-L-arabinosidyl cyanidin. Reduced tissue pH value and cell death results in a deep reddish purple colour. Following this, glycosidases split the anthocyanin into a sugar and an anthocyanidin. The resulting change in the electronic structure of the chromophore leads to development of a bleached violet colour. The anthocyanidin is subsequently oxidized by polyphenol oxidase (see below) to produce the brown colour of fully fermented beans. It is probable that different glucosidases exist with specificity for either the α- or β-linkages.

Unfermented coffee beans contain a series of polyphenols, including anthocyanidins and cyanidins, catechins, leucocyanidins and the flavonol quercetin. Polyphenols undergo reactions with proteins during fermentation and are also oxidized by polyphenol oxidases to form the brown colour associated with cocoa (enzymic browning). Lower molecular weight catechins undergo condensation reactions, which result in reduction of astringency. Under some circumstances the concentration of catechins remains relatively high and the resulting astringency is interpreted as a flavour defect.

Extending the length of fermentation can lead to deamination and, ultimately, decarboxylation of amino acids. These reactions result in formation of acids, such as butyric, propionic, valeric and isovaleric and a consequent 'hammy' or 'over-fermented' off-flavour. It is not usually possible to remove this off-flavour during subsequent processing.

The role of micro-organisms in producing precursors for desirable flavour compounds is obscure, although it is known that off-

---

* *Streptomyces* is a soilborne bacterium, which produces a stable, branched mycelium and which reproduces by exospores. The organism is well known for its role in antibiotic production, but is also responsible, in conjunction with the myxomycete *Nannocystis*, for the odour of newly dug soil. This is due to production of volatiles, notably geosmin and 2- methyl-isoborneol. Earthy taints in piped water supplies have been attributed to growth of *Streptomyces* in the distribution system. Where the water is used in food processing the volatiles can be transferred to the product.

flavours can be derived from metabolites of moulds and *Streptomyces*.

The testa is known to be permeable to acetic and lactic acids and to ethanol, but the extent of permeability to other substances is not known. It appears likely that permeability is determined by the molecular size and it has been suggested that a cut-off occurs at 17 000 Da for globular molecules and 6000 Da for linear molecules. In theory at least, this would permit a large number of precursors and flavour-active bacterial metabolites to enter the bean.

(b) Drying

Browning of the cotyledons as a consequence of polyphenol oxidase activity continues during drying and is the most important reaction at this stage. Drying at too high an initial temperature results in the testa adhering to the cotyledon and obstructing oxygen diffusion. Under these circumstances, pigment formation can be incomplete.

The acetic acid content of the beans falls during drying. This is partly due to evaporation, but oxidation by micro-organisms is also a significant factor. The relatively high operating temperature of some mechanical dryers inhibits micro-organisms and leads to high acetic acid levels in the dried beans.

Although drying has little direct role in flavour development, quality can be adversly affected by pick up of taints from smoke. This is usually a consequence of contact with combustion gases in the dryer, but may also result from contact with smoke from domestic fires in the vicinity of the dryer or during storage. Phenolic compounds are responsible for the resulting 'smoky' off-flavour, 2,6-dimethoxyphenol being of greatest significance.

(c) Roasting

Many of the reactions which occur during roasting are similar to those which occur during coffee roasting and result in the formation of a large number of pyrazines and other heterocyclic compounds (Table 6.2, see also pages 244–5). The unique flavour associated with cocoa products, however, is primarily derived from the aldol condensation product 5-methyl-2-phenyl-2-hexenal. The immediate precursors of this compound are phenylacetaldehyde

Table 6.2 Main heterocyclic compounds formed during roasting of cocoa beans

Maltol	Thiophene
Isomaltol	Furans
Furanones	Thiazoles
Pyrazines	Trithiolane
Pyridine	Pyrroles
1,3,5-trithiane	Oxazoles
Tetrahydrothiophene	3-thiazoline

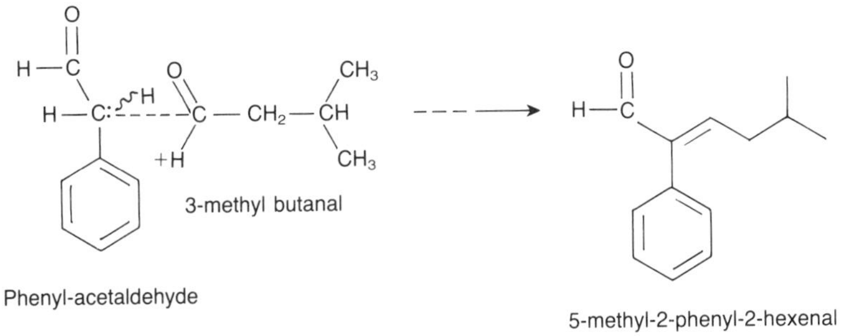

Figure 6.8 Formation of 5-methyl-2-phenyl-2-hexenal.

and 3-methylbutanal (Figure 6.8), formed from phenylalanine and leucine respectively by Strecker degradation. Strecker aldehydes, including phenylacetaldehyde, isobutyraldehyde and isovaleraldehyde, are themselves of importance in the flavour of cocoa. There are also increases in the concentration of pyrocatechaic acid and vanillic acid and a decrease in the concentration of caffeic acid during roasting. The consequences for flavour are not known.

6.6.2 Cocoa butter

Standard cocoa powder contains *ca*. 20% cocoa butter and reduced fat cocoa powder *ca*. 8% (legal minima vary on a national basis). Cocoa butter has little role in flavour, but contributes to mouthfeel and the overall character of the made beverage.

Cocoa butter is a relatively simple fat. The main fatty acids are palmitic, stearic and oleic, together with small quantities of myristic, linolenic and arachidic acids. Triacylglycerols comprise *ca*. 98% of cocoa butter. The major type of triacylglycerol contains two saturated fatty acids and one unsaturated, while the second in importance has one saturated and two unsaturated. Changes in levels of cocoa butter have been reported during fermentation, but these are very small and within experimental error.

6.7 THE MICROBIOLOGY OF COCOA

Cocoa is unusual amongst beverages in its potential role as a source of pathogenic micro-organisms, notably *Salmonella*, and of spoilage micro-organisms, notably *Bacillus*. For this reason, microbiological analysis of cocoa products is common practice.

6.7.1 Cocoa as a source of pathogenic micro-organisms

Salmonella is derived from handling at the country of origin and the incidence of *Salmonella* before roasting is high. The organism will not, however, survive roasting and its presence in the roasted product is a direct consequence of cross-contamination from unroast material. In most circumstances, no subsequent manufacturing stage for either cocoa beverage of confectionery is guaranteed to kill *Salmonella*. Heat treatment can be applied to the liquor in the case of cocoa beverage (see page 274), but this practice is considered to lead to poor taste in the case of confectionery. Prevention of contamination of roast cocoa requires rigorous control (see Table 6.1, page 272). In this context, concern has been expressed at the trend toward roasting in the country of production where hygiene standards and process control may be poor.

Chocolate confections have been implicated in a number of outbreaks of salmonellosis. These include international outbreaks in which a striking feature was an infective dose as low as 1×10^1 cells.

Cocoa beverage powder is known to have been involved in only one outbreak of salmonellosis. It has been proposed that the use of

* Although human adapted serovars, such as *Salmonella typhi* are known to cause illness at an infective dose of less than 10^3 cells, it is generally considered that serovars associated with food poisoning require between 10^5 and 10^7 cells to initiate infection. A number of cases are however known where the infective dose of food poisoning serovars was significantly lower than 10^3 cells. In many cases the foods involved have been of high fat content, such as cheese, chocolate and hamburger, and it has been assumed that this protected the cells from gastric acidity. One case, however, involved water where the infective dose of *S. typhimurium* was only 1.7×10^1 cells.

boiling milk or water in making up cocoa beverage acts as an additional safeguard. This may well be true in the case of traditional cocoa and drinking chocolate powders, but many products suitable for making up at lower temperatures are now available. The increasing complexity of some modern products means that a risk may also be introduced by non-cocoa ingredients. Dried milk powder is recognized as a potential source of *Salmonella* and control procedures must be implemented accordingly.

Although *Salmonella* is of over-riding concern with respect to the safety of cocoa-based products, handling conditions are such that a risk of contamination from other pathogens must exist. A cocoa-based syrup used for post-pasteurization flavouring of chocolate milk was believed to be the source of *Yersinia enterocolitica* in an outbreak in New York. Cocoa powder-containing chocolate milk has also been the cause of a large outbreak of Staphylococcal intoxication in California. In this case, however, the organism grew and elaborated toxin in the milk before pasteurization.

6.7.2 Cocoa as a source of spoilage organisms

Cocoa powder of poor microbiological quality can contain a range of spoilage micro-organisms. Moulds, yeasts and endospores of *Bacillus* species are generally numerically dominant. None of the micro-organisms are able to grow in cocoa powder, but *Bacillus* is of importance where the powder is used as an ingredient in chocolate milk. Cocoa powder is the source of highly heat resistant endospores of thermophilic *Bacillus* species, responsible for spoilage of UHT-processed chocolate milk exported from Europe to the Middle East.

6.7.3 Microbiological analysis

Analysis for *Salmonella* is required for roast beans, the end product and environmental samples, such as dust. Standard methods have been found to be inadequate on some occasions and there has been considerable effort applied to development of acceptable methodology.

Cells of *Salmonella* in cocoa products are stressed and pre-enrichment (resuscitation) is necessary. Buffered peptone water is used supplemented with malachite green oxalate, to reduce competition in environmental samples, and with casein to eliminate inhibitors present in some cocoa powders. Motility enrichment using a semi-solid version of Rappaport-Vassiliadis broth medium has been found to be superior to conventional cultural methods. Motility enrichments may be made direct from the pre-enrichment (direct motility enrichment) or after a eight hour incubation period in a conventional enrichment broth (indirect motility enrichment). Motility enrichment is particularly effective in detection of lactose fermenting strains of *Salmonella* and is also suitable for examination of milk powder and, if required, other ingredients.

Standard methods are suitable for the enumeration of other micro-organisms from cocoa, although in general only *Salmonella* is of concern.

* Papers describing the development of the motility enrichment method for isolation of *Salmonella* from cocoa powder contain a fundamental error, although this in no way affects the validity of the method itself. The error concerns the use of the selective mannitol-lysine-crystal violet-brilliant green agar for purifying cultures of presumptive *Salmonella*. Selective media, or selective incubation conditions, should never be used for purifying cultures, since contaminants may be suppressed and remain undetected. This may be demonstrated by the example of obligate anaerobes, such as *Clostridium*. Incubation under anaerobic conditions is effectively selective and during purification a streak plate incubated aerobically must always be included to permit the growth of aerobic contaminants. (de Smedt, J.M. *et al.* 1991. *International Journal of Food Microbiology*, **13**, 301-8).

EXERCISE 6.1.

During discussions at an international symposium a scientist stated that '. . . it is a convenient fiction that cocoa beans do not support mycotoxin production . . .'. The actual situation, however, is unclear. As a mycologist employed by a research institute, you are asked to assess the risk to consumers of cocoa and cocoa products made from mouldy beans. Draw up a proposal for submission to your senior scientist indicating how you would determine:

1. The mycotoxigenic potential of the spoilage moulds of cocoa beans.
2. Factors affecting mycotoxin production during mould growth on cocoa beans.
3. The fate of mycotoxins during subsequent storage and processing of the cocoa beans.

Include in your proposal a summary of methodology, staff requirements for an initial 3 year period and an assessment of cost, including capital purchases (assume your laboratory is equipped to basic mycological standards, but has no specialist equipment).

At the end of the first phase, it appears that both mould growth and mycotoxin production is determined by three factors: water activity (a_w) level, storage temperature and ratio of lactic:acetic acid concentration. You are then asked to develop a mathematical model to assist safety assessment. Your senior scientist considers that only mycotoxin production need be modelled. Do you consider that this view is valid, or should mould growth be modelled in addition to, or in place of, mycotoxin production? Which of the various types of model would you use in each case?

Additional information on mathematical modelling is available in Bazin, M.J. and Prosser, I.I. (1992). *Journal of Applied Bacteriology* (Supplement), **73**, 89–95; Farberi, J.M. (1986) in *Foodborne Micro-organisms and their Toxins: Developing Methodology* (eds M.D. Pierson and N.J. Stern), Marcel Dekker, New York; McMeekin, T.A. *et al.* (1993) *Predictive Microbiology: Theory and Applications*. John Wiley, Chichester.

EXERCISE 6.2.

You have been appointed as product development manager for a small, family-owned company manufacturing cocoa and drinking chocolate. The products are downmarket and sales have been falling for several years. A new and energetic managing director has been appointed and an injection of capital secured. It has been decided to develop a range of non-traditional chocolate beverages and 'brain-storming' sessions have been held which have identified four possible products and their target markets. You are asked to write short notes on each, indicating technical feasibility, potential major obstacles in development and the likelihood of market success.

1. A clear (non-turbid) chocolate drink to be sold from refrigerated cabinets and to be available in several flavours (e.g. orange, peppermint, almond). The target market is the more sophisticated 20–35 year old.
2. A high energy, high protein chocolate drink for the sports and health drinks market. The drink is to be stable at room temperature and to be packaged in individual packs and in bulk packs for sale through dispensers.
3. A 'winter warmer' based on brandy, spices and cocoa. The product is to be consumed when heated and must be suitable for use with microwave ovens. The target market is the older consumer.
4. A chocolate-flavoured, lightly carbonated white wine. The target market is the young drinker, especially females.

EXERCISE 6.3.

You are employed as microbiologist for a multi-national chocolate manufacturer. The company management is dominated by accountants, who are increasingly unwilling to take account of factors, such as safety, which cannot be quantified on the balance sheet. In an initiative to reduce costs it is planned to roast cocoa beans at point of production in an Asian country, rather than at plants in Europe and North America. You have visited the Asian plant and found that, while equipment is modern, operator training and process control are rudimentary at best. Prepare a report for company management, written with sufficient strength to persuade the directors to implement a major improvement programme, including operator training and the use of HACCP. (You are advised to stress the financial consequences alongside technical considerations.) *Assuming* that your report is accepted draw up a programme of basic hygiene training suitable for all staff, including management, at the Asian plant. Assume that education at a basic level is good, but there is little knowledge of any form of biology and a poor appreciation of basic hygiene.

Further information concerning the economic impact of food poisoning is available from: Roberts, J.A. *et al.* 1989. *British Medical Journal*, **298**, 1227–30; Todd, E.C.D. 1987a. *Journal of Food Protection*, **50**, 1048–57; Socket, P.N. (1991) *Journal of Applied Bacteriology*, **71**, 289–95.

7

ALCOHOLIC BEVERAGES: I. BEER

OBJECTIVES

When you have read this chapter you should understand

- The key role of yeast in the production of beer and other alcoholic beverages
- The difference between ale-type and lager-type beer
- The ingredients of beer, their processing and their technological role
- The technology of the brewing process
- Maturation
- Quality assurance and control
- The major chemical and biochemical reactions during the brewing of beer
- The nature of the important flavour compounds of beer and its ingredients
- Potential microbiological problems

7.1 INTRODUCTION

The production, and consumption, of alcoholic beverages is one of man's oldest activities. Today brewing, wine making and distilling are of major commercial importance in many non-Islamic countries and, through taxation, can be an important source of government revenue. At the same time it must be recognized that excess consumption of alcohol leads to serious social and medical problems for individuals as well as loss to the national economy through loss of productivity, cost of medical treatment, etc.

Over the years a vast range of alcoholic beverages have evolved although, in most cases, it is possible to place these in one of three categories - beer, wine or distilled spirits - according to ingredients and method of manufacture. These three categories are discussed in separate chapters, but in all the production of ethanol by fermentation of carbohydrates is the definitive stage, in which, in the majority of cases, species of *Saccharomyces* play a key role.

7.2 THE CENTRAL ROLE OF *SACCHAROMYCES*

7.2.1 Taxonomy and nomenclature

In common with other micro-organisms the taxonomy and nomenclature of *Saccharomyces* is subject to change and re-appraisal. Classically two species, *Sacch. cerevisiae* (ale-type beer) and *Sacch. carlsbergensis* (*Sacch. uvarum*; lager-type beer) have been used in production of alcoholic beverages. The separation of the two species is supported by recent work and while the name *Sacch. pastorianus* has been proposed for *Sacch. carlsbergensis*, the traditional nomenclature is retained in this book. Although the taxonomy of *Saccharomyces* is of importance to the science of microbiology, the prime concern in the production of alcoholic beverages is technological performance. Strains have been selected over a period of many years on the basis of this performance and differ from each other, from strains used in other industrial processes such as baking and from 'laboratory' strains. Strain differences are often considered to be minor and taxonomically insignificant but are of the utmost importance to the practical brewer's man and his colleagues involved in production of other alcoholic beverages.

7.2.2 The alcoholic fermentation

Saccharomyces is capable of fermenting a wide range of sugars including sucrose, glucose, fructose, galactose, mannose, maltose and maltotriose. Production of ethanol, the major fermentation end product, involves the aerobic formation of pyruvate *via* the Embden–Meyerhof pathway and the subsequent anaerobic decarboxylation of pyruvate to acetaldehyde. Finally acetaldehyde is reduced to ethanol with concomitant reoxidation of NADH (Figure 7.1).

Glycerol, the major fermentation end product after ethanol and CO_2 is formed by the glycerol-3-phosphate dehydrogenase mediated conversion of dihydroxyacetone phosphate to glycerol-3-phosphate, and significant amounts of acetic acid may also be formed through the action of pyruvate decarboxylase.

BOX 7.1 Burton ale

Louis Pasteur is well remembered for his work with beer and wine, but it is often forgotten that the brewery town of Burton on Trent housed, during the late 19th century, a remarkable collection of eminent scientists. Colourful characters by any standards, these men were responsible for laying the foundations of modern brewing science and introducing principles of quality control. Fundamental studies of yeasts were also made including a creditable attempt at elucidating the glycolytic pathway.

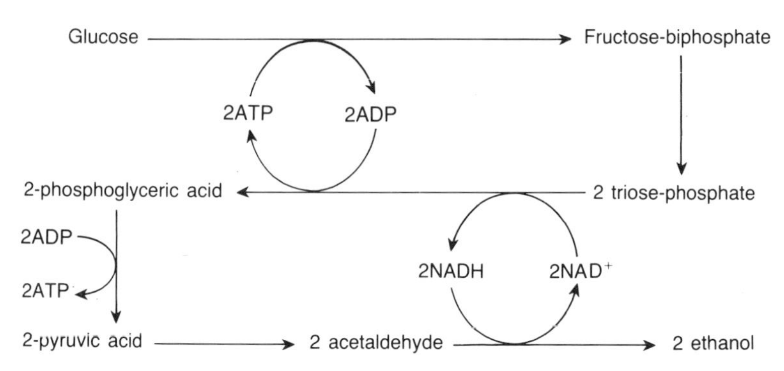

Figure 7.1 Simplified pathway of alcoholic fermentation.

7.2.3 Technologically important properties

(a) Ethanol tolerance

Most industrially used strains of *Saccharomyces* have an ethanol tolerance significantly higher than their wild counterparts. A high tolerance is required to ensure that the fermentation continues to the desired ethanol content and is of particular importance in the 'stronger' beverages. Ethanol tolerance is under polygenic control and involves a number of possible mechanisms of which end product inhibition of glycolytic enzymes and damage or modification of the cell membrane are considered most important.

(b) Flocculation

Flocculation involves the formation of an open agglomeration of cells and is thus distinguished from other forms of aggregation such as 'clumpy growth' and 'chain formation'. The property is required for ready separation of the yeast from the product at the end of fermentation and to minimize off-flavours due to an excessively long contact period. However a too rapid flocculation can result in failure to complete the fermentation. Flocculation is not dependent on cell division and involves divalent ions, usually Ca^{2+}, forming bridges between anionic groups at the cell surface. Several mechanisms have been proposed including cross-linking of cells by glycoproteins and the possession of a 'hairy' coat by flocculent strains, in comparison with the 'smooth' coat of non-flocculent. Flocculation is genetically determined but is also affected by the the environment in the immediate vicinity of the yeast cell wall. The phenomenon is affected by a number of factors, but it may be broadly stated that flocculation is inhibited by sugars and promoted by salts.

(c) Resistance to killer activity

Killer activity results from the production, by some strains of *Saccharomyces*, of extra-cellular toxins, zymocins, which are lethal for sensitive strains. The presence of killer strains can seriously disrupt fermentations and resistance to killer activity is an desirable attribute for any commercially used strain of *Saccharomyces*.

7.2.4 Genetics of *Saccharomyces*

Saccharomyces used in production of alcoholic beverages differ from the 'laboratory' strains on which knowledge of yeast genetics has been based. In contrast to the predominantly diploid 'laboratory' strains, commercial strains are usually polyploid or aneuploid. Multiple gene structures are highly stable and less prone to mutation, permitting the continued use of strains over many fermentation cycles. Such strains are not, however, amenable to genetic manipulation. Most industrially important characteristics of *Saccharomyces* are chromosomally mediated, but plasmids have been detected in *Sacch. cerevisiae* and both zymocin production and resistance are plasmid-encoded.

* Killer activity has been introduced to industrially used yeasts by genetic modification and is of potential benefit in control of wild yeasts during fermentation and storage of 'cold-sterilized' beer. The use of both killer and sensitive strains in the same premises may, however, lead to fermentation problems with due to inactivation of the sensitive strains.

Table 7.1 Examples of genetic modification of *Saccharomyces*

General
introduction of killer activity and killer resistance
Brewing
introduction of glucoamylase activity in brewing of low carbohydrate beer
elimination of phenolic off-flavour
reduction of vicinal diketone production

Despite the inherent problems associated with genetic manipulation, there is much interest in either breeding, or constructing, strains of yeast with improved characteristics for production of alcoholic beverages. A range of techniques are used although recent emphasis has been on recombinant DNA technology. It has proved possible to introduce, or eliminate, a wide range of properties (Table 7.1) although work is still largely at an experimental stage.

7.3 TECHNOLOGY OF BREWING

Although a vast number of beers is available, the majority may be classified as one of two basic types, the ale-type, brewed with a top-fermenting strain of *Sacch. cerevisiae*, and most popular in the UK and some former colonies, and the lager-type, brewed with a bottom-fermenting strain of *Sacch. carlsbergensis* and most popular in the rest of the world. The basic process for the two types is similar and is summarized in Figure 7.2.

7.3.1 The ingredients of beer

In addition to yeast, the main ingredients of beer are a cereal, usually malted barley, which serves as the source of fermentable carbohydrates, proteins, polypeptides, minerals, etc.; hops, which impart bitterness and the hop characteristic and which also have anti-microbial properties valuable in draught beer and water (liquor). In countries such as Germany, beer may be made only from barley, hops, yeast and water, but additional ingredients permitted elsewhere include adjuncts, which partly replace malted barley, caramel as colouring and various enzyme preparations.

Malting barley grown in Europe is usually of the hulled, two-row type; However, in the US, hulled, six-row varieties are more suited

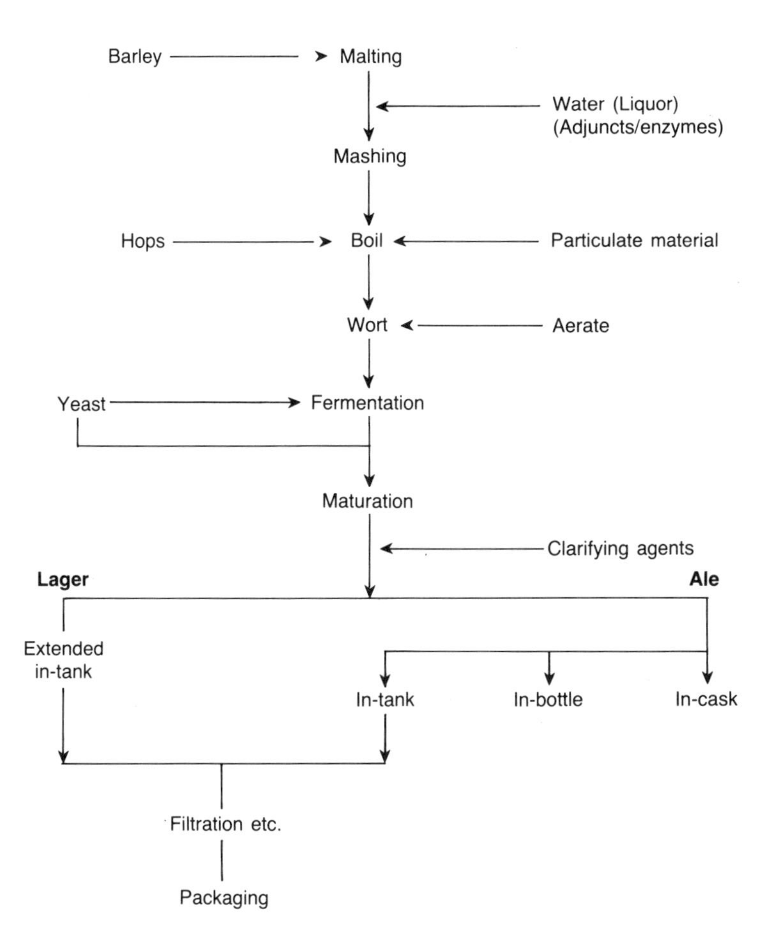

Figure 7.2 Simplified procedure for the brewing of beer.

BOX 7.2 Grains of paradise

Like most beverages, beer was subject to widespread adulteration. 'Dilution with water, restoration of bitterness with quassia or gentian and of 'strength' with *Cocculus indicus*, and the use of sulphate of iron for 'heading' were the stock-in-trade of most, if not all, publicans'. Narcotics, such as grains of paradise, were also widely used to adulterate hops and beer itself and were responsible for many cases of apparent drunkenness. (Burnett, J. 1989. *Plenty and Want*, 3rd edn. Routledge, London).

Table 7.2 Malt and corresponding malting barley quality parameters

Malt quality parameters
high extract yields
high proteolytic and cytolytic enzyme activities
Corresponding barley quality parameters
germinative capacity greater than 94%
fraction of plumper grains greater than 90%
husk content not greater than 11%
protein content not greater than 11.5%

to the environmental conditions. Six-row varieties are generally of lower malting quality, but have a high level of enzyme activity and are thus highly suitable for American brewing practices which involve a high proportion of unmalted cereals in the mash.

Malting barley has been the subject of long-term breeding programmes and varieties such as Triumph, which are both of acceptable malting quality (Table 7.2) and high yield have been developed. In addition the recognition of the role of malt-derived proanthocyanidins in promoting non-biological haze formation (see page 352) has led to a requirement for barley varieties which produce low levels of proanthocyanidins. Malt produced from low-proanthocyanidin barley has been successfully used with little effect on beer quality, but the dominant variety, Galant, is not entirely satisfactory with respect either to agronomy or malting qualities.

Malting is a controlled germination process involving 'modification' - the mobilization and development of enzymes and the liberation of starch granules from the endosperm (see pages 333-6 for a more detailed account of the chemistry of malting). Considerable progress has been made in recent years in refining the process which now takes a total of 8-9 days compared with 14 days during

* Extrinsic factors, such as such as dirt, mould growth, *etc.* also affect the quality of malt. Growth of *Fusarium* affects malting quality and plays a role in the phenomenon of 'gushing'. Concern has also been expressed over the possibility of mycotoxin production by species including *F. graminearum* and *F. moniliforme*. The risk to consumers of beer made with *Fusarium* contaminated malt is considered to be low, however, due to loss, or breakdown during malting and brewing.

the 1950s. At the same time, the weight of material lost in the process has been reduced from *ca.* 10 to 5% or less.

Following initial drying and cleaning the grain is steeped in water at 10–16°C to raise the water content of the grain to 42–46% when germination can begin. The distinction between steeping and germination is, however, artificial. Freshly harvested barley will not germinate due to dormancy, but this may be reduced by drying. A lack of oxygen in the steeping water also inhibits germination, but this may be overcome by periodically draining the steeping grains or by aerating the steep water. Most modern maltings employ combined steeping-germination vessels characterized by a shallow grain bed and a high rate of air flow.

The main advances in malting in recent years have stemmed from an improved understanding of the physiology of the barley grain. Germination may be accelerated and regularized by post-steeping application of gibberellic acid to initiate enzyme production by the aleurone. It is also possible to enhance the passage of gibberellic acid to the aleurone cells by mechanically abrading the husk and underlying layer. This process is used commercially to accelerate malting or to assist the malting of poor quality barley.

In modern practice conditions during germination are controlled by passing humidified, temperature controlled air through the germinating grain. It is desirable to reduce malting loss by limiting root growth and respiration and this may be achieved by re-steeping. A final steep at 40°C may also be employed to achieve root killing.

The degree of 'modification' required depends on the type of beer to be brewed and when this has been achieved malting is stopped by drying. Drying is a two-stage process, involving an initial low-temperature, followed by a temperature sufficiently high to suspend, but not to inactivate, enzyme activity. In the final stage temperatures of 100°C or higher may be used. High-temperature short-time extrusion cooking has been proposed as a cheaper alternative to kiln-drying. Special types of malt are produced by special kilning procedures. Caramel malt is produced by kilning green malt in a drum to hydrolyse and caramelize the starch in individual grains, while brown malt is made by rapid drying in a special draft kiln. Roasted malts are of various types including chocolate malt and the less severely heated amber malt, and are made by subjecting dried malt to further heating in a drum. Special types of malt are used as adjuncts and contribute to beer colour and stability.

Malt is milled before mixing with water at the mash stage of brewing. Four roll mills were previously common and give a good extract with well modified malt, but the increasing use of less well modified, lager-type malts and a demand for rapid processing has meant the increasing use of six roll equipment. Either a dry or wet process can be used. Dry milling gives a good extract but fragmentation of the husk leads to increased wort run-off time after mashing. This problem may be overcome, and extract increased, by steam conditioning which results in a rubbery husk which is less prone to fragmentation. Wet milling also increases the extraction rate and allows a greater weight of mash to be placed in the lauter tun. Wet milling also offers high throughput, but is unsuitable for use with poorly modified malt and involves high energy costs.

Steep conditioning avoids many of the disadvantages of steam conditioning and wet milling by limiting the contact time between malt and water to 60 s. Hot water at 75°C is used and while the water content of the husk is raised to *ca*. 20% the endosperm remains dry.

Unmalted adjuncts are now widely used in some brewery operations. The use of adjuncts is partly dictated by economics, but adjuncts can also be of importance in imparting particular characteristics to the end-product.

Adjuncts are of two types: (i) sugars and sugar syrups, and (ii) starch-rich materials. Sugars and syrups may be used to adjust beer characteristics, usage involving little additional capital cost. Cane and beet sugars are now less commonly used than glucose and other syrups produced by enzymic hydrolysis of corn starch. Use is also made of malt extracts and caramelized syrup which is

* The climate of some east European countries means that the barley produced is unsuitable for malting and, to avoid importation of malting barley, 'all barley' beer is produced by use of enzymes in a special mashing process. Problems have also arisen in countries such as Nigeria, which have a large consumption of European-type beer but where barley is not grown. An embargo on the importation of barley for brewing has led to a search for locally grown substitutes. Sorghum appears to be most suitable and has been used to brew both a lager-type beer and a stout.

produced by heating glucose syrup with ammonium hydroxide. Caramel syrup is important in imparting colour and flavour to dark beers but doubts have been raised concerning the safety of this and other types of caramel.

Starch-rich adjuncts are typically unmalted cereal products. Carbohydrate extraction requires three stages: gelatinization, solubilization and amylolysis. Extrusion cooking is of interest as a means of gelatinization and offers a cheap and high yielding source of extract relative to malt. Where nitrogenous materials are present in addition to starch, however, extrusion cooking can be associated with an 'artificial' burnt, coffee or caramel-like flavour due to formation of alkylpyrazines. Maize and rice starch are both widely used as adjuncts, pre-gelatinization being required in each case. Wheat and barley may be added to the mash without pre-treatment but problems of high wort viscosity are potentially greater. Adjuncts can affect beer flavour and their use leads to a high viscosity wort and a slight lowering of fermentability. The presence of high levels of glucose, fructose or sucrose may lead to repression of maltose metabolism (see page 344). Despite this the use of adjuncts is highly advantageous in many circumstances and 'fine-tuning' of the brewing process may be used to minimize difficulties.

Hops can be grown in many parts of the world and are essential for the production of beer. Varieties are generally considered to be of two types: high α-acid varieties, containing high levels of humulones which are responsible for the bitter flavour of beer, and aroma varieties, containing high levels of essential oils which impart the hop character ('kettle hop' or 'noble hop' flavour). Hop varieties have been considerably improved by breeding programmes directed both at husbandry and at brewing-associated properties. The major interest lies in improving the α-acid content which has been raised from 3.5 to 5% in traditional varieties, to 7–10% in varieties such as Brewer's Gold and Northern Brewer, and as much as 11–13% in Pride of Ringwood, Galena, Target, Yeoman and other very high α-acid varieties. The yield of modern varieties

* Unlike barley, both wheat and rye contain gluten. The use of these cereals as adjuncts has caused some concern to persons with coeliac disease, who are intolerant of gluten. The level of gluten in the final beer, however, is extremely low and the possibility of problems seems remote.

is also markedly higher than that of traditional and up to three times as much α-acid per hectare is produced. There has also been some success in breeding improved aroma varieties, but fear of altering the flavour of an established beer means that brewers tend to continue using traditional varieties. High α-acid hops, however, often also have good aroma properties.

Hops cones are dried on the farm and, in traditional handling, are pressed into bales or pockets. In each case the bulk densities are very low, and transportation and storage of hops is both inconvenient and expensive. It is generally agreed that hops should be subject to as little change as possible during storage, although flavour may benefit from a short ageing, and problems arise from the susceptibility of humulones to oxidation. Oxidation may be delayed by low temperature storage, at significantly increased cost, or by vacuum packing. The low bulk density of hops, however, restricts the use of vacuum packing. Further problems arise in usage where the yield of isohumulones in the finished beer is usually no more than 40%. For these reasons there has been a long-standing requirement for hop products which are convenient to handle, stable in storage and of high isohumulone yield. A number of hop products are commercially available (Table 7.3) and are gaining an increasing share of the hop market.

Hop powders or pellets are the simplest form of hop product. Their advantage lies primarily in convenience of handling although the disruption of lupulin glands during pelleting does make resins and essential oils more accessible during hop boiling. The susceptibility to oxidation is, however, markedly increased and vacuum packing, or gas packing in nitrogen or CO_2 is required. In most cases any increase in yield of isohumulones is minimal at best, but powders and pellets do have advantages in usage, minimizing wort loss in the kettle, eliminating the need to remove hops by filtration and obviating problems of disposal of spent hops.

Enriched powders and pellets are effectively high α-acid content hops produced by physical concentration rather than plant

* The introduction of low-proanthocyanidin barley varieties has increased the relative importance of hop proanthocyanidins in production of non-biological haze in beer. It is fortuitous, therefore, that high α-acid content hops tend also to contain low levels of proanthocyanidins.

Table 7.3 Summary of commercial hop products

Hop powders and pellets dried, hammer milled hop cones, packaged as powders or pelleted.
Enriched hop powders and pellets powders concentrated by mechanical sieving at a maximum temperature of −20°C. At such temperatures lupulin glands are rigid and easily separated from cellulosic material. Contain *ca*. 75% original weight of hops and 90–95% of α-acids.
Speciality hop powders and pellets (a) incorporate hop extract. (b) incorporate inorganic salts.
Hop extracts Extracted by single, or mixed, solvents with no, or minimal, change to the extracted components.
Speciality hop extracts (a) incorporate additional hop oil. (b) incorporate inorganic salts.
Isomerized hop extracts Isolated, isomerized α-acids produced by: (a) treatment of liquid extract with alkalis. (b) treating of metal salts of α-acids. (c) simultaneous isomerization and reduction with borohydrides.
Hop oils (a) traditional – essential oils steam distilled from hops either in concentrated form or a water-based emulsion. (b) late kettle hopping – fractionated oil-rich liquid CO_2 extract. (c) Dry hopping – fractionated oil prepared by low-temperature distillation under reduced pressure or liquid CO_2 extraction.

breeding. There is a decrease in polyphenol content, the consequences of this decrease being the subject of widely varying opinions amongst brewers.

Speciality hop powders and pellets are of two types. The first, produced by blending with hop extract, is effectively an extract standardized with milled hops and having the advantage of being more easily handled and metered into the kettle. The second type is produced by mixing hops with a metal salt such as calcium or magnesium hydroxide. This stabilizes the α-acids and may lead to more rapid isomerization of humulones during boiling. A degree of pre-isomerization also occurs as a consequence of the heat of pelleting. This may lead to higher utilization of α-acids in the hop kettle, but the mixture of humulones and isohumulones in the pellets makes control of bittering difficult.

Hop extracts have been available for many years, the underlying principle being the concentration, by physical means, of desirable hop components. Organic solvents such as hexane, methanol and methylene chloride were previously used in extraction, but liquid, or supercritical, CO_2 and ethanol are now preferred. Liquid CO_2 is a non-polar solvent and highly selective for the desirable components of hops. Essential oils and α-acids are extracted sequentially (Figure 7.3) resulting in two products and greater flexibility in use. The efficiency of extraction, 92–95%, is, however, relatively low.

The polarity of supercritical CO_2 varies according to temperature and pressure employed. Selectivity is always less than that of liquid CO_2 and 'hard resin components' consisting of oxidation and/or polymerization products of α-acids are present in the extract. The extract is of lower quality and more subject to batch-to-batch variation, but the efficiency of extraction is higher (95–98%). Ethanol is highly polar and while the efficiency of extraction is marginally greater, the selectivity is poor.

Hop extracts are highly stable and may be stored without significant deterioration. The extracts are usually standardized to a specific α-acid content in sugar syrup and are convenient to use.

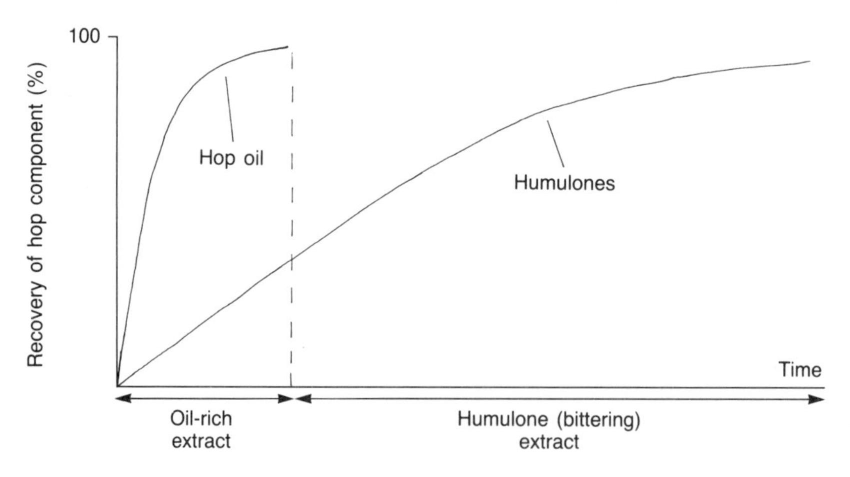

Figure 7.3 Differential extraction of hop components by liquid CO_2.

Dispersion and isomerization is more rapid but there is little benefit in terms of α-acid utilization. Speciality hop extracts have also been developed and contain either hop oil or inorganic additives. In neither case is there any obvious advantage over other hop products.

The underlying rationale for the development of pre-isomerized hop products is that isomerization of humulones can be achieved more efficiently under high pH conditions not possible during the brewing process. Early processes were unsatisfactory, but subsequently methods for the production of a high quality pre-isomerized extract were developed using high purity liquid CO_2 extracts as base material and carbonates as isomerizing agents. Pre-isomerized powders and pellets are also available. Powders are produced by heating magnesium humulate, formed by reacting humulone with magnesium oxide. Pellets are produced by heat treatment of calcium or magnesium humulones at 100°C for 40 min to achieve a *ca*. 95% conversion of humulones to isohumulones. Extrusion cooking has also been used to produce pre-isomerized hop extrudates. Hop powder and a calcium or magnesium salt are processed at 120–150°C, a *ca*. 90% conversion of humulones to isohumulones being achieved in less than 90 s.

A special type of isomerized hop extract is obtained by simultaneously reducing and isomerizing humulones by treatment with borohydride. This extract permits the use of clear glass bottles for

BOX 7.3 **Yellow is the colour**

Food and beverage companies are extremely sensitive to the possibility of product tampering. Concern amongst consumers, especially in the US, is such that totally unsubstantiated rumours rapidly gain credence. The effect on the company involved can be very serious indeed, involving direct loss of sales, expensive public relations exercises to recover brand credibility and, in some cases, significant falls in the share value. This situation has been faced by the American importers of the popular Mexican beer Corona™. Rumours circulated variously stating that the beer, a distinctive yellow colour and packed in clear glass bottles, contained urine as an ingredient or had been deliberately contaminated with urine. Counter-publicity by the importers was, fortunately, effective.

packaging without the development of 'sunstruck' flavour (see page 349).

The advantages of pre-isomerized extracts lie in a significant increase in utilization of α-acids, and improved control of bittering. Addition is usually made to the beer stream in passage from the fermentation vessel to storage or at filtration. The final beer has a quality of bittering markedly superior to that obtained by conventional kettle hopping and utilization of as much as 80–95% of added α-acids

Hop oil is traditionally prepared by steam distillation and is a complex mixture of some 450 compounds many of which, in combination, impart the hop character to the beer. Hops effectively undergo steam distillation during wort boiling and only the more water soluble components remain at the end of that process. For this reason hops may be added towards the end of wort boiling (late kettle hopping) or added to the beer at the start of conditioning (dry hopping). Each process produces in a distinct flavour in the final beer as a result of the different chemistry involved. At the same time changes in the hop oil during conventional distillation mean that it is not possible to match exactly the flavour of either late kettle hopped or dry hopped beer. Technological developments, however, mean that it is now possible to produce hop oil which closely matches the flavour of dry hops. This is achieved by fractionation of oil produced by ambient temperature distillation at reduced pressure or of the oil preferentially removed in the early stage of liquid CO_2 extraction. Some hop resins should also be present in the extract to enable selective extraction of oil components to occur on addition to the beer and thus to mimic the dry hopping process. The situation with late kettle hopping is more complex, but it has been possible to develop a late hop essence which produces a flavour in the final beer closely resembling that of traditional late hopping. This is produced by physical separation and fractionation of oil-rich liquid CO_2 extract and comprises two main fractions – ketones imparting a floral character and alcohols imparting an astringent mouth-feel.

The use of hop oils increases control over the consistency, quality and intensity of the hop character as well as offering a much wider range of flavours than traditional hopping. Problems of utilization lie in the extreme flavour intensity per unit weight and difficulties in dispersion and solution in the beer. Water-based emulsions have been used for distillates but tend to be unstable and injection of solutions in liquid CO_2 into the beer stream have been recommended.

Water or, as commonly known in brewing, liquor is an important constituent of beer and must be both of potable quality and of suitable chemical composition. Many centres of brewing developed where a ready supply of suitable water existed, well known examples being Burton on Trent in the UK where the water, rich in calcium and magnesium sulphate, is particularly suited to brewing bitter beer, and Pilsen in Czechoslovakia where the water is particularly suited to brewing pale lager.

Many new breweries are situated in green field sites where the chemical composition is only one of many factors deciding location. Further, some older breweries have lost the use of traditional sources such as artesian wells due to pollution. In addition to any treatment required to ensure potability, modifications may be made to the levels of inorganic ions present. This is most simply achieved by 'Burtonization', the addition of calcium sulphate and, sometimes NaCl, while bicarbonate may be removed by addition of lime and consequent conversion to insoluble carbonate. This process, however, is labour intensive and results in problems of sludge disposal and a more satisfactory method is de-alkalization by passage over acidic ion exchangers.

7.3.2 Mashing

The mashing process consists of two parts (i) conversion, during which malt enzymes are reactivated and enzymic processes continued, and (ii) lautering, in which soluble components are extracted from the malt and the spent malt finally separated from the wort. Wort is a complex and finely balanced solution of fermentable carbohydrates, amino acids and minerals which serves both as a substrate for yeast growth and ethanol production and as a source of flavour and aroma precursors. The mashing conditions, i.e. temperature, mash viscosity, pH value and water composition, may be varied to alter the amount of individual components present in the wort and to serve as a means of obtaining the correct wort composition.

Three distinct types of mashing procedure are in use: single-temperature infusion, temperature-programmed infusion and decoction

(Figure 7.4). Single-temperature infusion is traditionally used in ale brewing with well modified malts. The process is technically straightforward and both conversion and lautering may take place in a single vessel, the mash tun, although in modern practice greater operational flexibility may be obtained by using a separate converter and lauter tun. Coarse ground malt and hot water are mixed in a 'Steel's masher' and passed to the tun. Temperatures of 60–65°C are normally used, although 70°C may be used under some circumstances. Exo- and endo-β-glucanases and peptidases are inactivated at these temperatures, but starch is rapidly solubilized and hydrolysed. However, inactivation of β-amylase commences at *ca*. 65°C depending on the mash thickness and, in general, higher temperatures result in a less fermentable mash. Conversion is completed in 0.5–2 h after which lautering commences for up to 18 h. The traditional mash tun is simple to use, requiring no wort transfer or agitation during conversion and produces a very bright wort and well drained grains. Mash tuns are also flexible in terms of quantity of material added, but are expensive to operate due to the requirement for well modified malt and the slow run-off time resulting from the large bed depth. Extract recovery varies with run-off time from *ca*. 95% at 3 h to 100% at 18 h.

Temperature-programmed and decoction mashing are similar in that in each case the temperature is raised during mashing to enhance protein and β-glucan breakdown at lower temperatures and starch solubilization and hydrolysis at higher. They are thus suitable for use with less well modified malts. Both systems employ separate conversion and lautering vessels, but the means by which the temperature of the mash is raised differs.

Temperature-programmed mashing involves the use of a hot water jacket or internal heater and a mash mixer to raise the temperature in the conversion vessel through a pre-determined profile. The mash must be stirred sufficiently to ensure adequate heat transfer but mechanical damage resulting in release of higher molecular weight β-glucans must be avoided.

Two mash vessels, the conversion vessel and the mash copper, are employed in decoction mashing. Temperature is raised by withdrawing a portion of the mash from the conversion vessel, boiling it in the mash kettle and then returning it to the conversion vessel. Decoction mashing is most widely used in continental Europe and is able to handle poorly modified malts.

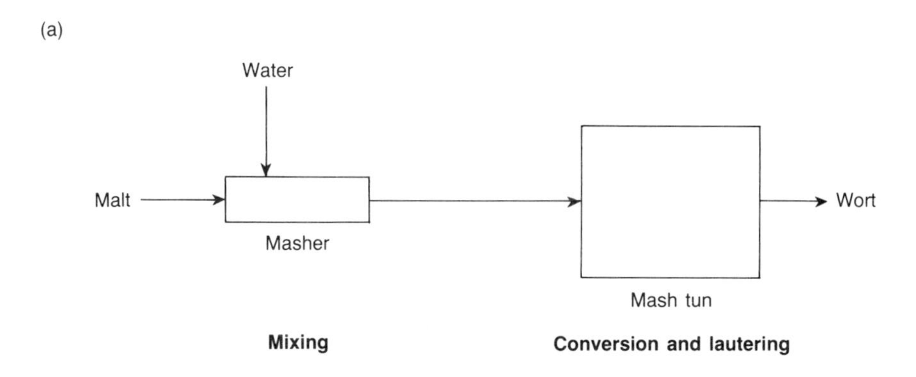

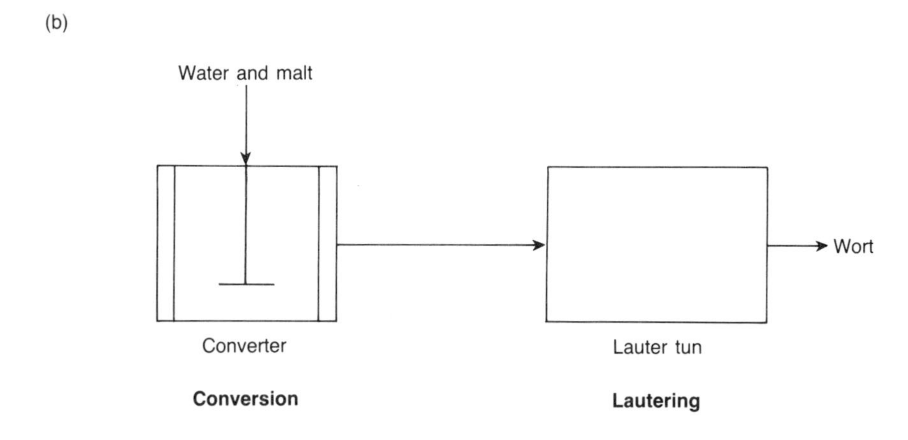

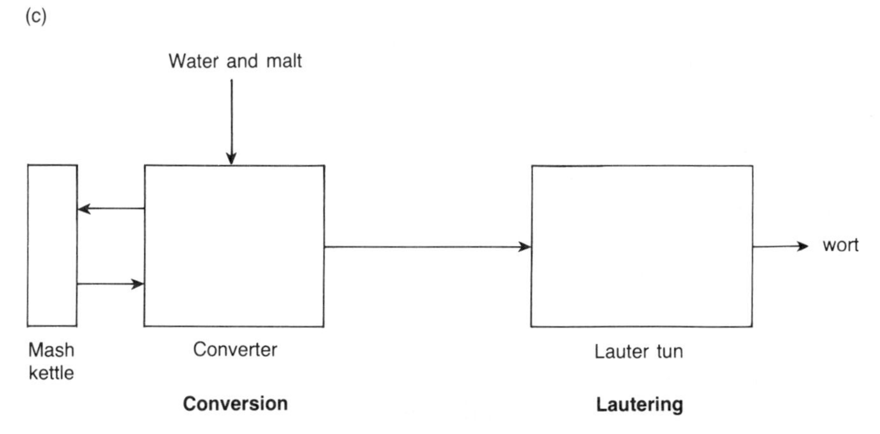

Figure 7.4 Alternative mashing procedures. (a) Traditional mash tun, (b) temperature programmed, (c) decoction.

A lauter tun is most commonly used in conjunction with temperature programmed or decoction mashing and is designed for rapid filtration of pre-converted mash. A shallow bed depth is used, giving a run-off time of 2–3 h and 98–99% extract recovery. The sparge liquid may be added in one operation and solids suspended by rotating knives (rakes). Lauter tuns are flexible in usage and spent grain well dried. Variants such as the Strainmaster™ have been developed which combine a higher filtration area with a lower floor space. The run-off time is reduced to 1–2 h at 98–99% extract recovery, but grains are very wet usually requiring further drying and the tuns are poorly suited to producing worts of specific gravities above 1.6.

Mash filters have been used for many years, but fluctuate in popularity. The construction is that of a filter press, cloth filter plates alternating with hollow plates. Run-off time is 1.5–2.5 h at 99.95% extract recovery, but the equipment is inflexible and produces notoriously cloudy worts. A modern variant, the high pressure mash filter overcomes the problem of cloudy worts and permits a reduction in sparging without loss of extract. This equipment is highly suitable for production of a high specific gravity wort and economical in terms of water use.

The use of adjuncts (see pages 304–5) is now widespread in brewing. The conventional mash tun can cope with only a small quantity of additional starchy material and sugars are normally used where adjuncts are required. Starchy adjuncts may be used in temperature programmed and decoction mashing without major difficulties. Depending on the level of enzymes present in the malt, however, supplementation with commercial enzymes is required when adjuncts are present at levels exceeding 25%. Enzymes may also be added to normal mashes to increase the amount of carbohydrate extracted or to reduce the β-glucan content and thus the viscosity of the wort. Commercial enzymes are usually of microbial origin and less heat-labile than those of malt and must be of sufficient purity to eliminate unwanted co-activity.

7.3.3 Wort boiling

Wort boiling satisfies seven technological requirements – evaporation of excess water, extraction of hop components and isomerization of humulones, destruction of malt, or added, enzymes, sterilization of wort, removal of undesirable volatiles, formation of flavour and colour compounds *via* the Maillard reaction and protein coagulation. Heating and evaporation is not sufficient and agitation is required to ensure complete protein coagulation. Wort boiling requires large amounts of energy and the potential for savings at this stage are the greatest in the brewing process.

In traditional brewing hops are added to the wort at the start of boiling (early hopping). Extraction of humulones by boiling is an inefficient process recovering at best *ca.* 65% of isomerized humulones after two hours boiling. Conditions favour oxidation of the humulones, but in practice this is minimized by the blanket of steam at the liquid surface. Oxidation may, however, be responsible for 'hard to explain' differences between breweries. Considerable loss of hop oil occurs during prolonged boiling, being replaced in some cases by addition of hops at the end of boiling (late kettle hopping) or to the beer before conditioning (see page 323).

The simplest means of wort boiling involves the use of open top vessels employing thermosiphon circulation as a means of agitation. Coppers of this type are operated for 1.5–2 h with an evaporation rate of *ca.* 8%/h. Various configurations of copper have been used but all are relatively energy-expensive. A more modern and energy-efficient type involves a separate heater (calandria) through which the wort is circulated. The higher operating temperature of 106–110°C results in better hop utilization (*ca.* 15%) and decreases the boiling time to 0.5–1.5 h. Agitation is provided by circulation through the external heater and excessive evaporation avoided. Further energy savings may be made by using mechanically recompressed vapour from the copper as a heat source. Other high thermal efficiency wort boiling systems operate at pressures above atmospheric. The 'low pressure' wort boiling system involves heating to boiling point, over a 10 min period, using a pressure-resistant copper fitted with an external heater. A second heating phase follows during which the wort is heated to 110–112°C over a further 10–15 min. The wort is then boiled

---

* Despite major improvements in efficiency, wort boiling remains an expensive procedure. Attempts have been made to obtain the effects of the process without actual boiling. Partial success has been achieved by filtration of wort through silica hydrogel and removal of unwanted volatiles by spray evaporation. This results in a beer which, although satisfactory, differs from that produced by conventional boiling and the process is yet to be adopted commercially.

under pressure for 15 min followed by pressure release and a 10 min post-boiling phase. Vapours are used to to heat wort in the initial heating phase and *ca.* 95% of the energy is recovered. The 'high-pressure' system requires more complex plant but is continuous in operation and achieves a steam saving of *ca.* 69% over conventional. Wort is collected in the kettle at 72°C and heated in a succession of three heat exchangers to a 140°C and held at that temperature for 3 min. The wort then undergoes stepwise cooling in two expansion vessels, the vapours being used in the heat exchangers.

The extent of change during boiling is usually dependent on the length and temperature of boiling and the constitution of the wort. Protein coagulation is an exception and the extent to which this reaction proceeds depends also on physical factors such as copper shape and flow and the heating system. Coagulation is an important reaction, inadequate coagulation leading both to a poor quality beer and adversely effecting the subsequent fermentation by interference with yeast–substrate exchange processes (membrane blocking). Only a proportion of the excess protein ('trub' or 'hot break') coagulates during boiling, some ('cold break') remaining to precipitate, complexed with polyphenols, at lower temperatures prior to or during fermentation. This results in a haze which is traditionally removed after fermentation (see page 324). It is now common practice in ale brewing to add copper finings, negatively-charged polysaccharides which complex with positively-charged beer proteins in the boiling wort to enhance post-fermentation removal. Alternatively, in lager brewing, some of the haze-forming proteins may be removed after conventional wort clarification procedures.

7.3.4 Wort clarification and cooling

Wort clarification is based either on sedimentation or filtration. Where whole hop cones are used it is necessary to employ either a hop back or a hop separator-filter. The hop back consists of a vessel fitted with a slotted false bottom, the hops themselves forming a filter bed. Hop backs are not suitable for use with hop powders or pellets, are labour intensive and unsuited to intensive brewing operations. Further the practice in the UK of sparging the hops to increase extract produces high concentrations of fats, polyphenols, etc. which have an adverse effect on beer quality. Hop separator-filters consist of a vessel with a conical perforated insert and are used in continental Europe and the US where sparging the hops is not practised, but losses of wort are significant. The reduction in hop usage and the widespread acceptance of pre-isomerized extracts means that hop backs and separator-filters are now much less widely used. In many breweries trub is removed by a whirlpool, a vertical cylinder into which the wort is introduced at high velocity at an angle to the vessel side. This induces sustainable circulation, the trub collecting as a compact cone in the base. Whirlpools are more suited to lager worts but may be used with ale. A number of modifications to the basic design exist to achieve maximum trub removal while minimizing wort loss and it is possible to combine the boiling copper with the whirlpool in a single vessel. In modern breweries centrifuges are effectively the only alternative to whirlpools. Centrifuges are efficient having high extract recovery and, especially when of the self-desludging type, requiring little supervision. Little floor space is required but the initial capital cost is considerable.

Wort is cooled after clarification, tubular or plate heat exchangers now being employed in virtually all breweries. In most cases the wort then passes direct to the fermentation vessel, although additional processing may be applied.

Hot wort aeration is a process designed to decrease haze by oxidizing and removing haze-forming protein. The process is used primarily in continental Europe and can impede the elimination of undesirable flavours during fermentation and conditioning by inactivating hop reductones. It is normally used in conjunction with cold wort filtration and may be supplemented by ultraviolet irradiation.

Cold wort filtration, using kieselguhr, reduces haze and shortens maturation time by removal of sulphur-bearing proteins. There is a significant increase in operating cost but the process is of some benefit when operating a Nathan-type fermentation-conditioning system for production of lager-type beer. Cold wort filtration may be detrimental in ale-type beer production since the absence of suspended protein to serve as nuclei for gas formation can lead to sluggish fermentation and premature termination. Wort stabilization (deep chilling) is used in New Zealand in conjunction with continuous fermentation. Wort is cooled to 0–5°C, held for up to 48 h and filtered. A major benefit stems from the ability to store unfermented wort from Friday to Monday and thus avoid weekend working.

7.3.5 Yeast handling

Under normal conditions yeast cells grown during one fermentation cycle are used as inoculum in subsequent cycles. It is necessary to maintain the yeast in a satisfactory physiological condition before re-use and, in this context, glycogen levels within the cell are of prime importance (see page 343). Yeast is usually stored in water or beer. Handling procedures should avoid inclusion of oxygen and yeast should be rapidly cooled to 4–6°C. A larger inoculum size (pitching rate) should be used for yeast of low glycogen content. Two tests, the acidification power test and the specific oxygen uptake rate have been developed, both of which correlate closely with fermentation performance. Either test may be used to determine optimum pitching rate or to reject a yeast of unsatisfactory performance.

Pitching yeast is now recognized as the major reservoir of bacterial contaminants, especially *Lactobacillus*, *Pediococcus* and *Obesumbacterium*. These organisms are involved in beer spoilage and, in the case of *Obesumbacterium*, reduce nitrate to nitrite with consequent formation of *N*-nitroso compounds. Yeast decontamination procedures have become widely used, current methods involving an acid treatment most commonly using dilute phosphoric acid or, more effectively, acidified ammonium persulphate. Nisin has been proposed as an alternative and is said to avoid adverse effects on yeast performance. Changes to the yeast cell surface and ATP leakage do occur, but acid washing is an effective procedure providing temperatures are maintained below 5°C and the yeast is pitched immediately. A solution of higher pH value and a shorter wash should be employed if the yeast is in poor physiological condition.

7.3.6 Wort aeration

Brewery fermentations differ from most others in that oxygen is supplied only once, usually with incoming wort, and in strictly limited amounts. The main role of wort aeration is not stimulation of yeast growth as such, but the promotion of the biosynthesis of lipids required for growth. Setting the level of aeration according to the dissolved oxygen concentration of the wort is misleading and a superior criterion is the lipid levels of the early yeast. An automatic on-line control system for wort aeration, based on indirect measurement of lipid levels by determination of specific oxygen uptake rate has been described.

7.3.7 Fermentation

Fermentation is recognized to play a central role in production of alcoholic beverages, although the full extent of processes taking place in the fermenter may not be appreciated. These are summarized in Table 7.4.

Brewery fermentations have traditionally been considered to be of two distinct types; top fermentations, used in production of ales and bottom fermentations, used in production of lagers, the difference in fermentation pattern tending to dictate the design of the fermenter. The increasing use of cylindro-conical and similar fermenters, however, means that differentiation on the basis of the behaviour of the yeast is becoming less distinct.

Ale-type beers are traditionally fermented in relatively shallow, circular or rectangular vessels from which the 'top' yeast is removed by skimming. The use of a 'bottom' yeast in lager fermentations obviates the need for skimming, but similar, although rather deeper vessels are used.

Table 7.4 Processes taking place within brewery fermenters

1. Aseptic collection of wort
2. Mixing of fermenter contents to avoid layering
3. Fermentation of wort under conditions of controlled temperature and pressure
4. Collection of surplus CO_2
5. Utilizing heat generated during later stages of fermentation to raise temperature of the contents to that desired for maturation
6. Controlled chilling of the contents to 3°C (temperature of maximum density for most beers), or above for maturation and/or yeast sedimentation
7. Dissolution of CO_2 in beer to desired level
8. Sedimentation of yeast and ready removal from vessel
9. Gas washing to remove unwanted volatiles
10. Propagation and storage of yeast
11. Cold conditioning at temperatures of 0°C or below[1]
12. Storage of finished beer[2]

[1] Not general practice
[2] Not recommended practice

BOX 7.4 **The power of tradition**

In the case of ales, many localized fermentation systems were developed, including the Burton Union and Yorkshire Square. Some such systems remain in use today and are associated by many beer aficionados with a product of high and traditional quality. It should be appreciated, however, that while beer produced by these systems has characteristic organoleptic properties, the 19th century brewers, responsible for their design and installation, were as much concerned with efficient fermentation, low costs and ease of operation as their late 20th century counterparts. Both the Burton Union and Yorkshire Square systems were, in fact, designed to permit use of yeast with generally excellent brewing performance, but whose mechanical properties precluded use in skimmed systems.

Since the 1960s most fermenters installed have been of low surface area to volume ratio. The Nathan cylindro-conical design has been predominant, although other configurations are used. Cylindro-conical fermenters are able to produce either ale-type or lager-type beers and have a number of advantages stemming from their structural geometry. These include a high ratio of fermenting capacity to total vessel capacity, greater hop utilization, due to the absence of a yeast top crop which adsorbs hop resins, low beer loss, efficient temperature control and ease of cleaning. Yeast handling is simplified by sedimentation into a compact plug in the cone and the ability to apply CO_2 top pressure.

Fermentation is rapid in cylindro-conical vessels as a consequence of the vigorous agitation arising from CO_2 production. Nucleation of CO_2 bubbles occurs at the lowest point of the fermenter, agitation being due to gas lift, the power of which increases logarithmically with fermenter height. A high rate of yeast growth, however, results in ester formation and a poor quality product. In lager brewing this has been prevented by brewing under moderate CO_2 pressures but, while effective in reducing flavour defects, there is an adverse effect on yeast crop viability and this process has been abandoned by some breweries. In many cases a simpler and more satisfactory solution is to control the fermentation by reducing the temperature.

The use of cylindro-conical fermenters for maturation has a number of advantages including convenience and reduced beer losses. Some brewers, however, consider this practice to be a retrograde step and prefer a two vessel system which offers advantages in relation to utilization of tank capacity, chilling, blending and post-fermentation additions. Problems due to yeast entering the product stream and interfering with filtering are also minimized.

Continuous fermentation systems have been available since the early part of the 20th century and offer a number of potential advantages including higher throughput, improved productivity and reduced costs. Interest in continuous fermentation revived during the 1960s, but the process is currently in commercial use in only two breweries both situated in New Zealand. The failure to exploit continuous fermentation may stem partly from technological inertia but more fundamental objections arise from inflexibility and the fact that the benefits fail to outweigh the disadvantages of greater complexity of plant and its operation. Further, fermentation is only one part of the brewing process and it seems likely that widescale adoption of a continuous process must await continuous wort production. Ultimately continuous fermentation using immobilized enzymes may be most acceptable, but for the moment advances in brewery fermentation technology would seem to lie primarily in the application of sophisticated process control.

Inexpensive microprocessor-based equipment now permits fermentations to be controlled using interactive systems. In the case of brewing, however, sensor technology has failed to match information handling technology and the use of automated control remains limited. Considerable efforts have been made to develop suitable in-line sensors to monitor the progress of fermentation, rate of oxygen uptake and evolution of CO_2 being of particular interest as control parameters.

Fermentation of wort in production of ale-type beer is conducted at 12–24°C for 4–7 days in traditional fermenting vessels or for 2–3 days in cylindro-conical vessels. Fermentation temperatures of 3–14°C are used in production of lager-type beer, 8–10 days being required when traditional vessels are used and 5–7 days in the case of cylindro-conical. Fermentations proceed more rapidly at higher temperatures but the risk of flavour defects is greater and the temperature used is a compromise between these factors.

A number of factors affect the course of fermentation, some of which, such as seasonal variation in malt quality, cannot be eliminated by control. The most serious situation faced by the brewer is a 'stuck' fermentation where the required ethanol level is not attained. Common causes include premature yeast flocculation, or failure of the yeast to metabolize maltotriose due to repression by glucose.

The use of high gravity wort (high gravity brewing) is particularly popular in North America and involves fermentation of a concentrated wort followed by dilution with deoxygenated water. The process has a number of technological advantages, although brewhouse material efficiency is reduced. Beer produced by high gravity brewing is of improved flavour stability and smoother taste but, while highly acceptable, can be difficult to match to that produced from a conventional wort. The use of high gravity wort also requires a yeast capable of tolerating high ethanol levels. The behaviour of the yeast changes, becoming more flocculent and sedimentary as the gravity is increased and producing markedly higher quantities of acetate esters. This can be controlled by introducing oxygen into the fermenting vessel as oxygen-saturated water, or by increasing the levels of fermentable sugars in the wort.

Low carbohydrate beers have become increasingly important due to their relatively low calorie content, high potential for intoxication and suitability, in some cases, for consumption by diabetics. Several methods of manufacture may be used but the most common involves the use of amyloglucosidase from *Aspergillus niger* to hydrolyse dextrins to sugars which are then fermented by brewery strains of *Saccharomyces*. A separate species, *Sacch. diastaticus* is capable of fermenting dextrins directly but is unsuitable for use in beer production since it possesses the gene for phenolic off-flavour production (*POF*1), a serious flavour defect. Genetically modified brewery yeast have been constructed containing the amyloglucosidase gene from either *Sacch. diastaticus* or, more effectively, *A. niger*.

---

* Tyramine is produced during fermentation and is hazardous to persons undergoing therapy with monoamine oxidase inhibiting drugs giving rise to a tyramine-pressor response. Levels of tyramine ranged from 0.98 to 3.21 mg/l in alcoholic beer and from 0.52 to 3.96 mg/l in 'non-alcoholic' beer produced by dealcoholization. Sensitive persons should thus avoid all types of beer.

The fermentation of lambic beer differs from that of other types in being carried out by yeasts and other micro-organisms derived from the environment. Micro-organisms considered as undesirable in most types of beer are necessary for development of the characteristic lambic flavour. A mixed lactic-alcoholic fermentation is involved, the barley and wheat mash being initially acidified by lactic acid bacteria. The alcoholic fermentation is initiated by maltose non-fermenting yeast species such as *Hanseniaspora* but the main fermentation becomes dominated by *Sacch. carlsbergensis* and other *Saccharomyces*. Lambic beer undergoes a lengthy maturation characterized initially by the successive growth of *Pediococcus* and *Dekkera*. Similar microbial successions occur during the manufacture of the sorghum-based 'kaffir' beer but a much more varied yeast microflora is involved.

7.3.8 Maturation (conditioning; secondary fermentation)

Fermentation and maturation are usually considered to be two separate stages in brewing, but in modern practice the distinction between the two is increasingly blurred. Maturation is considered to include all transformations between the end of primary fermentation and the final filtration or, in the case of cask conditioned, dispensing of the beer. These include carbonation by fermentation of residual sugars, removal of excess yeast, adsorption of various non-volatiles onto the surface of the yeast and consequent removal, precipitation of haze-forming complexes and progressive changes in aroma and flavour.

Ale-type beer is traditionally conditioned for 1–2 weeks at 12–20°C, either in bulk tanks or, especially in the case of high quality beers, in individual casks. In-bottle maturation is also used for a few types of beer including premium stout and high alcohol light ale. Priming sugar is added or amyloglucosidase used to hydrolyse dextrins to provide a substrate for the secondary fermentation, but sugars are not fully metabolized, that remaining determining the sweetness of the beer. Hops may also be added at this stage, the so-called dry-hopping.

Traditional lagers are conditioned in bulk at *ca.* 4°C for several months, although the process is occasionally completed in-bottle. Partly fermented wort may be added to initiate the secondary fermentation (krausening). The length of maturation can be significantly reduced by producing green beer, which contains

> BOX 7.5 **Cask lager?**
>
> Although cask lager is really a contradiction in terms, draught cask lagers, such as *Gold Cross*, were produced during the 1980s. None made any significant impact and all have been withdrawn from the market. It is suggested, however, that products of this type may re-emerge during the 1990s. (*The Cask Ale Report* 1993. Carlsberg-Tetley Brewing Ltd, Burton on Trent).

minimal levels of undesirable flavour compounds. Currently this is achieved by control of fermentation, although the genetic modification of yeasts to minimize production of vicinal diketones is a future possibility. Removal of volatiles by CO_2 washing also permits maturation to be shortened. In some breweries fermentation and maturation are both conducted in cylindro-conical fermenters by introducing a warm period at the end of the main fermentation. This system can lead to defects due to the development of autolysed flavours and it is necessary to implement the programmed removal of sedimented yeast.

Clarification of beer during maturation is aided by addition of finings. The role has changed somewhat over the years and finings are now used as a pre-filtration treatment rather than to remove large quantities of sediment. Isinglass is the main fining agent used at the maturation stage and consists of collagen of net positive charge which reacts with yeast, polyphenols and other negatively-charged material. Auxiliary finings, negatively-charged polysaccharides, are also added to remove positively charged material.

Fining is not sufficient to remove the polypeptide–polyphenol complexes responsible for 'chill-haze' and, while progress has been made in preventing complex formation through use of low-proanthocyanidin malt and hops, a significant problem remains. For many years proteolytic enzymes, especially papain, have been added to beer during maturation, but while effective in limiting haze formation, papain has an adverse effect on head retention, causes allergic reactions amongst some consumers and is expensive. Higher purity proteases from yeast such as *Candida olea* and *Pichia pini* overcome the disadvantages of papain and, if required, may be added during fermentation. The possibility exists of incorporating the appropriate genes into *Saccharomyces*, although the use of proteolytic enzymes is now less extensive than in previous years.

7.3.9 Post-maturation treatment

In the case of cask-conditioned beer and beer conditioned by an in-bottle secondary fermentation, post-maturation treatment is usually neither desirable, nor possible. In contrast, beers matured in bulk can undergo a variety of post-maturation processes, depending on the type of beer and the operating practices at different breweries. Filtration has been used in brewing for many years and can be used for various degrees of clarification as well as final polishing and removal of micro-organisms (filter sterilization). Filtration systems are of two basic types according to the underlying physical principles. Surface filtration, typified by the concurrent filtration of particulate material and kieselguhr bodyfeed, involves the bridging of pores on the surface of the filter by suspended solids to form a constantly changing filter surface. Depth filtration, in contrast, involves many factors of which adsorption of particulate material during flow through the filter is probably of greatest importance in relation to beer processing.

Kieselguhr has been used as a filter aid for many years and is highly effective in use. Kieselguhr, however, presents a health risk when handled as a dry powder and alternatives have been sought. These include perlite, silicates derived from volcanic rock and silica hydrogels. Perlite has been disappointing in performance in some breweries but silica hydrogels have a performance similar to kieselguhr.

Various types of filtration equipment are available including plate and frame filters, filter candles and leaf filters. Filter sheets consisting of cotton/cellulose impregnated with asbestos are effective, but have been replaced by various grades of kieselguhr filter in response to concerns over the safety of asbestos.

Cartridge filters, consisting of small, enclosed units containing a filter element, are convenient to use and, according to purpose, can contain either surface or depth filter elements. Cartridge filters are widely used for cold sterilization in conjunction with aseptic filling of PET bottles. The requirements are more demanding than for rough beer clarification, nylon and teflon membranes being used following rigorous pre-filtration. The use of regenerable fine sand beds to replace existing filtration has been developed to pilot plant level and offers a number of potential advantages including the elimination of a requirement for bodyfeed. Cross-flow membrane filtration however, offers more radical changes in technology including the ability to filter rough beer and beer recovered from tank bottoms in a single operation.

Beer is chilled during at least some of the filtering operations to precipitate protein–polyphenol complexes and tannic acid may be added to encourage precipitate formation. Synthetic resins such as polyamides and polyvinylpyrrolidone preferentially remove polyphenols and thus limit haze formation, while maintaining most of the polypeptides of importance in head stability. Greatest efficiency is obtained by use of insoluble, cross-linked polyvinylpyrrolidone.

Centrifugation has been used in place of filtration and is suitable for long, continuous process runs. Centrifugation may also be used at an earlier stage to separate yeast from beer at the end of filtration. Application, however, has been limited by the high capital expenditure.

Pasteurization is widely used for prevention of spoilage of canned and bottled beers and some types of keg beer. Heat treatment is measured in pasteurization units, the equivalent of 1 min at 60°C and a treatment of three to four units is common. Cans and bottles are pasteurized in-bottle, by passage through hot water, but plate heat exchangers are used for keg beer and some bottled beer. Hot filling of bottles at 65°C is sometimes used to prevent contamination, but flawless bottles are required if a satisfactory level of carbonation is to be attained. Pasteurization is now being replaced by sterile filtration for heat-sensitive polyethyleneterephthalate (PET) bottles and some canned beer is also sterile filtered.

7.3.10 Packaging and dispensing

Bulk beer is packaged in barrels of various size although road tankers have been used for transportation, the beer being piped into tanks at the point of sale. Aluminium has now largely replaced wood, but stainless steel or glass fibre are used for large tanks. Bulk beer is dispensed either by tapping the barrel directly (from the wood), by hand or electrically operated pumps, or by gas (CO_2 or nitrogen) pressure. The method of dispensing the beer can affect its acceptability; the use of gas pressure imparts a 'fizziness', especially when CO_2 is used, which is generally disliked in ale-type beer, although acceptable in lager-type. In reality, there is little difference between electric and hand operated pumps, although in each case the design of the dispensing equipment may affect perceived quality by affecting the head on the beer. Production of a stable head depends on, amongst other factors, the production of stable bubbles to act as nucleation sites to give the characteristic 'surge' when the beer is drawn. Dispensing equipment is usually fitted with orifice plates, or inserts, which provide sufficient shear force to produce a large number of small, stable, bubbles. The role of dispensing equipment in producing a stable head is particularly important in draught stouts such as Guinness™ which are characterized by a thick, creamy head. This is obtained by dispensing, through an orifice, under combined CO_2 and nitrogen pressure.

BOX 7.6 **Like unto crystal**

In the UK there has been a trend towards reviving traditional dark beers, such as mild and porter. This has been reflected, to some extent, in the US where a number of micro-breweries now produce an all-barley beer significantly darker than the average American brew. At least one of the large brewers, Miller's, have followed the small brewers lead. In contrast, US brewers are also attempting to exploit the fashion for 'new age' clear drinks, such as clear colas, by developing clear beer. The lead has been taken by Coor's, a brewer often considered to have lost ground to competitors in recent years. Coor's are producing a clear malt liquor aimed at women and young drinkers. Critics have pointed out that a clear appearance is not consistent with beer and, more seriously, fault the brewer for targeting the young. It appears, however, that the reception for clear beer in the licensed trade is less than enthusiastic.

Beer is packaged in both cans and bottles. In the vast majority of cases the beer is carbonated before filling. Glass bottles, sealed with non-replaceable crown caps, are still widely used, but larger sizes of bottle are commonly made of blow-moulded PET. Early usage of PET was not entirely satisfactory due to the material permitting the entry of oxygen and light of a wavelength leading to 'sunstruck' flavour, but this has been overcome by application of a coating with suitable barrier properties.

The rapid release of CO_2 on opening cans or bottles of ale-type beer leads to organoleptic differences to bulk dispensed. This is of no consequence in the case of beers such as light or brown ales, which are brewed for bottling, and may even be considered desirable in lager-type beer drunk, according to the dictates of fashion, direct from the bottle. Acceptability of canned bitter and stouts was, however, limited until the development of an in-can device which provides the shear force necessary for the 'initiation' of the head. The device was initially developed for use with Guinness™ stout, but similar systems are used for other stouts and for bitter ales. The device consists of a small plastic chamber communicating with the remainder of the can volume by a small hole. The filled can is pressurized to *ca.* 25 lb/in^2 and sealed. On opening, beer and gas are forced through the hole in the in-can device creating stable bubbles which act as nucleation sites.

7.3.11 Reduced-alcohol and alcohol-free beers

Reduced-alcohol and alcohol-free beers were developed partly in view of concern over the effects of alcohol consumption on health and partly in an attempt to provide an acceptable social drink for motorists. The latter factor appears to be of greatest importance in consumption, demand being greatest where strict blood alcohol levels for drivers are enforced.

A number of approaches have been taken to production of beer of reduced alcohol content. There has been considerable interest in the use of yeast strains which are unable to complete the alcoholic fermentation, but which produce other flavour components and thus a 'brewed-taste'. The most common means of production, however, involves post-brewing removal of alcohol from the beer. Thermal evaporation, under vacuum, is currently used, but alternative methods, which avoid heating have been proposed. These include removal of alcohol by selective freezing and contraflow dialysis using cellulose membranes.

Modern thermal evaporation de-alcoholizing equipment is highly energy efficient and of low running cost, although initially requiring a high level of capital expenditure. Such equipment is typically capable of producing an end-product with an alcohol content of *ca*. 0.03% to just below the original content. Thermal de-alcoholization of beer was originally associated with the development of 'heated' flavours and the loss of some volatile aroma compounds. These problems have been largely overcome in modern equipment by use of evaporators with a very short residence time and an evaporation temperature of less than 45°C. Volatile aroma compounds may be recovered from the vapours by use of a distillation column, which also recovers alcohol. CO_2 is lost during evaporation and must be replaced at a later stage.

7.3.12 Quality assurance and control

It is a common misconception that brewing is a craft process, relying on skilled judgement and empiricism to obtain an end-product of satisfactory quality. Brewing certainly requires skill and experience, to overcome the problems inherent in a process of this type. Production of a beer of consistent quality and character, however, is the aim of all brewers and, while the level of technical input obviously varies according to the size of the operation, control at all stages is essential to success.

Quality assurance and control in breweries involves four main areas; adequacy of plant cleaning, acceptability of materials, the status of wort and beer in production and properties of the finished beer. A high standard of plant cleaning is required to prevent contamination of the beer with spoilage micro-organisms. A thorough clean is required after each production cycle and this should be verified by visual inspection and microbiological analysis of swabs. A thorough knowledge of plant lay-out is required to identify areas where cleaning is most difficult, and to eliminate these where necessary. In many breweries, it is common practice to incorporate a regular 'alkali-brew', in which the brewing process is followed using a solution of NaOH. Particular care is required to ensure that re-usable bottles and casks are adequately cleaned. In this context, it is of note that problems due to acetic acid bacteria are often markedly less where wooden casks have been replaced by aluminium, or stainless steel.

Selection of raw materials is usually the responsibility of the brewer, with technical support as required. Laboratory analyses are required for all ingredients, including water. Large breweries may be capable of their own analysis, but in many cases the specification and quality of malt, hops and adjuncts is guaranteed by the supplier. Quality assurance of yeast is the responsibility of the microbiologist (see page 359).

It is necessary for the brewer to be aware of the progress of each batch of beer through production of wort, fermentation and maturation. Much information is available from production records, but this must be supported by chemical and microbiological analysis (Table 7.5).

The finished beer is subject to chemical, microbiolological and organoleptic analysis (Table 7.6). This is required to verify that the beer is within its specification range. It should be appreciated that, while laboratory analysis is required, organoleptic analysis of the finished beer is the single most important analytical procedure in brewing. In smaller breweries, assessing the flavour of the beer is the responsibility of the brewer who, usually together with other personnel, has the onerous responsibility of tasting every batch before release. In larger breweries, various types of taste panel are used, often in conjunction with expert tasting. In all cases, the beer must be tasted from the viewpoint of the critical and informed consumer and organoleptic analysis must determine not only that the beer is free of overt defects, but that its flavour and character conforms with that expected by its consumers. In large breweries flavour-wheels and similar techniques are used to ensure conformity and for early detection of changes in flavour and character. Organoleptic assessment is also valuable in detection of faults. Specific problems, such as excessive diacetyl levels, are

Table 7.5 Analysis of wort and beer during brewing

Wort	Beer
Specific gravity	Specific gravity
pH value	Degree of fermentation
Total acidity	pH value
Reducing sugars	Total acidity
Free amino acids	Fermentable sugars
Protein	Free amino acids
Starch	Protein
Colour	Colour
Viscosity	Dissolved oxygen
	'Wild' yeast
	Bacterial contaminants

Table 7.6 Analysis of finished beer

Alcohol content by volume	Specific gravity
Original extract	Final extract
Degree of fermentation	Head (foam) stability
Dissolved oxygen	Protein
Starch	Iso-α-acids
pH value	Diacetyl
Sulphur dioxide	Copper
Sodium	Calcium/oxalate
Flavour	Flavour stability[1]
Carbon dioxide	Clarity (new beer)
Clarity (after abuse)	Bacterial contaminants
Pitching yeast[2]	Wild yeast

[1] Only applicable to canned and bottled beer.
[2] Not applicable to cask-conditioned beer, or beer finished by an in-bottle fermentation.

easily recognized and problems identified before laboratory analysis is completed.

7.4 THE CHEMISTRY OF BREWING AND OF BEER

7.4.1 The chemical basis for the organoleptic properties of beer

(a) Flavour and aroma

A very large number of compounds, derived from a number of sources, contribute to the flavour and aroma of beer (Table 7.7). In many cases the compounds involved contribute to the overall flavour spectrum rather than having a direct flavour impact. With the exception of ethanol, for example, the alcoholic fermentation does not yield character impact compounds, although many compounds are produced which contribute to the overall flavour and aroma.

The situation is complicated by the fact that compounds such as acetic acid, which is usually associated with spoilage by acetic acid bacteria but which is also produced during fermentation, may be considered a desirable flavour component when balanced by other compounds. The quantity of a compound present may also affect the desirability as a flavour component. Over-production of esters,

Table 7.7 Summary of desirable flavour and aroma compounds in beer

Source	Compound
Malt	volatile Maillard products[1] e.g. maltol
	dimethyl sulphide[2]
Hops	iso-humulones
	essential oils and oxidation products e.g. linalool
Fermentation	ethanol
	esters, e.g. ethyl acetate
	Organic acids e.g. acetic acid

[1] Production commences during kilning and proceeds during wort boiling (see page 343).
[2] Produced from S-methyl methionine in malt during kilning and wort boiling and from dimethyl sulphoxide during fermentation (see page 348). Important in lager only.

for example, results in flavour defects in high alcohol beers, while under-production leads to lack of flavour in low alcohol. There is also an interaction between flavour compounds and the type of beer. Diacetyl and pentane-2,3-dione are readily detectable in lager and light Canadian ales and invariably considered undesirable, whereas in English ales the taste threshold concentration is higher and these compounds considered less objectionable and even, in the case of some brown ales, desirable.

The importance of aftertaste is often overlooked in discussion of sensory characteristics. Beer has a bitter aftertaste which is ameliorated by the sugars present in sweeter types of beer. In other types, particularly some heavily hopped traditional bitter ales, however, the aftertaste may be considered unpleasantly harsh. In some cases degradation of dextrins by salivary α-amylase may produce sufficient sugar to balance the bitterness without a detectable sweet taste.

(b) Mouth-feel

Mouth-feel is an important property of beer which can modify the sensory appreciation. For a number of years it was thought that dextrins, while essentially tasteless, are major contributors to body and viscosity. More recent work, however, suggests that the importance of dextrins in determining mouth-feel characteristics is overstated. Beta-glucans, especially those of molecular weight greater than 3×10^5, ethanol, glycerol, glycoproteins and melanoidins, being of greater significance. Level of carbonation may also be of importance in determining perception of mouth-feel.

(c) Head

Head (the foam layer) is an important factor in determining acceptability of beer. Although small stable bubbles are necessary to act as nucleation sites during head formation, these result from mechanical forces applied during dispensing rather than from properties of the beer itself. The bubbles form an initial 'gas emulsion' which breaks down by bubble rise and drainage and an upper layer of 'true foam' separates. In all foams, the gas bubbles are separated by a continuous phase of thin liquid layers, the lamellae. Mechanical energy is required for creation of this interface and surface active agents are usually required to prevent the coalescence of gas bubbles. A high viscosity in the bulk liquid phase also increases foam stability.

It is not known if foam stability in beer depends on one or two key compounds, or is the balance of a number of different compounds. Polypeptides have been associated with foam stability for many years, although it appears that while large polypeptides (molecular weight greater than 10 000) stabilize foams, small polypeptides interfere with stability. Glycoproteins have also been thought to have stabilizing properties, but their role is now questioned. In addition to the apparently major role of polypeptides, it is likely that many other compounds play at least some part. These include ethanol, isohumulones and melanoidins. The continuing release of CO_2 in highly carbonated beer also serves to maintain a head. Additives such as propylene glycol alginate may be used to aid beer foam stability, but are ineffective in the absence of stabilizing polypeptides.

7.4.2 The chemistry of the malting process

(a) Mobilization of endospermal reserves

The germinating barley seed is a complex physiological system, the major processes involving the synthesis of amylolytic and proteolytic enzymes and degradation of endospermal structures. The barley grain consists of two main parts: the endosperm, which has a storage function, and the embryo, from which the new plant is derived. The starchy endosperm accounts for *ca*. 90% of the total endosperm mass and consists of large, thin walled dead cells packed with starch granules and storage proteins. The living aleurone layer, composed of small thick walled cells, surrounds the starchy endosperm. The embryo consists of the 'embryo proper' which contains the leaf and root initials and a scutellum which serves as an absorptive organ. The scutellum is attached to the embryonic axis without a distinct border and is probably homologous with the cotyledon.

A number of plant hormones are produced in the embryo and involved in the regulation of metabolism in the germinating grain. The gibberellins, a structurally related group of which A_1 and A_3 (gibberellic acid) are most important, have been most widely studied and are known to induce α-amylase synthesis by the aleurone layer. Enzyme synthesis is also influenced by indolyl acetic acid, abscisic acid and cytokinins and control depends on a complex interplay between these hormones, the gibberellins and inorganic ions.

Degradation of the cell walls surrounding the starch and storage proteins of the starchy endosperm, the process of modification, is necessary to permit the movement of hydrolytic enzymes and mobilization of reserves. Beta-glucan comprises *ca*. 75% of the cell walls together with *ca*. 15% arabinoxylan, *ca*. 5% protein and *ca*. 3% glucomannan. The β-glucan and protein constituents complex to form an insoluble matrix.

The first stage of β-glucan degradation is solubilization by the enzyme β-glucan solubilase. It is probable that several enzymes are capable of solubilizing β-glucan by various pathways, but carboxypeptidase activity is of greatest importance under brewery conditions.

At least three enzymes are capable of the hydrolysis of solubilized β-glucans; endo-(1–3)-β-glucanase, endo-(1–4)-β-glucanase and endo-(1–3:1–4)-β-glucanase (barley-endo-β-glucanase), although endo-(1–4)-β-glucanase may be derived from contaminating fungi.

Various patterns of modification may be observed, enzyme activity frequently extending around cracks in the starchy endosperm. In completely modified regions the cell walls become fragmentary, although some physical integrity may remain, possibly due to residual middle lamellar tissue containing fibrillar material. Insufficient modification was a common problem in the early days of accelerated malting and may still occur, contributory factors being a low level of β-glucanase activity or a high level of β-glucan in the barley. Excessive levels of high molecular weight β-glucans in the wort lead to slow filtration and, possibly, gel precipitation during lagering and in bottled beer. The addition of fungal β-glucanases to the wort is necessary when a high level of unmalted starchy adjuncts are present and in some breweries is added on a routine basis.

Protein accounts for virtually all of the nitrogen content of the barley grain. Degradation of the protein matrix surrounding the starch granules is necessary to permit access of α-amylase, the resulting amino acids being used as the nitrogen source for yeast growth in brewing.

Germinating barley contains three groups of endopeptidases (proteinases) and three groups of exopeptidases (peptidases). Endopeptidase activity increases 20-fold during germination and at the same time inhibitors disappear. Approximately 66% of endospermal reserve proteins are present in the starchy endosperm, especially in the tissue below the aleurone layer. The remainder is localized in living aleurone layer cells in the form of protein bodies (aleurone grains). Similar protein bodies are present in the scutellum. Mobilization of the protein during germination commences with degradation of the protein bodies of both the aleurone layer and scutellum. The free amino acids and peptides then appear to pass to the cytosol where peptides are hydrolysed by neutral and/or alkaline exopeptidases. Amino acids are involved in the synthesis of the enzymes subsequently responsible for hydrolytic activity in the starchy endosperm.

Hydrolysis of the proteins of the starchy endosperm appears to be initiated by acid endopeptidases. Enzymes are probably first secreted from the scutellum followed by gibberellin stimulated synthesis in the aleurone layer. The resulting peptides are, in turn, hydrolysed by carboxypeptidases. This reaction does not go to completion but results in a mixture of free amino acids and small peptides. Under brewery conditions carboxypeptidases are in excess and the rate of overall protein degradation is governed by endopeptidase activity. The final stage of protein mobilization involves uptake of amino acids into the scutellum where peptides are hydrolysed by neutral and/or alkaline exopeptidases.

Starch comprises *ca.* 66% of the total dry weight of grain, and is composed of two types of glucose polymer. The major constituent of barley starch, amylopectin, is a branched molecule formed from glucose units joined by α-(1–4)-glucosidic links and a smaller number of α-(1–6)-glucosidic links. The minor constituent, amylose, is a linear molecule formed from glucose units joined primarily by α-(1–4)-glucosidic links. The relative arrangement of the two types in the starch granule is not known, but distribution appears to be fairly uniform.

Germinating barley contains a number of amylolytic enzymes of which α- and β-amylase are of greatest importance. Alpha-amylases are calcium-dependent endo-enzymes which catalyse hydrolysis of α-(1–4)-glucosidic links in the central region of the chain, leaving the terminal sugar in the α-configuration. The low molecular weight products are maltose, maltotriose and relatively small quantities of glucose.

Beta-amylases are exo-enzymes, which catalyse the hydrolysis of the penultimate glycosidic bond from the non-reducing end of an α-glucan chain. The enzyme is synthesized in the developing grain, but during ripening *ca.* 66% becomes complexed with proteins to produce an insoluble 'latent' form. During germination the 'latent' form is progressively decomplexed releasing the free, active enzyme.

In addition to the amylases, germinating barley contains at least two 'debranching enzymes', pullanases, which catalyse the hydrolysis of the α-(1–6)-glucosidic links of amylopectin and limit dextrins to liberate maltotriose. Maltose is hydrolysed to glucose by α-glucosidase (maltase), levels of which rise rapidly during germination.

(b) Chemistry of kilning

Enzyme activity increases during the early stages of kilning and, in continental Europe, modification may be completed in the kiln. A high malting loss may, however, be anticipated if the malt is kept warm without drying, although in the case of crystal malt, metabolic activity is allowed to continue until virtually all starch is converted to sugar.

During the higher temperature part of the kilning cycle some enzymes are partly, or completely, inactivated. The extent of inactivation depends on the heat stability of the enzymes, the temperature and duration of heating and the moisture content at any given temperature. Temperatures within the kilning cycle are controlled to ensure sufficient remaining enzyme activity to produce a wort of the desired composition, although in some speciality malts virtually no enzyme activity remains after heating.

Malt enzymes vary in heat stability, the most labile being both isoenzymes of endo-(1–3:1–4)-β-glucanase. Endo- and exo-peptidase activity is also reduced while, in general terms, α-amylases are more heat stable than β-amylases.

The Maillard reaction between amino compounds, usually proteins, and reducing sugars commences during kilning producing melanoidins which are important flavour and colouring compounds. Flavour is also improved by the removal of 'raw grain' flavours. The extent to which the Maillard reaction progresses during kilning depends largely on the degree of heating and, in most cases, the reaction continues during wort boiling.

S-methylmethionine (SMM), the probable precursor of dimethyl sulphide (DMS), an important flavour compound in lager is formed during malting. Higher levels of SMM are present in malt from high nitrogen barley and formation is also increased by addition of gibberellic acid and abrasion treatment. Dimethyl sulphide and its oxidation product dimethyl sulphoxide (DMSO) are formed during kilning, but at the high temperatures used in malts for ale production, most of the DMS is lost by volatalization. Maintaining the kilning temperature at below 60°C until the moisture content is 4–5% minimizes conversion of SMM to DMS, but also favours DMSO production. S-methylmethionine remaining after kilning is converted to DMS during wort boiling, the most important stage in DMS production (see page 342).

7.4.3 The chemistry of hops

(a) Humulones (α-acids)

The humulone fraction comprises a mixture of homologues and analogues of humulone itself, the relative percentages present being:

humulone	35–70%
cohumulone	20–55%
adhumulone	10–15%
prehumulone	1–10%
posthumulone	1–5%

The structure of humulones is illustrated in Figure 7.5.

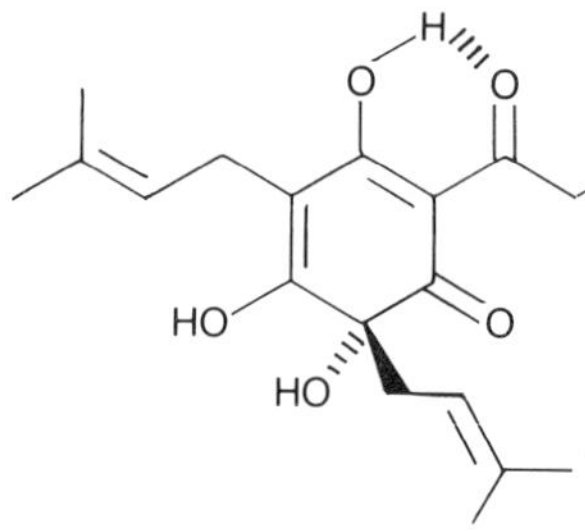

Figure 7.5 Structure of humulones.

Humulones, are not present in beer as such but, in the isomerized form, isohumulones, which are considerably more soluble and have a more bitter taste. The isomerization of humulones is the most important reaction in hop chemistry and occurs during wort boiling. Isomerization involves a change from a six-membered to a five-membered ring structure (Figure 7.6) and is unusual in that, while the reaction proceeds in boiling wort at pH values of 5.0–5.5, bases such as sodium bicarbonate or potassium carbonate are needed to catalyse isomerization of extracted humulones. Potassium and magnesium salts of humulones are also isomerized by heating, but boiling with water at a pH value of *ca*. 7.0 yields virtually no isohumulones.

Isomerization during wort boiling is an inefficient process due to the low solubility of humulones in wort and losses of isohumulones on precipitated protein, yeast and filter material. Yield in the finished beer is no more than 30–40% resulting in a content of *ca*. 15–35 mg/l.

Humulones are highly susceptible to oxidation, the sensitive parts of the molecule being double bonds, which can be epoxidized into very reactive oxygen intermediates. Tertiary carbon centres are also oxygen sensitive. The products of oxidation are complex, most having a five-membered ring system and thus resulting from oxida-

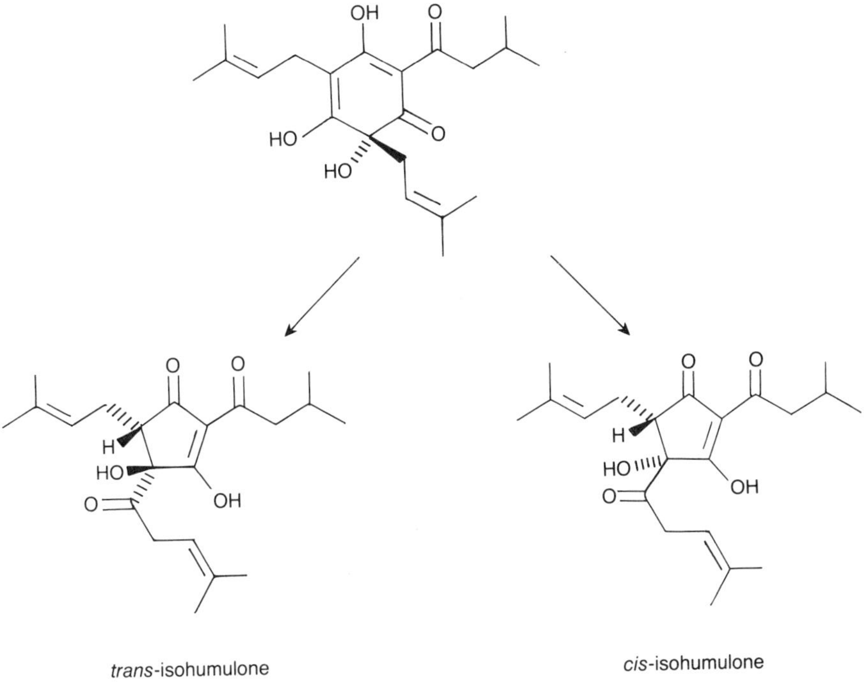

trans-isohumulone

cis-isohumulone

Figure 7.6 Isomerization of humulones.

tion of isohumulones. A very large number of oxidation compounds are formed (Figure 7.7), but individual yields are low. None of the oxidation products have significant bitterness and oxidation is inevit-ably accompanied by loss of hop quality.

In the presence of borohydrides, alkaline-catalysed isomerization of humulones is accompanied by specific reduction of the carbonyl group in the isohexonyl side chain. Resulting structures have three chiral centres and are of technological importance due to the inability to participate in the formation of 3-methyl-2-butenethiol, the cause of the 'sunstruck' off-flavour in bottled beer (see page 349).

(b) Lupulones (β-acids)

The lupulones consist of a parallel series to the humulones. The relative quantities present in hops are:

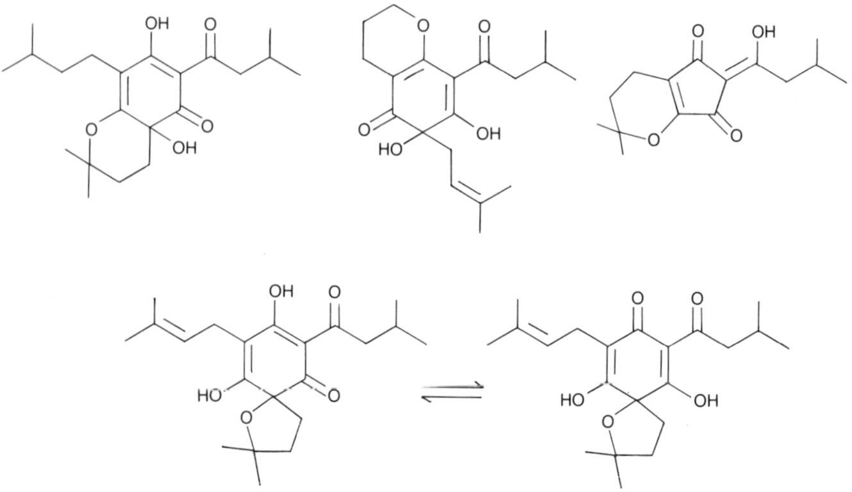

Figure 7.7 Humulone derivatives formed during hop boiling.

lupulone	30–55%
colupulone	20–55%
adlupulone	5–10%
prelupulone	1–3%
postlupulone	?

Lupulones are not utilized in the brewing process, but chemical treatment such as catalytic hydrogenation of colupulone yields tetrahydro-isocohumulones which have industrial potential as bittering compounds. It appears unlikely, however, that such compounds will find significant application.

(c) Essential oils

The main components of essential oils are the terpene hydrocarbons, myrcene, humulene and caryophyllene (Figure 7.8), which together account for 80 to 90% of the total essential oils. These components are not, however, directly responsible for the contribution of hop essential oils to beer flavour. Many compounds are potentially involved, usually at sub-threshold level, but humulenol II, humulene diepoxides and, to a lesser extent, humulene monoepoxides and α-terpineol are responsible for the herbal, spicy flavours and the terpene alcohols linalool, geraniol and citronellol for the floral, citrus flavours. Unidentified compounds are responsible for grapefruit flavours in beer made from very aged hops. Oxi-

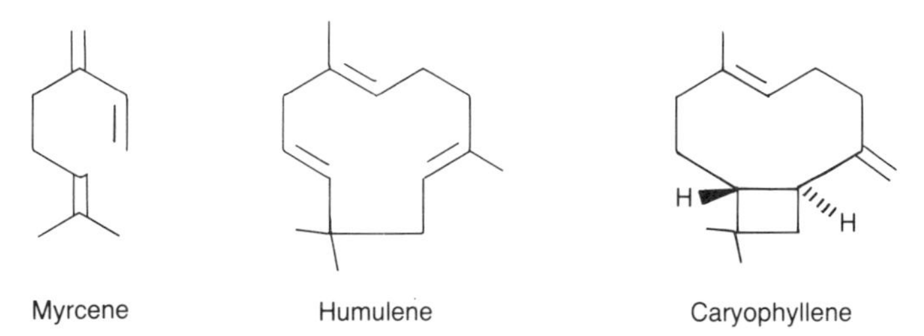

Figure 7.8 Structure of major precursors of flavour-active compounds derived from hop essential oils.

dation of essential oil components commences during ageing of the hops and continues during wort boiling. Hop ageing is accompanied by an increase in the levels of humulene epoxides. Humulene epoxide II serves as a precursor of humulenol II, humuladienone and and humulol and accumulates to a lesser extent than the other humulene epoxide isomers. Humulene diepoxides are formed by further oxidation of monoepoxides during both ageing and wort boiling.

Myrcene is highly labile and largely disappears from the fresh hops by oxidation and evaporation, none surviving wort boiling. Survival of humulene and caryophyllene during wort boiling is variable, but a high level of extraction of the oxidation products occurs. Levels of α-terpineol increase during boiling, some being formed by oxidation of linolene.

Although sometimes overlooked in this context, important changes to essential oil constituents occur during fermentation. Loss of the oxidation products of humulene and caryophyllene occurs possibly as a result of their lipophilicity causing adsorption to yeast cells. Levels of linalool increase in the early stages of fermentation, while geraniol and geraniol-type compounds undergo a number of yeast-mediated transformations. The net result is the total loss of geraniol-type compounds such as geranyl acetate and geranyl isobutyrate and the production of citronellol, primarily by reduction of geraniol.

7.4.5 Chemical changes during wort boiling

Wort boiling is recognized as serving a number of technologically important functions including the extraction and transformation of hop components discussed above.

(a) Inactivation of residual enzyme activity

In most cases the enzyme activity of the wort is low after mash separation. Distinct residual activity is usually associated only with polyphenol oxidases and total inactivation occurs relatively early in boiling.

(b) Protein coagulation

The mechanism of protein coagulation during wort boiling is still not fully understood. The process is generally thought to involve two stages: chemical denaturation followed by colloidal coagulation. It remains a common misconception that polyphenols play an important role in coagulation during wort boiling. Protein–polyphenol complexes are based on hydrogen bonds with a very low binding energy and are thus unstable at high temperature, although polyphenols are, of course, involved in complex formation at lower temperatures.

The key reaction during protein coagulation appears to be the destruction of disulphide bridges. This leads to the formation of free thiol-groups which react with the thiol-groups of other proteins and peptides. Formation of complexes between the ε-amino-group of lysine and hop acids is also involved. Wort pH value is of considerable influence on the course of protein coagulation and the process is readily completed at the isoelectric points of the individual proteins. The isoelectric point of wort proteins such as β-globulin and hordein is, however, below the pH value of wort and coagulation is often incomplete.

(c) Production of dimethyl sulphide

Wort boiling is the most important stage for production of DMS from the precursor SMM, although a smaller quantity is formed during fermentation by reduction of dimethyl sulphoxide. In some, but not all lagers, DMS is an important flavour compound when present at concentrations between the threshold level of *ca.* 30–100 μg/l. At higher concentrations DMS is associated with 'cooked sweet corn' and 'blackcurrant' off-flavours. Dimethyl sulphide is not present in significant quantities in ales unless microbial spoilage has occurred. Conversion of SMM to DMS occurs at a high rate during the actual boil but the DMS formed is lost by volatalization. Conversion continues during hot holding before cooling and the majority of DMS entering the final beer is formed at this stage.

(d) The Maillard reaction

Although the Maillard reaction commences during malt kilning it proceeds to completion during wort boiling. The end-products are melanoidin pigments, high molecular weight structures containing pyrazine and imidazole rings and volatile compounds. Volatile compounds are mainly heterocyclic compounds including carbonyls such as 2- and 3-methyl butanol, *O*-heterocyclics such as furanes and pyranes such as maltol and derivatives. These compounds provide malty flavours. *N*-heterocyclic compounds including pyrazines and pyrroles are also present and provide roast or bready flavour notes at very low thresholds. Sulphur–nitrogen heterocyclics also provide bready notes. Proline derivatives such as pyrrolizines, azepines and 'malt-oxazine' are of importance as the consequence of their extreme bitterness. Intermediate products such as 5-hydroxymethyl-2-furaldehyde, methyl reductone and α-dicarbonyls are also likely to be present.

Maillard end-products are important determinants of the colour and flavour of beer. Each type of beer has an optimal level above which the products are undesirable, these are signifcantly higher in dark than in light beers.

7.4.6 Fermentation and maturation

(a) Glycogen and growth of yeast

Glycogen is the major reserve material in yeast and has a similar function to amylopectin in higher plants. Glycogen serves as a carbon and energy source during non-growth or starvation and provides an endogenous supply of energy during the lag phase at commencement of the growth cycle. At this stage the main energy requirement is for the synthesis of such compounds as sterols and fatty acids and there is a corresponding demand for oxygen. Oxygen uptake is virtually immediate and very rapid, but there is no uptake of wort glucose for *ca.* 5 h during which time oxygen is almost depleted. Glycogen mobilization, however is immediate, ATP produced by respiration activating the phosphorylase system required for the hydrolysis of glycogen to glucose. During the first 6 h of fermentation glycogen levels fall from *ca.* 27 to 5% of the dry cell weight while lipid levels rise from *ca.* 5 to 11.5%. The glycogen reserve is restored, however, towards the end of the fermentation.

Lipids permit yeast growth in deoxygenated wort and stimulate growth and fermentation, in air-saturated wort, by yeast strains of high oxygen requirement. Yeasts are able to incorporate unsaturated fatty acids and sitosterol from spent malt lipids. In practice, the quantity of lipid present in wort is small, only 1.4% of malt lipids surviving after mashing. While it is generally considered desirable, however, to remove lipids as completely as possible to minimize potential adverse effects on flavour and foaming properties, the beneficial role in stimulating yeast growth should not be forgotten.

(b) Uptake of sugars, amino acids and peptides

Although there is some degree of overlap, wort sugar utilization by yeasts usually follows the sequence sucrose, glucose, fructose, maltose and maltotriose, dextrins such as maltotetraose remaining unfermented. The means of entry into the cell varies, maltose and maltotriose, the major sugars present in wort, are transported intact across the cell membrane by maltose and maltotriose permease respectively and subsequently hydrolysed to glucose units by the α-glucosidase system. In contrast other sugars such as sucrose are hydrolysed by extra-cellular enzymes before entering the cell.

Uptake of maltose and maltotriose is subject to catabolite repression by glucose and does not commence until *ca*. 50% of the glucose has been metabolized. This is a major limiting factor in wort fermentation rate and the use of derepressed mutants offers considerable economic potential through increased control of fermentation rate. At present, however, failure of yeast to take up and metabolize maltotriose remains an important cause of a 'stuck' fermentation. The uptake of fructose is also repressed by glucose and its presence at the end of fermentation can lead to a sweet off-flavour. Conversely, however, the presence of fructose is desirable in some types of sweet beer.

Amino acids also tend to be taken up from wort according to a fixed sequence (Table 7.8). The pattern is fairly consistent and probably results from the interplay of amino acid permeases, competition for permease sites and transinhibition exerted by the internal amino acid pools. Total absorption of amino acids is governed more by the total assimilable nitrogen level of the wort than by the concentration of specific amino acids.

Table 7.8 Uptake of wort amino acids during fermentation

Group A
Absorbed rapidly in an almost linear fashion from the medium resulting in their elimination early in the growth cycle:
arginine, asparagine, aspartic acid, glutamic acid, glutamine, lysine, serine, threonine

Group B1
Absorbed gradually
histidine, isoleucine, leucine, methionine, valine

Group C
Absorbed after a lag which corresponds to the complete removal of Group A amino acids
alanine, ammonia, glycine, phenylalanine, tryptophan, tyrosine

Group D
Slowly absorbed under anaerobic conditions, high concentrations remain unabsorbed
proline

In contrast to amino acids, relatively little is known of peptide uptake. Wort contains a variety of peptides of different chain lengths which account for *ca*. 33% of the total nitrogen content. Yeast contain a general peptide-transport system whereby *ca*. 40% of the peptides are taken up during fermentation, but most strains of *Sacch. cerevisiae* transport only di- or tri-peptides. Removal of polypeptides commences at the beginning of fermentation without a significant lag phase, removal being faster in all-malt than in adjunct-containing worts. Linoleic acid increases the uptake of peptides from adjunct-containing, but not all-malt worts.

(c) Yeast excretion products

In addition to ethanol varying amounts of higher alcohols (fusel oils) are produced during fermentation and are both flavour-active compounds and substrates for ester formation by reaction with acyl coenzyme A derivatives. Higher alcohols are formed either by deamination or decarboxylation of wort amino acids or by synthesis from carbohydrates. Different strains of yeast vary considerably in formation of higher alcohols and significantly higher quantities are formed at higher temperatures. Temperature has a greater effect on formation of aromatic alcohols such as 2-phenylethanol than on aliphatic alcohols such as butanol, hexanol and propanol.

$$CH_3COSCoA + C_2H_5OH \rightleftharpoons CH_3CO_2C_2H_5 + CoASH$$

acetyl coenzyme A, ethanol, ethyl acetate, coenzyme A

Figure 7.9 Production of ethyl acetate during fermentation.

Esters, which have strong fruity flavours, are formed as secondary products during the anaerobic metabolism of sugars. Potentially a large number of esters may be produced, but in practice ethyl acetate, isoamyl acetate, isobutyl acetatc, cthyl caproate and 2-phenylethyl acetate are of greatest significance. Synthesis may involve esterases working in reverse, or alcohol acetyl transferases mediating the reaction between acetyl coenzyme A and alcohols (Figure 7.9). The latter route is probably the more important. In addition to strain variation, ester production is affected by many factors, being increased by high fermentation temperature, and the use of all-malt and high gravity worts and decreased by aeration and a high pitching rate. Ester production can be a particular problem in high gravity brewing, but may be controlled by oxygenation during fermentation to channel cell energy for growth, rather than ester production, or by directing acetyl coenzyme A molecules to lipid synthesis, rather than ester formation by ensuring a high level of fermentable carbohydrates in the wort.

Acetaldehyde is the most important of the high flavour-potential carbonyl compounds associated with beer. Acetaldehyde is a metabolic branch point in the yeast alcoholic fermentation and may either be reduced to ethanol or oxidized to acetate. Concentration reaches a maximum during fermentation then decreases and there is considerable variation in the quantity present in beer from brewery to brewery. Acetaldehyde production is relatively unaffected by strain of yeast or fermentation temperature, but is greater when a high pitching rate is used and when air is present during fermentation.

* Esters can also be produced in a non-enzyme mediated reaction. This involves an equilibrium reaction between an acid and an alcohol. The rate of reaction is, however, slow and the quantity formed too small to account for the levels of esters present in many types of beer.

Acetic acid is the major fatty acid of beer comprising 40 to 80% of the total. The levels present vary considerably and have a strong effect on the pH value of the finished beer. Aeration of the pitching yeast markedly increases production of acetate through stimulation of pyruvate decarboxylase activity. This enzyme utilizes Fe^{3+} ions in the direct oxidative decarboxylation of pyruvate to acetate.

Both diacetyl and pentane-2,3-dione (vicinal diketones) produce buttery, honey, or toffee-like aromas. Both compounds are formed outside the cell by spontaneous oxidative decarboxylation of the α-acetohydroxy acids, α-acetolactate (diacetyl) and α-acetohydroxybutyrate (pentane-2,3-dione). These acids are intermediates in the synthesis of leucine and valine (α-acetolactate) and isoleucine (α-acetohydroxybutyrate) and leak from the cell during fermentation. Actively metabolizing yeast produce reductases which reduce diacetyl to acetoin and butanediol and pentane-2,3-dione to pentane-2,3-diol. Under most circumstances, therefore, diacetyl and pentane-2,3-dione do not accumulate and are present only in sub-threshold quantities. Accumulation does however occur in circumstances where decomposition of α-acetohydroxy acids continues in the absence of yeast reductase activity.

Although small quantities of sulphur compounds are desirable in beer, an excess causes unpleasant off-tastes and aromas. Volatile sulphur compounds may, however, be of less importance than thought previously, since SO_2 levels are usually below the taste threshold and since beer free from bacterial infection does not normally contain free H_2S. However, SO_2 is an important anti-oxidant and also forms complexes with carbonyl compounds which, in the free state, generate stale flavours during storage. Hops, malt and adjuncts all contribute volatile sulphur compounds, but the major source is fermentation. The pathways of H_2S production are only poorly understood, but it is known that pantothenate-requiring strains form sulphides from sulphite or sulphate when a deficiency of the vitamin exists. Hydrogen sulphide production is also stimulated by amino acids such as threonine and glycine but

---

* One of the long term aims of genetic modification of *Sacch. cerevisiae* is development of strains, which produce insignificant levels of vicinal diketones. Such strains are unlikely to be available for some time, and various alternatives have been suggested. One, which has attracted interest is the use of bacterial acetolactate decarboxylase to shunt α-acetolactate direct to acetoin.

retarded by methionine, the lag before production of H_2S during fermentation corresponding with the removal of methionine. A relationship has also been established between H_2S production and the bud index (the ratio of the number of daughter cells to the number of mother cells), a low bud index correlating with a high level of H_2S production.

Hot holding of wort after boiling is the most important stage in production of dimethyl sulphide, but small quantities are produced during fermentation by reduction of dimethyl sulphoxide. The ability to reduce DMSO is widespread amongst *Sacch. cerevisiae* and spoilage bacteria and involves three enzymes: thioredoxin, thioredoxin reductase and sulphoxide reductase, with NADPH as hydrogen donor. The main purpose of thioredoxin, however, is to generate reducing power during yeast cell growth, while sulphoxide reductase is subject to catabolite repression in high nitrogen worts, methionine sulphoxide being of particular importance. These factors are likely to limit the extent of reduction of DMSO during fermentation.

(d) Maturation

Chemical changes during maturation are complex and, in some cases remain poorly understood. There are major differences between ales and lagers in that secondary fermentation of carbohydrates and removal of undesirable flavour-active compounds such as acetaldehyde proceeds rapidly in ales during maturation at 12–20°C, but follows a different pattern during low temperature maturation of lagers.

A considerable metabolic stress is placed on yeast by cooling to the *ca.* 4°C temperature employed in lager maturation. Problems with excess levels of residual sugars can occur due to the inability to ferment maltotriose which results from both catabolite repression and catabolite inhibition. Similar problems can result from the presence of excess fructose, especially in high-gravity brewing.

It is generally accepted that the removal of acetaldehyde, vicinal diketones such as diacetyl and sulphur compounds such as H_2S is of major importance in the maturation of beer, although in lager brewing the actual participation of yeast in secondary fermentation has been doubted. In contrast to the general concept of maturation as a process of continuing improvement of beer flavour and quality, levels of H_2S and vicinal diketones may rise rather than fall during the first week of maturation, especially where the krausening process is used. This may be attributed to the creation of growth promoting conditions by agitation and aeration during transfer. This in turn leads to depression of biosynthetic pathways to balance the demand for valine, isoleucine and methionine under the prevailing conditions of nitrogen limitation. Non-growth conditions also favour production of DMS especially in situations where levels of methionine sulphoxide are low.

Although the biochemical role of yeast in removal of unwanted flavour-active compounds may be questioned, taste improvements can be related to yeast sedimentation and the removal of polyphenol–protein complexes by adsorption on the yeast cell, or by precipitation. The length of time needed for complete stabilization is, however, commercially unacceptable.

Other changes result from the release of compounds from the yeast during the maturation process. Compounds involved include amino acids, peptides, nucleotides and both inorganic and organic phosphates. Participation of these compounds in flavour improvement depends upon their intracellular accumulation, changes in cell permeability and consequent exchange reactions between the yeast and the beer. These factors in turn depend on many variables including the yeast, its physiological condition and behaviour, length and temperature of maturation and the physical nature of the maturation vessel. The release of compounds from yeast may also have a deleterious effect on quality, medium chain length fatty acids, for example, can impart autolytic and yeasty flavours to beer.

7.4.7 The deterioration of beer quality

(a) Sunstruck flavour

Sunstruck (lightstruck) beer develops a characteristic and objectionable 'skunky' flavour. This results from the formation of 3-methyl-2-butene-1-thiol (prenylmercaptan) following the photocatalysed α-scission of the 4-methyl-3-pentenoyl side-chain of isohumulones. The 3-methylbutenyl radical formed by this reaction combines with H_2S, produced during fermentation, to form 3-methyl-2-butene-1-thiol. Selective reduction of the carbonyl group of preisomerized hop extract by treatment with borohydride prevents this reaction (see page 309) and permits the bottling of beer in clear glass bottles. In other cases the development of sunstruck flavour may be markedly delayed by the traditional means of bottling in brown bottles. This does not entirely eliminate the problem, however, and sunstruck flavour can develop during the very long storage lives demanded by supermarket chains.

(b) Development of flavour-active carbonyls

Carbonyl compounds, especially aldehydes, have very low detection thresholds and are a major factor in the deterioration of beer flavour during storage. Production of carbonyls is usually associated with extended storage, but under some circumstances detectable levels are produced before beer has left the warehouse.

A number of pathways are potentially involved in the formation of carbonyls (Table 7.9). Oxygen plays a key role, especially in bottled beer, by promoting some reactions and accelerating others. Free radicals, especially the hydroxyl radical are involved.

The most important source of oxidized flavours in beer is considered to be carbonyls, especially unsaturated, with more than six carbon atoms, derived from linoleic acid, although a further important source involves aldolic condensation reactions between carbonyls of one to four carbon atoms arising from oxidative degradation of isohumulones. Degradation of isohumulones also results in a loss of bitterness which may be recognized as the first stage in quality deterioration. *Trans*-2-nonanal, which imparts a 'cardboard' flavour is generally considered to be the most important individual carbonyl with respect to beer deterioration. This is formed by degradation of a precursor, trihydroxyoctadecanoic acid, formed by oxidation of unsaturated C_{18} fatty acids, especially linoleic. Degradation of trihydroxyoctadecanoic acid is accelerated

Table 7.9. Main reactions leading to carbonyl formation during flavour deterioration in beer

Strecker degradation of amino acids
Oxidative degradation of isohumulones
Oxidation of alcohols to aldehydes
Autoxidation of fatty acids
Enzymatic degradation of lipids
Aldol condensation of aldehydes
Secondary autoxidation of long-chain unsaturated aldehydes

by higher temperatures and light levels and the presence of metal ions.

The relationship between oxygen content of beer and formation of staling radicals may readily be demonstrated. Beer of high oxygen content contains significantly higher levels of carbonyls after storage, including *trans*-2-nonenal, octanal, *iso*-butenal and 2-phenylacetaldehyde, although levels of *trans*-2-*trans*-4-octadienal and *trans*-2-*trans*-4-nonadienal decrease. In beer of very high oxygen content, *trans*-2-nonenal is further degraded to other flavour-active compounds of unknown nature.

Although a degree of protection is obtained by minimizing the oxygen content of beer after packaging, a high level of oxidation during earlier stages of brewing also results in flavour instability in the final beer. Oxidation during wort production, for example, decreases the concentration of reducing compounds and the reductive capacity of polyphenols, etc.

There is an obvious relationship between the development of oxidized flavours and the concentration of precursors in the wort and beer which, to a considerable extent, can be controlled by good brewery practice. Turbid worts, for example, resulting from a deficient lauter system have a high level of long chain fatty acids. Beer prepared from turbid wort may contain high levels of carbonyls such as *trans*-2-butenol, *iso*-butanal, *trans*-2-nonenal, *iso*-valeral and 2-phenylacetaldehyde shortly after manufacture. Similarly, the amino acid level content of wort should be as low as compatible with good fermentation to minimize the formation of carbonyls through Strecker degradation and other reactions. Proline, which is present in significant quantities in beer, is of particular importance in promoting the aldol condensation of aldehydes.

The role of Maillard products in flavour deterioration of beer is sometimes overlooked. Excess production of Maillard products leads directly to flavour defects, but in addition, intermediates include α-dicarbonyls which react with amino acids in the Strecker degradation reaction. Melanoidins themselves also reduce flavour stability and in aged beer the extent of the Maillard reaction is more important in determining levels of some carbonyls (*trans*-2-*trans*-4-decadienal, *trans*-2-*trans*-4-undecadienal) than oxygen content.

(c) Non-biological haze formation

Non-biological haze formation remains a problem in brewing despite the various preventative and corrective measures taken. Haze formation during storage of finished beer also occurs, in association with the development of oxidized flavours and results from an increased affinity of proanthocyanidins for protein. The reason for this is not fully understood, but earlier theories involving the oxidative polymerization of proanthocyanidins are being superseded by the concept of direct or indirect hydroxyl radical attack producing proanthocyanidin oxidation products with a large affinity for proteins.

7.5 MICROBIOLOGICAL PROBLEMS ASSOCIATED WITH BEER

Microbiological problems associated with beer are almost entirely those of off-taints and visible spoilage caused by growth of non-culture micro-organisms during the brewing process or in the finished beer. Improvements in the general standard of hygiene and in the construction and design of equipment, together with widespread application of refrigeration, pasteurization and sterile filtration have reduced spoilage problems in recent years. The potential for spoilage, however, always exists and it must be emphasized that the financial consequences, direct and indirect, can be sufficiently serious to threaten the survival of smaller brewery companies.

7.5.1 Beer as an environment for micro-organisms

Beer is a poor environment for micro-organisms and this is reflected in the limited number of micro-organisms able to grow in beer. Five intrinsic factors, i.e. low pH value, low redox potential, relatively low nutrient level, presence of ethanol and other metabolites and presence of hop isohumulones, control microbial growth and must be tolerated, in combination, by any micro-organisms developing in beer. In some cases tolerance of the stress factors reflects a general characteristic of the type of micro-organism, but in others

* Beer does not support the growth of pathogenic micro-organisms, but allegations of 'food poisoning' due to beer are surprisingly common. Such allegations are rarely fully investigated and are often due to overindulgence. Mild gastrointestinal disturbances can, however, result from consumption of overtly spoiled beer, while the presence of mould pellicles, due to inefficient cleaning of re-usable bottles, can result in revulsion and consequent vomiting.

tolerance is a associated only with brewery strains. Ability to grow at low pH values, for example, is a generic characteristic of *Lactobacillus*, but tolerance of isohumulones is restricted to brewery strains.

The response of beer spoilage micro-organisms to the different stress factors varies. The growth of lactic acid bacteria, for example, is limited by the low level of amino acid growth factors, while the low redox potential restricts growth of acetic acid bacteria. To some extent these factors may be manipulated during brewing to limit the spoilage potential of the final beer. Production of low alcohol beer presents a special case in that the effective total removal of a stress factor not only increases the growth of recognized beer spoilage micro-organisms but permits other micro-organisms to grow. Low alcohol beer requires pasteurization for stability, but the absence of ethanol also means that pasteurization is less effective than with 'normal' beer. Increasing the hopping rate presents a possible solution, but many low alcohol beers require refrigeration for stability.

Extrinsic stress factors also limit microbial growth in beer. The importance of refrigeration has been recognized for many years and natural ice was employed in central Europe before the advent of mechanical means of cooling. Until recently, refrigeration was largely restricted to brewery operations, but the widespread application of public house cellar cooling has significantly reduced the warm weather spoilage rates of cask conditioned ales.

Extrinsic stress may also be applied by the addition of preservatives. In many countries, preservatives are not permitted, but in the UK SO_2, benzoic acid and various benzoates are permitted at levels of 70 mg/kg. Sorbic acid and sorbates are also permitted in some countries.

It should be appreciated that manipulation of stress factors to suppress one type of micro-organism may create the conditions necessary for growth of another type. Thus while problems caused

* The use of the polypeptide antibiotic, nisin, has been proposed for control of lactic acid bacteria in beer, but is not currently permitted. Nisin has no medical uses, but the climate of consumer opinion is opposed to any wider application of preservatives and a more acceptable future alternative may be the introduction of the *nis* gene of *Lactococcus lactis* into brewery strains of *Saccharomyces*.

by acetic acid bacteria have been considerably reduced by the creation of more highly anaerobic conditions in beer during bulk storage, conditions have been created which permit the growth of obligately anaerobic Gram-negative bacteria such as *Pectinatus*.

7.5.2 The main types of beer spoilage micro-organisms

(a) Acetic acid bacteria (Acetobacteraceae)

The acetic acid bacteria comprise two main genera, *Acetobacter* and *Gluconobacter*. Both *Acetobacter* and *Gluconobacter* are able to oxidize ethanol to acetic acid, although there are differences in that *Acetobacter* can further oxidize acetic acid to CO_2 and H_2O. In the past both genera have been associated with spoilage of beer, but *Acetobacter*, which favours alcohol-enriched niches, is of considerably greater importance than *Gluconobacter*. Four species of *Acetobacter* are currently recognized of which *A. pasteurianus* and *A. aceti*, are of greatest importance.

The usual spoilage pattern of *Acetobacter* is acetification, rope formation, turbidity and discolouration. 'Strange' off-flavours may develop before detectable acetification and may be due to the oxidation of polyalcohols, such as glycerol, to dihydroxyacetone. *Acetobacter* may be isolated from pitching yeast and survives fermentation, but the usual brewery sources are storage, filtration and filling equipment. As noted above the importance of *Acetobacter* as a spoilage organism has been much reduced and where problems do occur, these tend to result from poor public house cellar hygiene and the consequent growth and colonization of inadequately cleaned and maintained dispensing equipment.

(b) Anaerobic, Gram-negative, bacteria

Obligately anaerobic Gram-negative, rod-shaped bacteria have only recently been implicated in the spoilage of beer and are restricted, as far as is known, to lager-type beer. Original isolates were assigned to a new genus, *Pectinatus* (family Bacteroidaceae), as *Pect. cerevisiiphilus*. More recently a systematic study has been made of *Pectinatus* and similar bacteria from pitching yeast and spoiled beer. Isolates were assigned to *Pect. cerevisiiphilus* and a new species of *Pectinatus*, *Pect. frisingensis*; a new species of the existing genus *Selenomonas*, *Sel. lacticifex* and to a new genus *Zymophilus*. In addition obligately anaerobic, Gram-negative cocci, assigned to the genus *Megasphaera* as *M. cerevisiae* have also been isolated from spoiled beer.

Spoilage by anaerobic, Gram-negative bacteria involves turbidity and off-flavours, including sourness, and rotting egg odours. The bacteria have been isolated from pitching yeast which acts as a continual source of infection.

(c) The Enterobacteriaceae

Members of the Enterobacteriaceae may be isolated from wort in significant numbers. Such bacteria are referred to as 'wort bacteria' and normally only survive the initial stages of fermentation. A number of genera have been isolated (some names used earlier may not conform to modern taxonomic usage) including *Citrobacter*, *Enterobacter*, *Escherichia*, *Hafnia* and *Klebsiella*. These may be accompanied by non-Enterobacteriaceae described as *Pseudomonas* and *Achromobacter*.

Growth of Enterobacteriaceae in wort can result in various off-flavours, including 'vegetable' and 'phenolic' taints. A number of strains produce DMSO reductase and may be involved in production of excess dimethyl sulphide in lager-type beer. The original source is believed to be malt, hops, etc., but problems rarely arise unless cooled wort is stored for long periods before pitching.

Obesumbacterium proteus (*Hafnia protea*, *Flavobacterium proteus*) is of considerably greater significance than other members of the Enterobacteriaceae. The organism is a common contam-inant of pitching yeast, in which it may be recognized microscopically as short, fat rods, and grows in wort during the alcoholic fermentation. In the past it has been thought that *O. proteus* was unable to grow in beer and of no spoilage significance. It is now recognized, however, that growth occurs in beer of high pH value (above 4.5) and that the presence in pitching yeast can be associated with increased levels of *n*-propanol, *iso*pentanol, 2,3-butanediol and 2-acetolactate in the finished beer. Some strains also possess DMSO reductase. The most significant role of *O. proteus* may, however, be nitrate reduction to nitrite and the consequent formation of *N*-nitroso compounds.

Obesumbacterium proteus has two distinct biogroups which are effectively different species. The organism is closely related to

Hafnia alvei, itself a brewery organism which can grow during the first 25 h of fermentation, and which may have a more significant role than other 'wort bacteria'.

Rahnella aquatilis, a relatively recently described member of the Enterobacteriaceae has also been proposed as a beer spoilage organism and is able to grow in pitching yeast and during lager fermentations.

(d) Lactic acid bacteria

Lactic acid bacteria are common brewery spoilage organisms and the most prevalent during fermentations. Two genera, the rod-shaped *Lactobacillus* and the coccoid *Pediococcus* are involved in beer spoilage. A third genus, *Leuconostoc*, is a common spoilage organism in wine, but rare in beer. Species of beer spoilage lactobacilli differ from their counterparts isolated from other habitats by their resistance to isohumulones.

A number of species of *Lactobacillus* have been implicated in beer spoilage including the heterofermentative *L. brevis*, *L. buchneri* and *L. fermentum* and the homofermentative *L. casei*, *L. plantarum* and *L. delbrueckii* ssp. *lactis* (*L. leichmannii*). In general spoilage by heterofermentative species is more common, but the spoilage pattern is similar in all cases and involves excess acidity, a silky turbidity and 'buttery' and 'dirty' off-flavours resulting from diacetyl and 2,3-butanediol production. Slime and rope may also be present. Lactobacilli are common contaminants of pitching yeast and grow throughout fermentation, causing particular problems in tower fermenters. Growth in the finished beer is limited only by deficiency of amino acids.

Pediococcus is rarely involved in spoilage of top fermented beer but is a common problem in bottom fermented. This may be a consequence of its optimum growth temperature being lower than that of *Lactobacillus*. Several species of *Pediococcus* have been

---

* For many years resistance to isohumulones was thought to be an unstable character. More recently however, the trait has been shown to be both phenotypically and genetically stable. Ability to grow in beer may be induced in hop-resistant, but not hop-sensitive, lactic acid bacteria by growth in the presence of a sub-inhibitory concentration of isohumulone.

described, but 90% of spoilage has been attributed to a single species *P. damnosus*.

Pediococcus is responsible for the classic beer spoilage condition 'sarcina sickness' which, in its extreme form, involves excess acidity, turbidity, a granular sediment, rope formation and off-flavours due to diacetyl formation. Such spoilage, however, occurs only when growth has been extensive and in other situations diacetyl production alone is of greatest importance. The threshold level of diacetyl is very low in lager (*ca.* 0.12 mg/l) and as few as 2×10^4 cells/ml can produce readily detectable levels of diacetyl at up to 0.36 mg/l. Spoilage may therefore occur in the absence of visible turbidity. Pediococci are contaminants of pitching yeast and grow throughout fermentation and storage and in the finished beer. The organism has a high level of resistance to commonly used sanitizers and is able to colonize equipment which then serves as a focus of infection.

(e) Wild yeast

Wild yeast may be defined as '. . . any yeast not deliberately used and under full control'. This is of necessity a broad definition and includes both 'harmful' yeasts which have undesirable effects on beer and 'harmless' yeasts which have no detectable effects. Further non-culture strains of *Sacch. cerevisiae* and *Sacch. carlsbergensis* are included in the definition of wild yeasts. A flocculent strain of *Sacch. cerevisiae*, for example, is considered as a spoilage organism in a fermentation carried out by a non-flocculent strain. Equally, *Sacch. cerevisiae* may cause spoilage during lager fermentation and *Sacch. carlsbergensis* during ale fermentation.

A wide range of yeasts derived from the environment have been implicated in beer spoilage including a number of species of *Saccharomyces*, most of which are currently classified as synonyms of *Sacch. cerevisiae*. Other common wild yeasts include *Pichia membranaefaciens*, *Hansenula anaerobia*, *Mycoderma cerevisiae* and some species of *Torulopsis*.

Spoilage by wild yeasts varies according to the type responsible, but typically involves unfinable turbidity, excess gas, excess acidity and off-flavours. Some wild yeasts produce high levels of C_4 to C_{10} fatty acids such as isobutyrate and isovalerate, while many non-culture species of *Saccharomyces* contain the *POF*1 gene and produce phenolic off-flavours by decarboxylation of hydroxycinnamic acids such as ferulic and *p*-coumaric. Wild yeasts are able to grow during fermentation and in the finished beer, spoilage being a particular problem in cask-conditioned ales. In breweries situated in rural areas, contamination may follow seasonal patterns and be associated with particular agricultural activities.

(f) Zymomonas mobilis

Spoilage by *Zymomonas* is unknown in lager-type beer, but although uncommon, is often considered the most serious form of spoilagc in ale-type beer. Infection with *Zymomonas* can lead to total and prolonged loss of production and the economic consequences have forced the closure of at least one brewery. *Zymomonas* is a facultative anaerobe, although some strains are unable to grow aerobically. The organism is unusual in that glucose is fermented anaerobically *via* the Entner–Doudoroff pathway. *Zymomonas* is related phenotypically, genetically and ecologically to the acetic acid bacteria and may be derived from a common aerobic ancestor. *Zymomonas mobilis* has two sub-species, but only one, *Z. mobilis* ssp. *mobilis*, is a problem in beer.

Spoilage by *Zymomonas* is characterized by a heavy turbidity and a highly objectionable stench due to the production of acetaldehyde and H_2S. *Zymomonas* is not known to be present in pitching yeast, but grows rapidly under anaerobic conditions in highly hopped beer of low pH value. Infection has been associated with building work in the vicinity of the brewery.

7.5.3 Microbiological analysis

Microbiological analysis of beer still tends to be based on use of conventional plate count or membrane filtration techniques, although there is interest in rapid methods, including impediometry, the direct immunofluorescence technique and ATP measurement. Multi-purpose media, such as universal beer medium and WL nutrient medium, are widely used. Universal beer medium supports the growth of acetic acid bacteria, lactic acid bacteria, wild yeasts and *Zymomonas* and can be used for examination of the pitching yeast, cooled wort and finished beer. Anaerobic incubation is required for lactic acid bacteria and *Zymomonas*. WL nutrient medium can be used for enumerating and differentiating, brewer's and wild yeast and enumerating bacterial contaminants. In the latter case, actidione is often added to suppress yeast.

Multi-purpose media are not always fully satisfactory and additional specialist media are often employed. Rogosa or MRS medium is effective for lactic acid bacteria, while lysine medium is often used for detection of wild yeast in the pitching yeast.

Zymomonas is present in only very small numbers and a forcing test, involving incubation of a large quantity of beer and examination for heavy turbidity and smell of H_2S, is widely used. A shorter method based on impediometry has also been developed.

Detection of Gram-negative anaerobes, such as *Pectinatus*, requires special media. A method using a selective and differential medium, lactate-lead actetate medium, has been developed especially for isolation of *Pectinatus* from beer. Beer is introduced into molten medium in a special screw-capped tube, which is incubated at 30-32°C for 5-10 days. Colonies of *Pectinatus* are black in colour.

Members of the Enterobacteriaceae can be enumerated on standard media, such as violet red-bile-glucose agar. Species such as *Obesumbacterium*, however, are slow growing and many strains require an incubation temperature of 25-30°C.

The microbiologist is usually also responsible for providing active yeast cultures. In addition to microbiological analysis for wild yeast and bacterial contaminants, this requires an assessment of suitability for use. Widescale use is made of viability stains and the percentage of yeast budding, although it is now considered that measurement of the physiological status of the yeast provides more accurate information (see page 318). A predictive test based on impediometry has also been developed, but is not widely used.

* Lysine medium differentiates brewer's strains of *Sacch. cerevisiae* and *Sacch. carlsbergensis* from wild yeasts on the ability to use lysine as sole nitrogen source. Most, but not all, wild yeasts are able to utilize lysine, but the property has never been found in a brewer's strain.

EXERCISE 7.1.

Lager-type beer from all over the world is readily available in the UK. The *Cask Ale Report*, 1993 (Carlsberg-Tetley Brewing Ltd, Burton on Trent), has suggested that it may be possible to develop a market for English cask ale in Germany (as *frische englische Bier*). You are employed as sales director of a small English ale brewery, which has pre-empted its larger competitors and negotiated a marketing agreement for cask-conditioned beer with a company owning beer cellars in large German towns. These range from university towns to industrial towns in the Ruhr and the former East Germany. Success of this initiative is considered essential to the future of your company. Draw up a plan for a marketing initiative designed to introduce cask ale to the German drinking public. It is anticipated that there will be initial problems due to unfamiliarity and in the first instance emphasis should be placed on encouraging the consumer to try the cask ale. *Assuming* your initial campaign was successful, decide the most cost-effective means of maintaining momentum after the initial novelty has worn off.

EXERCISE 7.2.

Having tired of corporate life, as senior technical manager of a large food retailer, you have purchased a public house in a large industrial town. The premises includes a micro-brewery, which is currently unused. This stems from failure of the previous owners to produce a beer suited to the taste of the major customer groups. The public house currently has a reputation as 'rough' and has been threatened with closure by the licensing authorities. You consider that successful establishment of the micro-brewery is essential for commercial success. Aware of the previous failure you decide to apply your knowledge of market research and sensory evaluation to develop a commercially successful range of beers for your micro-brewery. Draw up plans for an exercise to determine, firstly, the types of beer and their proposed markets and, secondly, to optimize the organoleptic properties with respect to local taste. Include details of the organization of consumer panels, comparative taste trials, etc. What, if any, are the particular difficulties with this approach in the context of a single, albeit large, public house?

EXERCISE 7.3.

You are employed as a microbiologist in a large lager brewery, producing a number of brands of lager. One brand, a premium, heavily hopped, high alcohol (5.6%) canned lager has been the subject of intermittent complaints due to sediment and off-odours. The causative organism has been presumptively identified as *Pectinatus*. The problem is restricted to certain batches, but no common features have been identified. *Pectinatus* has been isolated from pitching yeast and from lagers during maturation, but spoilage is restricted to the premium brand. You are instructed, as a matter of urgency, to investigate the factors leading to spoilage. A certain amount of preliminary work has been done and this has suggested that *Pectinatus* is stimulated by metabolites of the pitching yeast used for the premium lager. Draw up a plan for your investigation, bearing in mind the need to obtain a valid result as quickly as possible.

8

ALCOHOLIC BEVERAGES: II. WINES AND RELATED DRINKS

OBJECTIVES

After reading this chapter you should understand

- The nature of the different types of wine and related beverages
- The basic technology of wine making
- The involvement of various genera of yeasts in fermentation
- The importance of the malo-lactic fermentation
- Quality assurance and control
- The chemistry of compounds contributing to the flavour, aroma and character of wine
- The pigments of red wine and their changes during maturation
- Microbiological spoilage of wine

8.1 INTRODUCTION

Wine making is an important economic activity, not only in the traditional countries of the 'old world' such as France and Germany, but also in 'new world' countries such as Australia, the US and Argentina. Wine production is also increasing in the UK.

Wine may be made from any fruit, which contains sufficient fermentable carbohydrate. The grape (*Vitis vinifera* or, less commonly, *V. rotundifolia*) is of overwheming commercial importance, although wine is also made on a limited commercial scale from fruits such as strawberry, gooseberry and peach. Cider, produced by fermentation of apple juice, is not legally a wine, but shares a similar technology and, together with the less common perry, is also discussed in this chapter. Mention should also be made of mead, a type of wine now little produced, made from fermented honey and water, and palm wine, made by mixed lactic-alcoholic fermentation of the sap of various species of palm tree.

BOX 8.1 When the wine cup glistens

Consumer tastes have changed over the years, with a trend towards light wines of relatively low alcohol content and some types of heavy, fortified wine such as Madeira are now only rarely consumed. New wine-based products such as coolers, a blend of wine with fruit juice or soda water, have achieved some popularity, while low-alcohol and dealcoholized wines are also available.

8.2 TECHNOLOGY OF WINE MAKING

The manufacture of wine is summarized in Figure 8.1.

8.2.1 Production of the must and pressing

A large number of grape cultivars have been used in commercial wine making, it being recognized that the type of grape, together with factors such as the soil type and overall climate, is important

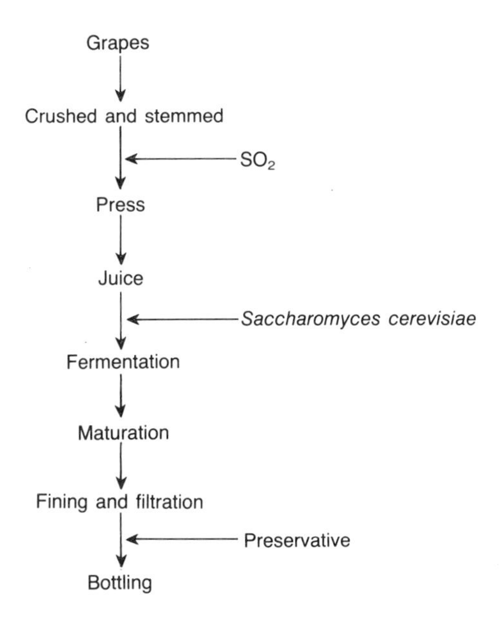

Figure 8.1 Basic technology of wine making.

Table 8.1 Classification of the flavours of wine

Varietal	Derived from compounds present in grapes
Pre-fermentation	Derived from compounds developing in the extraction of must
Fermentative	Derived from compounds produced during the primary fermentation by yeast and during the malo-lactic fermentation
Post-fermentative	Derived from compounds produced during ageing

in determining the character of the finished wine. Flavours imparted by the grape are usually referred to as varietal and combine with flavours arising during other stages of wine making (Table 8.1) to determine the overall character of the wine. In traditional wine-making countries such as France, the cultivar of grape used for each specific type of wine is controlled. Thus only Chardonnay and Pinot noir are authorized in Burgundy, and only Gamay in Beaujolais. Although such control serves to establish and protect the identity of existing wine regions, it also inhibits innovation and attempts to improve the quality of wine produced elsewhere. Control of this nature would not be acceptable in non-traditional countries such as the US or Australia.

The quality of the grape and its suitability for fermentation is affected by the climate during each individual growing season, cold wet weather leading to a grape of low sugar content and high volatile acidity and an increased possibility of mould infection. The composition of the natural yeast micro-flora is also likely to be affected (see below). Deciding the optimum time of ripening often presents problems to the wine maker. In all cases grapes should be of an acceptably high sugar content and, in some countries, picking cannot commence until a predetermined sugar content is reached. In the case of rich, sweet wines such as Tokay the high sugar content of the must is dependent on the growth of the

* *Botrytis cinerea* is not desirable on grapes destined for most other types of wine and the mould can be the cause of serious economic loss. Control is by use of fungicides and concern has been expressed in the US over the continued use of procymidone in Europe. A detention order is automatically placed on any wine entering the US which contains more than 0.02 mg/l procymidone.

Table 8.2 Effect of high phenolic compound and solids content of grape juice on the quality of wine

Phenolic compounds (tannins):	astringency, possible bitterness and browning
Solids (lees):	'mousey' and other off-flavours

mould *Botrytis cinerea* (the 'noble rot') on the grapes before harvesting.

The addition of sugar to must of low natural sugar content (chapitalization) is practised in some countries, but prohibited in others. The extent of ripeness, however, also affects the extent to which varietal characteristics are expressed as well as the development of flavour compounds during fermentation. The modern trend to lighter, low alcohol and rather acid wines has led to increased use of immature grapes. In the case of Chardonnay grapes, differences in maturity led to a major difference in alcohol content, but had little effect on production of esters or higher alcohols, or on acceptability. In contrast a study of Pinot Noir wines made using gas chromatography-olfactometry showed not only that odour profiles vary from year to year, but that profiles of wine from older grapes showed more odour-active peaks.

Grapes are picked mechanically or by hand, while stemming, crushing and pressing is usually mechanical, the colour of the wine being determined at this stage by the colour of the grape and the length of contact between the juice and the skin. For economic reasons it is desirable to obtain as high a yield of juice as posssible, but it is recognized that, in general, higher yields lead to poor quality wine due to the high levels of phenolic compounds and solids, both of which can result in faults (Table 8.2). Mechanical harvesting can also lead to high levels of phenolics as the result of material other than grape derived from stems, leaves and petioles. Pectinolytic enzymes may be added to facilitate pressing and, in the case of red wines, colour extraction. Commercial wine manufacture involves multiple pressing cycles, but it is now current

* Concentration of free monoterpenes, polyols and glycosides, and their distribution in the skin and pulp has been recommended as a guide to the suitability of grapes for harvesting and to establish the most suitable conditions for pressing.

practice to restrict premium wine production to juice from the first two pressings, third and subsequent pressings (press fractions), being used to produce lower quality wines and wine for distilling. Both quality of juice and yield vary according to the type of press, continuous augur presses achieving high yields, but producing a juice of very high solids and phenolic content. Conveyor belt presses have a lower, but acceptable yield and produce a juice of relatively low solids content. Conveyor belt presses, however, have disadvantages in terms of difficulty in operation and cleaning and initial cost. A more recent development, the tank press permits maximum juice yields with minimum solids and, despite the high cost, is becoming widely adopted. In modern commercial practice pressing precedes the fermentation, but post-fermentation pressing using batch presses is still used in small-scale traditional wineries.

It is usual practice to add SO_2 to the must to discourage growth of indigenous (wild) yeasts and spoilage bacteria. SO_2 is effective in this usage, but concerns have been expressed over safety. The use of pasteurization of juice in place of SO_2 treatment has been re-proposed but seems unlikely to be accepted.

Clarifying agents are sometimes added to the must before or in the early stages of fermentation. Clarifying agents remove protein and improve the colour of some types of wine. Powdered bentonite is still widely used, but is now being replaced by bentonite of higher activity such as filamentous and granular types. Other highy effective clarifying agents are available, based on albumins, micronized caseinates and charcoal. Care must be taken in the use of clarifying agents to avoid the removal of flavour components.

Carbonic maceration is an alternative procedure to crushing and pressing, which involves holding intact grapes for several days under an atmosphere of high CO_2 content. This results in cellular death and loss of semi-permeability of cell walls followed by enzymic reactions including alcohol production. There is also a rapid release of anthocyanin pigments and the process is normally used in manufacture of red wine. Carbonic maceration was prac-

* There can be considerable variation in SO_2 sensitivity within a given yeast species and a strain of *Candida stellata* which differs phenotypically from classical strains is able to tolerate 50 mg/l SO_2. This strain is strongly fermentative, producing 8% ethanol and being tolerant of 10%. Spoilage potential is high due to production of detectable quantities of 2-methyl-1-propanol.

tised in antiquity, but interest in the process has recently increased. Claimed advantages include early drinkability, fast maturation and a characteristic 'soft' taste and a distinct flavour and aroma favoured by modern tastes in red wine. Vinification does, however, tend to be slow and probably more expensive than in conventional wine making. There is also a tendency to produce wines of high volatile acidity. The success, or failure, of carbonic maceration often depends on the use of sound grapes and a very short time between picking and establishment of the CO_2 atmosphere.

8.3.2 Fermentation

(a) Properties of Saccharomyces cerevisiae

The general properties of *Sacch. cerevisiae* involved in wine making are the same as those of brewery strains (see Chapter 7, pages 297-300). Strains of importance in wine making are distinguishable from those used in brewing, although the older nomenclature for wine yeasts, *Sacch. ellipsoideus* (*Sacch. cerevisiae* var. *ellipsoideus*) is no longer considered valid. As with brewery yeasts there has been interest in improving performance through genetic modification (Table 8.3).

(b) Micro-organisms present during fermentation

In traditional European practice, fermentation depends on the presence of naturally occurring yeasts in the must. Initially, *Saccharomyces* is not numerically significant; dominant yeasts at this stage include *Aureobasidium pullulans*, *Candida stellata*, *Hanseniaspora uvarum*, *Issatchenkia orientalis*, *Kloeckera javanica*, *Metschnikowia pulcherrima* and *Pichia anomala*. The geographical location of the vineyard influences the micro-flora present to

Table 8.3 Genetic modification of wine yeasts

Introduction of flocculation and reduction of H_2S production.
Reduction of higher alcohol production
Improvement of fermentation efficiency
Reduction of foaming
Increased glycerol yield
Introduction of ability to effect malo-lactic fermentation
Elimination of urea (and thus ethyl carbamate) production (Sake yeast)

some extent, *H. uvarum* being dominant in the early stages of fermentation in California and mid-Europe, whilst *H. osmophila* dominates elsewhere. *Metschnikowia pulcherrima* is of particular prevalence in Europe. Climatic conditions during the growth of vines will also affect the yeast microflora of a particular vineyard. In a study of a Spanish vineyard, for example, cold, wet years led to a higher incidence of damaged grapes and significantly increased numbers of oxidative yeasts.

The addition of SO_2 to must reduces the total numbers of yeast present. The use of SO_2 also results in qualitative changes in the yeast microflora and favours the resistant *Sacch. cerevisiae* and *Saccharomycodes ludwigii* at the expense of more sensitive species such as *K. apiculata*.

Saccharomyces cerevisiae is primarily derived from winery equipment and in general becomes dominant as the fermentation proceeds and the increasing ethanol content inhibits many indigenous yeasts. Patterns of microbial succession and dominance are, however modified by temperature and at 25°C, *Sacch. exiguus* and *Zygosaccharomyces bailii* may predominate. At 10°C, *Sacch. cerevisiae* is dominant, but *K. apiculata*, *C. stellata* and *C. krusei* are also present in large numbers. Ideally, primary fermentation at *ca.* 24°C for 3–5 days should be used for red wines and 7–21°C for 7 days to several weeks for white. Over-heating can lead to a low quality wine and while traditional fermentation vats are uncooled, the introduction of cooling vats in the Californian wine industry has been considered to have a greater beneficial effect on quality than the use of pure yeast cultures.

Few wine makers in non-traditional areas rely on adventitious inoculation of the must with *Sacch. cerevisiae* and it is now common practice to add pure cultures at the start of fermentation. The use of pure cultures is also increasing in traditional regions. In some large wineries, the yeast is propagated from master cultures

---

* Histamine is produced in wine and, in association with a potentiator, can lead to headaches, flushing and other symptoms in sensitive persons. There is considerable variation in the quantity of histamine produced according to grape variety, geographical location and method of vinification. Red wines contain significantly more histamine than white, but white wines contain more putrescine, cadaverine and tyramine. Putrescine and cadaverine may act as potentiators to histamine reactions and the mixed drinking of red and white wines may be a hazardous pastime for sensitive persons.

held in-house, but liquid or dried cultures are more convenient in most cases. Dried cultures, which can be added direct to the vat, without any propagation are particularly useful in small wineries. The use of pure cultures of *Sacch. cerevisiae* is generally recognized to minimize problems of controlling the fermentation and to produce a wine of more consistent quality. The addition of *Sacch. cerevisiae*, however, does not affect the presence of indigenous yeast, nor to a great extent the pattern of fermentation. It appears that each stage of the natural fermentation is characterized by the development of different strains of *Sacch. cerevisiae*, and that the main consequence of adding pure cultures is to influence the development of *Sacch. cerevisiae* rather than to inhibit non-*Saccharomyces*. Until recently, however, there has been little information available on the population dynamics of inoculated *Sacch. cerevisiae* or the actual role of wild strains. Molecular methodology, mitochondrial DNA restriction analysis, has now been used to study the situation in both inoculated and natural fermentations. This has shown that the inoculated strain is primarily responsible for fermentation, but that there is no suppression of the naturally occuring strains in the early stages of fermentation. A great diversity of strains are present and, although only a few persist throughout the process, the same strains tend to be dominant in both natural and inoculated fermentations.

Although it is recognized that inoculated wines are generally of higher quality, there is some concern that special characteristics associated with the natural micro-flora may be lost. An re-evaluation has shown that some yeasts of low fermentation power, such as *K. apiculata*, produce significant quantities of volatile aroma compounds. Significant quantities of volatiles are also produced by some strains of *Sacch. cerevisiae*, but other strains, including some used for inoculation of the must, are markedly less effective. Despite this it is likely that there is sufficient development of natural aroma-producing strains in inoculated fermentations to permit development of typical characteristics. It has been suggested, however, that the emphasis on alcohol production in choosing strains of *Sacch. cerevisiae* for inoculation may result in a low quality wine and that mixed starter cultures should be used and should include yeasts which have a high potential for production of volatile aroma compounds.

Moulds and lactic acid bacteria are the most important micro-organisms, other than yeast, present at the commencement of fermentation. Moulds die in the early stages, while lactic acid bacteria, which are unable to compete effectively for fermentable sugars, decline during the alcoholic fermentation but increase in numbers during any subsequent malo-lactic fermentation.

(c) Design of fermentation vats

Traditionally, open, wooden fermentation vats were used, but these have been superseded by cylindrical, closed vessels, mounted either horizontally or vertically and constructed of stainless steel or fibreglass. Modern vats may be equipped with cooling coils or wall panels and incorporate the necessary fittings for cleaning-in-place.

Continuous fermentation systems have been developed for wine making, but none appears to have been fully successful.

(d) Premature death of Saccharomyces cerevisiae *during fermentation*

Although *Sacch. cerevisiae* multiplies rapidly during the exponential growth phase, non-growing cells are responsible for completing the alcoholic fermentation. Problems can arise, especially in natural fermentations, from premature death of cells as a result of intolerance of ethanol or other fermentation products. Such intolerance is a reflection of the physiological state of the yeast and may be prevented by adding sterol, a readily assimilable nitrogen source, such as ammonium salts, possibly in combination with thiamin, or yeast cell walls to the must in the early stages of fermentation. Commercially produced 'activators' are available which combine several approaches to the avoidance of premature yeast death. Bioactivator DC™, for example, is a blend of ammonium salts and thiamine with yeast cell walls and polysaccharide complexes which adsorb inhibitory fermentation products. Polysaccharide complexes also entrain air bubbles and thus are an effective alternative to oxygenation of the must which has been practised with some success in California. Urea was previously widely used as an 'activator' but is now recognized as a precursor of the potential carcinogen ethyl carbamate.

8.2.3 The malo-lactic fermentation

The malo-lactic fermentation (MLF) involves the decarboxylation of malic to lactic acid by strains of lactic acid bacteria (Figure 8.2).

Leuconostoc oenos is most commonly involved but some strains of *Lactobacillus* and *Pediococcus* are also capable of the conversion. In the majority of wines, the MLF follows the alcoholic fermentation, but during wine making using carbonic maceration (see pages 366–7), the MLF and alcoholic fermentation procede simultaneously.

Lactic acid is a weaker acid than malic acid and the main practical consequence of the MLF is deacidification of the wine, the extent depending on the proportion of the grape acids malic and tartaric present. The removal of malic acid reduces the likelihood of microbial spoilage, while the MLF also increases the complexity of the wine. The MLF is considered desirable during manufacture of highly acidic red wines such as Italian Piedmontese types, but undesirable in the low acid wines of California and other warm climates. The MLF is also considered undesirable in most types of high acid white wines, due to the resulting aroma and flavour modification and, with the exception of wines fermented from Chardonnay juice, chemical means of acidity reduction are employed.

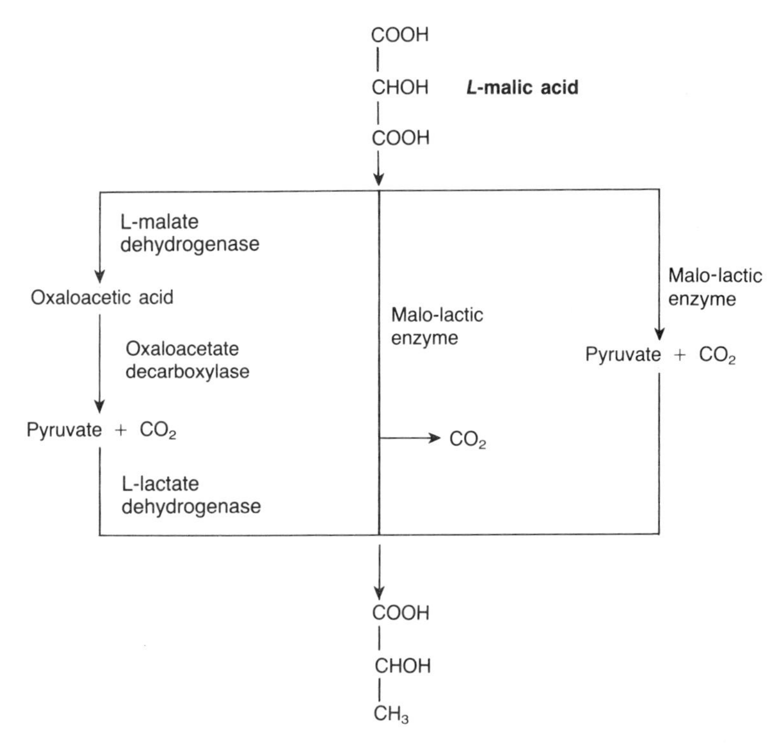

Figure 8.2 The malo-lactic fermentation.

Wine-making techniques can be manipulated to favour the induction, or suppression, of the MLF as required. Paradoxically and, for the wine maker infuriatingly, the MLF is often most difficult to induce when desirable and most difficult to inhibit when undesirable. Some success has been achieved by using pure cultures of *Leuc. oenos* to initiate the MLF and reduce the quality problems associated with some indigenous lactic acid bacteria. The use of pure cultures, however, introduces potential problems with bacteriophage, while the suitability of strains differs according to the nature of the wine. Pure cultures of *Leuc. oenos* have not entirely eliminated problems due to indigenous lactic acid bacteria and the addition of nisin has been proposed to control such strains. This would require the selection of nisin-resistant strains of *Leuc. oenos*. Immobilized micro-organisms are a potential alternative means of improving the reliability and efficiency of the malo-lactic fermentation, but further work is required before industrial application is possible.

Attempts have been made to genetically modify *Sacch. cerevisiae* by introduction of the malo-lactic gene. The malo-lactic gene of *Lb. delbrueckii* has been successfully transferred into *Sacch. cerevisiae*, but the gene was expressed to only a limited extent and practical application of this procedure is not yet possible.

8.2.4 Maturation

Maturation of wine initially takes place in the cask or, especially in large scale production in non-traditional areas, in large tanks. The

BOX 8.2 **Hearts of oak**

The wood of casks used for maturation contributes to the flavour and character of wine and in recent years the use of new oak casks has become very common in maturation of light Californian wines. The more conservative French wine makers have observed that wines of this type taste of 'fruit juice and sawdust'. Despite this the use of new oak casks for maturation has spread to many countries, including France. When properly balanced by other components, oak can make a positive contribution to the character of wine. In some cases, however, the use of new oak casks represents nothing more than the following of a fashion and the flavour imparted by the wood may be considered intrusive.

length of maturation varies considerably and is only a very short period in the case of 'young' wines, such as *Beaujolais nouveau*, which have a fruity character derived from the grape. White wine is generally matured in the cask for *ca*. 6 months, although some sweet types require maturation for over a year. In the case of 'young' wines, there is little further beneficial change after bottling and the wine is ready for immediate consumption. In other cases, however, maturation continues after bottling and many years may elapse before the wine is in optimum condition.

8.2.4 Filtration, clarification and bottling

(a) Filtration and clarification

Wine may undergo filtration at many stages during its production and filtration treatments range from the coarse to a full sterilizing treatment. Methods and equipment are similar to those used for beer (Chapter 7, pages 325-6), although the more sophisticated requirements of wine have led to the development of new filtration materials including cellulose fibrils used alone or in combination with diatomaceous earth.

Clarification and filtration sometimes results in loss of aroma. This has been attributed to binding of aroma compounds to macromolecules such as proteins derived from grape skin and pips, or to yeast cell walls and their subsequent removal. This can be a particular problem with removal of haze-forming proteins by bentonite. Proteins are derived from the grape and are thus a particular problem in high protein varieties. The possibility of using peptidases to hydrolyse the proteins has been explored, but at temperatures of *ca*. 16°C, proteins are highly resistant to active peptidases. Hydrolysis occurs at temperatures above 30°C, but heating to this extent is not acceptable in high quality wine.

In recent years microfiltration has been increasingly applied as the final process before bottling. Microfiltration membranes are usually in a tubular configuration for use with wine. Pre-filtration is not required, but clarifying and stabilizing agents such as bentonite are still necessary to maintain a sufficiently high product flow. The capital cost of microfiltration systems is relatively high, but this is offset by the operating efficiency, reliability and versatility. Maintenance and cleaning costs are also low.

(b) Bottling

Following filtration and clarification the wine passes to storage tanks prior to bottling. In the case of filter sterilized wine, final filtration usually takes place directly before filling. Some use is made of inert atmospheres (CO_2 and/or nitrogen) to protect wine from oxidation. Additions may also be made before bottling to stabilize the wine against microbiological and chemical deterioration. Anti-microbial agents are commonly added, and while national legislation varies, SO_2 and sorbic acid are most commonly used. Allyl isothiocyanate is also used as tablets in conjunction with an inert paraffin support. The tablets are floated on the surface of wine in bulk storage to prevent growth of film yeasts and moulds.

The acidity of wines can be reduced before bottling by addition of salts, usually potassium bicarbonate and calcium carbonate.

The use of glass bottles is universal for high quality wine and in some cases, such as chianti, the type of bottle used is specific to a particular type of wine. The cork is the traditional means of closing the bottle, and this is protected from dehydration and mould growth by a lead foil or, in recent years, a plastic outer cap. Wine is sometimes bottled under an inert atmosphere, but the effectiveness of this procedure depends on the success of earlier measures to protect the wine from oxidation.

Screw caps are now used for sealing cheaper types of wine, and some types of lightly carbonated wine such as Lambrusco. Some cheap wine has been retailed in bottles made from various plastics, but use is limited, except in specialist situations such as aircraft, where weight is critical.

---

* Microfiltration is a membrane process operated at a low pressure of *ca.* 10 lb/in^2. The membrane operates as a molecular sieve which permits the passage of water, ethanol, flavour compounds, selected macromolecules and other dissolved species, but retains suspended material such as colloids and microbial cells. In contrast to conventional filtering media, fluid flow is across the membrane rather than perpendicular to it.

* Corks can be a source of taints in wine and 'cork taint' has been attributed to chloroanisoles arising by methylation of chlorophenols used to treat the cork. Methylation is the result of the metabolic activity of moulds such as *Aspergillus* and *Penicillium*.

Hand filling and corking is still employed in small-scale production, but in large wineries bottling usually a large-scale, highly automated operation carried out under clean conditions to minimize the risk of aerial contamination. Medium sized wineries have, for many years, been faced with problems due to the high capital expenditure involved in installing efficient bottling plant. The use of sophisticated mobile bottling plant is becoming common in some areas, notably California and Germany and leads to substantial cost savings.

Large containers of polyethylene or other plastic have been used for some years for bulk packaging of wine for distribution at point-of-sale. This practice has, however, been restricted to cheap types, including fruit wines made in the UK from imported juice. The 'wine box' is a sophisticated extension of this principle into the retail market. Wines is filled into a 5 l capacity polyester:polythene:foil:paper laminate bag fitted with an integral tap, which is supported by an external cardboard box. As wine is drawn, the bag collapses thus minimizing the air space and oxidative deterioration. The wine box has achieved some popularity for parties, etc., but is seen as a relatively expensive means of buying wine. Use of plastic-based packaging, together with the substitution of plastic for stainless steel in construction of pipework, etc., also increases the likelihood of oxidative deterioration during storage.

Small quantities of wine are retailed in rip-top aluminium cans and more recently the Tetra Brik™ aseptic packaging system has been employed.

* In recent years concern has been expressed, especially in the US, over the levels of lead in foods, including wine. Investigations have suggested that the main source of lead in decanted wines is the foil used to seal the cork. This leaves corrosion deposits which contaminate the wine on pouring. Some lead may also be derived from decanters made from crystal glass, used in the home or, less commonly, for packaging wine for retail sale. Lead in wine may also be derived from environmental sources during grape growing and wine making. Maximum limits of 200 to 300 ppb have been proposed.

8.2.6 Special types of wine

(a) Sparkling wines

A wide range of sparkling wines are produced ranging from champagne and similar wines, in which carbonation is the result of a secondary in-bottle fermentation, to cheap substitutes carbonated by injection of CO_2.

> BOX 8.3 **Not a champagne teetoller**
>
> For more than 100 years the interests of the largest producers of the Champagne region have been promoted by the *Syndicat des Grandes Marques*, which functions as ' . . part club, part trade union and part public relations machine'. Little attention, however, has been given to quality standards and it is recognized that some of the champagnes permitted to use the cherished descriptor *Grande Marque* are of lower quality than others which are not. Against a background of falling sales, increasing competition from New World producers and an increasingly tarnished reputation, the *Syndicat*, in its current form, is to be disbanded. A replacement, as yet to be named, will be formed and enforce stricter regulations concerning vineyard and cellar practice. It also appears possible that the number of champagnes entitled to the descriptor *Grande Marque* will be reduced from the current 28 to around 20. (*The Guardian Weekend*, **June 12, 1993**, 56).

Champagne itself is a blended wine and both black grapes, such as Pinot Noir, and white grapes, such as Chardonnay, are used as starting material. Blending is based on sensory analysis and can involve wines from several years production. Secondary fermentation involves the addition of yeast and a small quantity of sugar to the blended wine, in a cask before transfer to bottles. A small quantity of tannin may also be added at this stage to inhibit bacterial growth. The bottles are sealed with a thick, wired-on cork. The yeast used must be capable of initiating secondary fermentation of a substrate that has been deprived of some of its growth factors, under increased pressure, at a relatively low temperature of *ca.* 10°C and in the presence of 10–12% alcohol. The yeast must also settle out rapidly and completely and die, or become inactive, before the 'dosage', the final addition of sugar syrup in Champagne brandy. In the case of champagne itself, the wine remains in contact with the secondary yeast growth for several years, the bottles being manipulated during storage so that the yeast forms a plug of sediment in the neck of the bottle from where it is removed by freezing. The long period of contact with the yeast cells is considered an important factor in development of the champagne character.

The quantity of sugar syrup added at 'dosing' depends on the type of champagne, the wine being classified as very dry (*brut*), extra dry, dry, semi-dry, semi-sweet or sweet. 'Dosage' varies from *ca.* 0.5% in very dry to *ca.* 10% in sweet champagne. After 'dosage' additional wine is added, if required, to fill bottles to the correct volume and the bottle finally corked. A special champagne cork made from criss-cross layers of cork is necessary to retain CO_2 and this is wired into place on a strengthened bottle neck.

The full champagne process is lengthy and labour-intensive and this is reflected in the price of the finished product. A number of attempts have been made to shorten the maturation and in some accelerated processes autolysed yeast cells or yeast extract is added at the end of the secondary fermentation.

Manufacture of other sparkling wines involves secondary fermentation either in the bottle, in casks or in large tanks. In each case the contact time with the yeast is limited both for economic reasons and because prolonged contact is considered to introduce off-flavours.

Artificially carbonated wines are generally considered to be low quality substitutes for champagne. CO_2 evolved during fermentation may be used or the gas obtained from an external source. Equipment is similar to that used for carbonating soft drinks (see Chapter 3, pages 92–4).

(b) Sherry and similar wines

Sherry proper is made only in a small area of southern Spain around Jerez de la Frontera, but similar wines are made world-wide. Either the Palomino or Pedro Ximinez grape cultivars are used. Before pulping and pressing the grapes are sun-dried to concentrate the juice and lightly dusted with gypsum. This lowers the pH value of the juice by converting potassium bitartrate to tartaric acid and reduces problems resulting from bacterial infection. In traditional practice, still widely applied in Spain, the juice is fermented in oak casks, the casks being only partly filled to encourage the development of a film (*flor*) of yeast. This consists largely of strains of *Sacch. cerevisiae*, although other yeasts are likely to be present and may contribute to flavour. The extent to which the *flor* develops varies from cask to cask and determines the organoleptic properties of the wine. At the end of fermentation the wine from each cask is sampled to determine its suitability for making one of the various types of sherry. The wine is then drawn off into a new barrel, fortified to 15–16% alcohol by addition of either grape brandy or neutral grain spirit and passed to the *solera* for maturation. A *solera* consists of a series of tiered oak casks containing sherry of the same type, but of different age, the oldest being at the bottom and the youngest at the top. Mature wine drawn off from the bottom tier is replaced by wine from the second tier which, in turn is replaced by wine from the third tier, etc. Many *soleras* have been in continuous operation for over 100 years and sherry thus consists of a blend of wines of many different ages. On leaving the *solera*, the sherry is clarified, blended and, if necessary, sweetened with a sweetening wine prepared from Pedro Ximinez grapes.

Although traditional methods can produce sherry of superb quality there may be problems due to lack of consistency. A modern method is to inoculate the young, fortified wine with a pure culture of a film forming strain of *Sacch. cerevisiae*. The *flor* then develops during maturation and usually includes other film yeasts. Sherry made in this way is highly consistent but, according to traditionalists, not of the very highest quality. In either case the *flor* yeast is of major importance in determining the sherry characteristic and in distinguishing the real product from imitations made by different processes. Metabolism of the *flor* yeast involves considerable utilization of acetic acid and, to a lesser extent, ethanol and production of acetaldehyde, acetal and other flavour compounds, including esters, isopentanol and hexanol.

Madeira is similar to sherry and is made using a *solera* maturation process but the characteristic flavour is achieved by heating the wine to *ca*. 41°C for up to 3 months before maturation. Specially constructed rooms, *estufas*, are used for heating. A wide variety of madeiras are made with an alcohol content varying from 17 to 21%

and are usually named after the grape used as starting material. A similar wine, baked sherry, is made in California.

(c) Port

Port is usually made from traditional black grape cultivars such as Tinta francisca, Bastardo and Rufeta, although white port is also made, common grape cultivars being Rabigato and Malvasia. Port is a fortified wine, grape brandy being used to stop the fermentation at the desired sugar content. The fermentation of sweet port is stopped in its early stages, while that of very dry port proceeds almost to completion. The highest quality vintage port is bottled young and matures in the bottle, in some cases for over 50 years. Most port, however, is blended and matured in wooden barrels (pipes), being bottled when ready for consumption. The character of the port is determined by the length of maturation, the colour of red becoming lighter and of brownish tinge as maturation proceeds. In contrast white port darkens with increasing maturity.

(d) Wine coolers, aperitifs and other wine-based products

A number of wine-based products are available in which wine is blended with other beverages, whether alcoholic or non-alcoholic. The simplest of these are 'coolers' comprising a mixture of wine with fruit juice, soda water or even spring water.

> BOX 8.4 **California dreamin'**
>
> Coolers are added-value convenience foods and there is nothing remotely difficult about mixing them at home, at a very considerably lower price. Demand is driven by the general trend towards lower alcohol products, reinforced by connotations of overseas holidays and memories of sun, sea, surf, etc. Specific names may be given to different combinations of wine and juice. One UK company used the same sparkling white wine base to produce 'pina colada' (pineapple juice), 'sangria' (mixed citrus juice) and 'Buck's fizz' (orange juice).

Wine-based aperitifs are often slightly more sophisticated and frequently contain herb extracts. Vermouth is a type of Italian wine widely drunk as an aperitif and, in combination with gin, forms the Martini. This may be mixed at point of consumption or retailed in a premixed form under a proprietary name such as Martini™ itself or Cinzano™. Vermouth itself is a herb-flavoured wine made with muscat grapes, originally produced in Turin, Italy which originated with the attempts of an 18th century wine maker to recreate the *absinthianum vinum* of ancient Rome. The name itself is derived from the German *vermut*, wormwood.

(e) Dealcoholized and low-alcohol wines

The market for dealcoholized and low-alcohol wines is largely driven by restrictions on drinking and driving, although such wines are also in demand for teetotal weddings, etc. In current commercial practice, wine is dealcoholized by thermal evaporation, an aroma-recovery plant being incorporated in equipment (see Chapter 7, pages 328–9). The use of SO_2 during wine making means that it may also be necessary to incorporate a desulphurization column into the evaporator.

8.2.7 Manufacture of other types of non-distilled alcoholic beverages related to wine

(a) Cider and perry

Cider and perry are similar to wine both in terms of manufacture and microbiology. Cider is made from the juice of the apple (*Malus pumila*) and perry from that of pears (*Pyrus communis*). Produc-

Box 8.5 Cider with Rosie

Commercially produced cider, perhaps fortunately, has been replacing farmhouse cider for many years and in some tourist areas the latter is reduced almost to the level of a novelty souvenir. Until recently, however, large-scale cider makers have continued to emphasize their rural origins even, in some cases, contriving to imply that a nationally marketed brand is still made in the scullery of a farmhouse. The bucolic approach has, however, been abandoned in the marketing of a new generation of ciders with brand names such as Red Rock™ and Desert Gold™, which appear positioned to compete with premium lagers. Cider 'coolers' containing fruit juice have also appeared, and a mix of cider and lager is also retailed, the notorious 'snakebite'.

tion of perry, with the exception of 'champagne-perry' marketed under brand names such as Babycham™, has fallen to very low levels and has ceased entirely in many traditional producing areas. In the UK cider is effectively of two types: farmhouse cider, often referred to as 'scrumpy', which is made on a small-scale, by a variety of primitive techniques, for purely local consumption, and commercially produced cider, manufactured on a large-scale and marketed on a regional or national basis.

Cider is made from a mixture of special, high phenolic-content cultivars (cider apples) together with culinary and dessert apples. Cider apples are classified into four categories (Table 8.4) and this classification may be extended to culinary and dessert types. Apples are harvested, taking care to minimize damage, and milled to a fine pulp in a grater or hammer mill. A higher yield of juice is obtained when particles are small, but excessively small particles are impossible to press. Various types of press are used including hydraulically operated plate presses and continuous screw presses (see Chapter 2, pages 47–9). Traditional hand operated presses are used only in small-scale farmhouse manufacture. Cider and culinary apples are usually of low soluble pectin content and treatment of the pulp with pectinase is unnecessary. In contrast, ripe, dessert apples are of high soluble pectin content and yield is increased by treatment with pectinase. In some cases additional steps such as heating pulp to 40°C are used to increase juice yield.

In traditional farmhouse manufacture fresh juice is used without further treatment. Procedures can vary widely from district to district and, indeed, from farm to farm. All, however, rely on a spontaneous fermentation by indigenous yeasts and can result in a product which is highly variable in its properties and highly unpredictable in its effect on the consumer. At the simplest, draught cider is fermented, matured and dispensed from a single barrel. A

Table 8.4 Classification of cider apples

Sweet	low in tannins and acidity. Bland and most suitable for blending
Bittersweet	low in acidity, high tannin content; the typical cider apple of south-west England and parts of France
Bittersharp	high in tannins and acidity; may be used as a single juice or in blends
Sharps	high in acidity, low tannin content

more common process, however involves fermentation in open barrels (keeves). A thick brown jelly consisting of a calcium pectate clot commonly develops and ultimately breaks up, removing yeast cells and other debris in a process of natural clarification (keeving). The clarified cider may then be transferred to clean barrels for sale and dispensing or to bottles where a secondary fermentation develops. Natural clarification, the process of defecation, is also a feature of some large-scale cider manufacture in France.

In large-scale manufacture the juice almost invariably undergoes further treatment. Concentration was previously used for convenience of storing surplus juice, but it is now common practice to base part or all of production on concentrate to minimize volumetric capacity. Several means of concentrating the juice have been used, but thermal evaporation in rising or falling film evaporators (see Chapter 2, pages 53–4) is most common.

It is also common practice for juice to receive treatment to reduce the indigenous microbial population. Practice varies somewhat according to the producing country. In northern France, for example, high speed centrifugation is used, while elsewhere in France sterile filtration is more common. Flash pasteurization is applied in Switzerland and some other countries, pasteurized juice often requiring supplementation with pectinases. The most common treatment, however, is sulphiting (*cf.* page 366).

The development of branded cider requires a high level of standardization. Base juices are standardized to a predetermined percentage of sugar and the solids content may also be adjusted and malic or lactic acid are added, if necessary, to increase the acidity. Water may be added to reduce acidity, produce a lighter coloured cider and to limit the final alcohol content. The use of concentrated juices, however, means that the cider may be diluted to the required alcohol content by addition of deaerated water to the fermented product.

In traditional manufacture and in countries such as Spain, the apple juice is fermented by naturally occurring micro-organisms. The general pattern of microbial succession and the course of the fermentation is similar to that of wine, although there tends to be considerably more variation both between plants and between batches produced by the same plant. In France the use of closed metal tanks and the consequent production of a CO_2 atmosphere, enabled natural fermentations to be used with minimal risk of problems due to undesirable micro-organisms, although the use of pure culture inoculation is now widespread.

In the UK, addition of pure cultures of yeast has been practised for many years, although in the past relatively little attention has been paid to yeast performance. In most cases a wine-making strain of *Sacch. cerevisiae* is used. Growth and fermentation characteristics have been considered to be of prime importance, but more recently increasing attention has been paid to the production of desirable flavours. Cider yeasts should also produce polygalacturonase to degrade de-esterified pectin to galacturonic acid and permit clarification of the cider. Fermenter technology remains fairly primitive in comparison with that applied in breweries and, while temperature control is well established in continental Europe, its application is not universal in the UK, even in some large-scale operations. Cider fermentations are allowed to proceed to completion, the sugar and, where necessary, the alcohol content being adjusted before packaging.

A malo-lactic fermentation often occurs concurrently with the alcoholic fermentation. This is generally considered to be undesirable and necessitates the addition of malic acid to restore the characteristic organoleptic properties. The malo-lactic fermentation may, however, be desirable in some harsh Spanish ciders.

Cider may be retailed either uncarbonated ('still') or carbonated. A small quantity of uncarbonated cider is still distributed in traditional wooden casks, but these are now rare and have largely been replaced by large glass jars or small plastic barrels. The cider may not be heat treated and in this case the storage life is no more than 2–3 weeks. More commonly the cider is heat treated by pasteurization at *ca.* 82°C for 30 s or by a variety of hot filling techniques. In-bottle pasteurization is common in continental Europe, although the prolonged heating can have an adverse effect on flavour. Sterile filtration is also used especially for premium 'vintage' ciders.

---

* In earlier practice, sweet cider was produced by addition of SO_2 to stop the fermentation when the required amount of sugar remained. The high sugar content favoured the growth of the sulphite-resistant spoilage organism *Zymomonas mobilis*, the resulting outbreaks of 'cider sickness' having severe economic consequences for small producers.

A small quantity of cider is still carbonated in-bottle but in the vast majority of cases the Charmat system is used, in which the secondary fermentation takes place under pressure in a temperature controlled tank. In some cases artificial carbonation is used. Sterile filtration or flash pasteurization at *ca.* 82°C for 30 s, followed by sterile filling is virtually universal, the cider being packaged for retail sale in 0.5 l aluminium cans, glass bottles of 0.75–1 l capacity and polyethyleneterephthalate (PET) bottles of up to 3 l capacity.

(b) Sake

Sake is a traditional Japanese drink, but similar products are made in China and other countries of south-east Asia. The raw material for fermentation, the *koji*, is steamed rice, which provides a substrate for the mould *Aspergillus oryzae*. Fermentation requires a strain of *Sacch. cerevisiae* which is tolerant of high concentrations of ethanol, acid and sugar and which can initiate fermentation at temperatures of below 10°C. The yeast is prepared as a starter culture, the *moto*. The *moto* contains lactic acid, which may either be added or produced by indigenous lactic acid bacteria, and which favours the sake yeast and inhibits undesirable micro-organisms. The main fermentation, the *moromi*, takes 20–25 days and is typified by simultaneous saccharification by *A. oryzae* amylase and fermentation by yeast. Sake is fermented in an open system and no attempt is made to exclude non-starter micro-organisms. The non-sterile culture, however, means that conditions must favour the growth of the starter strain of *Sacch. cerevisiae*. This is achieved by preparing the *moromi* in three successive stages during which the temperature is gradually lowered to less than 10°C. At the end of fermentation the ethanol content is 17.5–19.5%, the *moromi* is filtered and held for *ca.* 4 weeks before pasteurization and bottling.

(c) Palm wine

Palm wine is produced throughout tropical Africa, similar products being made elsewhere in the tropics. Manufacture involves fermentation of the sap of palm trees, the species of palm and the method of collecting the sap varying according to locality. In Nigeria, for example, the sap is obtained by tapping the *Raphia* palm and the oil palm (*Elaies* spp.).

Palm sap is a sweet, clear and colourless liquid containing *ca.* 10–12% sugar, predominantly sucrose. Fermentation is by a mixture of naturally occurring yeasts and bacteria, which remain in the product imparting a milky white appearance and active effervescence. Palm wine thus differs from most other alcoholic beverages in that in the latter, deliberate consumption of fermentation-associated micro-organisms is usually avoided. The continuing presence of metabolically active micro-organisms means that undesirable changes in taste due to production of organic acids occurs shortly after completion of the alcoholic fermentation and the shelf life is usually no more than 2–3 days.

Palm wine is produced by individual farmers and pooled before sale. Accurate figures for production are not available, but in Nigeria daily production is probably about 30 000 gallons. Palm wine is thus an economically important product and loss due to short shelf life is a matter of considerable concern. In the more rapidly developing tropical countries problems are exacerbated by increasing industrialization and urbanization. A number of attempts have been made to increase the shelf life while retaining the characteristic organoleptic properties. A shelf life of 6 months is considered ideal since this also overcomes problems caused by production being concentrated in the rainy season. The most promising method is to centrifuge the wine, followed by addition of 0.05% sodium metabisulphite and pasteurization at 60°C for 30 min. A clouding agent must be added to compensate for loss of turbidity during centrifugation.

Mention should also be made of pulque, a Mexican drink made by fermenting the sap of the undeveloped flowering stem of agave. Pulque is unique in that the fermentation is carried out by *Zymomonas* rather than by yeast.

8.2.8 Quality assurance and control

Unlike beer, wine and similar beverages are made from an inherently variable raw material and the quality can vary significantly from one year to another. Wine making, especially in traditional producing areas is often thought of as being a craft process, in which the skill of the wine maker is of overwhelming importance. This, however, is true only to a certain extent and, while skill and experience is of obvious importance, expert judgement must be supported by scientific analysis. This is most obvious in the determination of suitability of grapes for fermentation. Sugar content is most commonly used as criterion, but other determinations may be more suitable. Increasingly, especially in non-traditional producing areas, analysis is being applied to monitor the whole wine-making process from selection of raw materials to maturation. This is accompanied by much more rigorous process control than has previously been used and results in a much higher level of consistency than has previously been possible.

> BOX 8.6 **One gets so bored with good wine**
>
> It is part of the modern mythology of wine, that wines from non-traditional regions, such as California, made under a high level of control, are drinkable and consistent, but never truly great. In contrast the traditional European wine producing regions are perceived as possessing the flair to produce great wines, even if there is greater year to year variation. With some honourable exceptions, this is a delusion and wines from non-traditional regions often equal, or surpass, their traditional counterparts in terms of quality. This is reflected in the increasing market share of Californian, Australian, etc., wines at the expense of French and, to a lesser extent, German.

End-product testing is required to ensure wine is within its specification. Tasting is seen as being of prime importance, but this is supported by chemical analysis. Microbiological analysis is also required.

8.3 CHEMISTRY OF WINE

8.3.1 Flavour and aroma

The flavour and aroma of wine is the result of a complex synthesis of a very large number of compounds. Various estimates have been made of the number of compounds involved, but it is probably more important to appreciate that there is still considerable debate concerning the nature of the compounds and their role in the various different wines. As noted above (Table 8.1, page 364), it is possible to classify flavour active compounds according to their origin either in the grape, or in the various stages of wine making. The actual situation is rather more complex since some compounds may be derived by more than one route. Further, there is a dynamic process of chemical change and interactions which continue from the harvesting of the grape up to the point of consumption of the wine.

(a) Acids

A very large number of acids have been identified in wines and include aliphatic saturated and unsaturated, di- and tricarboxylic, aromatic carboxylic, hydroxy and keto carboxylic and sugar acids as well as miscellaneous types (Table 8.5). Quantitatively the grape

Table 8.5 Examples of acids which may contribute to flavour of wines

Aliphatic saturated and unsaturated
- acetic (produced by yeasts and bacteria, usually in white wine)
- pentanoic (derived from grapes, usually in white wine)
- decanoic (produced by yeasts, usually in white wine and sherry)

Di- and tricarboxylic
- succinic (produced by yeasts, in white and red wine)
- glutaric (produced by yeasts, usually in white wine and sherry)
- adipic (derived from grapes, usually in white wine)

Aromatic carboxylic
- cinnamic[1] (derived from grapes, usually in white wine)

Hydroxycarboxylic
- 2-methyl-2,3-dihydroxybutanoic (produced by yeasts, usually in white wine)
- malic (derived from grapes, in red and white wine)
- tartaric (derived from grapes, in red and white wine)
- citric (derived from grapes, usually in white wine)

Ketocarboxylic
- pyruvic (produced by yeasts, in red and white wine)
- 2-ketoglutaric (produced by yeasts, in red and white wine)
- levulinic (produced by yeasts, usually in white wine)

Sugar acids
- gluconic (various sources, in red and white wine)
- glucuronic (produced by yeasts, usually in white wine)

Miscellaneous
- vanillic[2] (derived from grape, usually in white wine and sherry)
- shikimic[3] (derived from grape, in red and white wine)
- caffeic[4] (derived from grape, in red and white wine)

[1] Cinnamic acid – 3-phenypropenoic acid
[2] Vanillic acid – 4-hydroxy-3-methoxy-benzoic acid
[3] Shikimic acid – 3,4,5-trihydroxycyclohexen(1)- carboxyllic acid
[4] Caffeic – 3,4-dihydroxycinnamic acid

is the major source of acids, the hydroxycarboxylic acids malic and tartaric accounting for over 90% of the total acid content of the grape. The mould *B. cinerea* is responsible for both quantitative and qualitative pre-fermentation changes in the composition of grape acids. These include the synthesis of gluconic, glucuronic and citric acids and degradation of tartaric acid. Structural modifications largely affect polar constituents and can vary considerably according to environmental conditions.

Modification and formation of acids during fermentation has a profound influence on the flavour of the wine. In general the acids of greatest importance with respect to the flavour and aroma of wine are formed during this stage, although the actual quantities synthesized may be small.

(b) Hydrocarbons

Hydrocarbons are among the neutral volatile substances and mainly originate in the grape, although some are derived from the oak wood used for making casks and some may be produced by microorganisms. Relatively few of the alkane hydrocarbon constituents of grapes are present in wine, the majority being precipitated with the must slurry. Conversely the majority of the terpenes, which are usually present only in small quantities in grapes, are too insoluble. Monoterpenes, including limonene, linalool and myrcene, are, however, detectable in wine and in the Riesling-type are the major volatile flavour compounds contributing floral flavours.

(c) Esters

A large number of esters have been identified in grapes, although some may be derived from microbial activity. Methyl anthranilate is of importance in imparting the characteristic aroma of *V. rotundifolia* and is not present in *V. vinifera*. Esters formed during yeast fermentation are of greatest importance in determining the aroma of wines. Ethyl, isobutyl and isopentyl esters are predominant (Table 8.6), reflecting the high concentrations of ethanol, isobutanol and isopentanol in wine. Esters as a group are responsible for much of the typical wine aroma.

(d) Alcohols

In addition to terpene alcohols, a series of C_4 to C_{10} aliphatic alcohols are present in grapes and are also detectable in wine. The

Table 8.6 Some esters contributing to the aroma of wine

Ethyl formate	Ethyl acetate
Isobutyl acetate	Isopentyl acetate
Hexyl acetate	2-phenylethyl acetate
Ethyl propanoate	Ethyl 2-hydroxy propanoate
Isobutyl 2-hydroxy propanoate	Isopentyl 2-hydroxypropanoate
Ethyl hexanoate	Ethyl octanoate
Diethyl succinate	

overwhelming importance of yeast fermentation means, however, that contribution to aroma of grape-derived alcohols is limited. Those alcohols of importance in determining the aroma of wine are derived from amino acids during fermentation. A contribution is also made by higher alcohols, which may be derived during fermentation by reduction of corresponding aldehydes. Butanol may be derived from 2,3-butandiol (Table 8.7).

Table 8.7 Some alcohols contributing to the aroma of wine

Ethanol	Methanol
1-Propanol	2-Methyl-1-propanol
1-Butanol	2-Butanol
2-Methyl-1-butanol	3-Methyl-1-butanol
2,3-Butandiol	2,3-Pentandiol
1-Hexanol	1-Octanol
2-Phenylethanol	

(e) Aldehydes and ketones

Grapes contain a number of aldehydes, although the importance in the aroma of wine appears limited. Acetaldehyde is the major aldehyde and may be derived both from the grape and, more importantly, by decarboxylation of pyruvate during fermentation. Sherry contains particularly large quantities of acetaldehyde as a consequence of its production under oxidative conditions. Benzaldehyde, 2-phenylacetaldehyde, furfural and 5-methylfurfural appear

* It is known that grape-derived aldehydes are lost during fermentation by reduction to the corresponding alcohols. It is also likely that, in the presence of SO_2, bisulphite addition compounds are formed. These are highly water-soluble and may escape detection due to their loss during extraction.

to be the most important aldehydes with respect to aroma, but analytical problems make the true situation difficult to assess.

Grapes contain a number of methyl and ethyl ketones as well as β-ionone and damascenone. In most cases these compounds are present in wine and contribute towards the aroma. Acetoin and vicinal diketones are not present in grapes, but are formed during fermentation. The importance of diacetyl as an aroma compound in wine is sometimes overlooked. In dry red wines, levels of up to 4 mg/l enhance quality by increasing complexity, but higher levels lower perceived quality. Smaller quantities of diacetyl have an adverse effect on white wines and levels in excess of 0.9 mg/l lead to a sweetish,'dairy' flavour in Riesling-type wines. Diacetyl is, however, considered to make a positive contribution at levels of up to *ca.* 0.7 mg/l. In general wines which have undergone a malo-lactic fermentation contain higher levels of diacetyl. Acetoin is usually present at higher levels than diacetyl, but has a significantly higher detection threshold. Levels of *ca.* 30 mg/l are considered desirable in fruity wines, but at levels of *ca.* 100 mg/l an undesirable musty flavour is imparted. Acetoin appears to be of importance in the characteristic flavour of sherry, which can contain as much as 350 mg/l.

(f) Lactones

Lactones are now recognized as being of considerable importance as aroma constituents of all types of wine. Gamma-lactones are of prime importance and only one δ-lactone has been identified (Table 8.8). Gamma-butyrolactone, produced from glutamic acid, is a constituent of the aroma complex of all types of wine. Other

Table 8.8 Some lactones contributing to the aroma of wine

4-Hydroxybutanoic acid γ-lactone
4-Acetyl-4-hydroxy-butanoic acid γ-lactone
4-Carboethoxy-4-hydroxybutanoic acid γ-lactone
4-Hydroxypentanoic acid γ-lactone
5-Keto-4-hydroxyhexanoic acid γ-lactone
trans-4-hydroxy-3-methyloctanoic acid γ-lactone
4-Hydroxydecanoic acid γ-lactone
5-Hydroxydecanoic acid δ-lactone
2-Vinyl-2-methyl-tetrahydrofuran-5-one

lactones, including pantolactone, appear only to be of significance in sherry.

Most lactones are the products of yeast metabolism, but 4-hydroxy-3-methyloctanoic acid γ-lactones are derived from oak wood casks during maturation, while 2-vinyl-2-methyl-tetrahydrofuran-1-one is derived from grapes.

(g) Volatile sulphur compounds

Little is known of the formation of volatile sulphur compounds. Sulphur-containing amino acids appear to be precursors, although volatile sulphur compounds may also be the product of yeast and bacterial metabolism (*cf.* beer; Chapter 7, pages 347–8). The flavour impact of volatile sulphur compounds is usually negative, although very low concentrations of some have a beneficial effect in certain types of wine. Dimethyl sulphoxide, for example, contributes flavours to neutral white wine variously described as asparagus, cooked corn or molasses. In contrast, the presence of thiols is invariably a defect, ethanethiol contributing undesirable onion and rubbery flavours. Defects of this nature may develop during the storage of wine due to the reduction of disulphides to thiols by sulphite. This reaction is very slow at the pH value of wine, but detectable amounts of thiols accumulate when storage is extended.

(h) Volatile phenolic compounds

Volatile phenolic compounds, especially vinyl phenols, can strongly influence the flavour and aroma of wine. Only one volatile phenolic compound, acetovanillone, has been identified in grapes and all others are formed as metabolites of micro-organisms or by cleavage of grape derived higher phenols. Both 4-vinylphenol and 4-vinylguaiacol have been detected in the aroma-complex of a number of different wines and are formed by decarboxylation of ferulic acid and *p*-coumaric acid, respectively. Decarboxylation may be mediated by enzymes from yeast or bacteria. Vinyl phenols may

---

* The production of ethanethiol is highly dependent on pH value. At pH value 3.5 and a temperature of 20°C, 700 days elapsed before detectable quantities of ethanethiol were produced by reduction of diethyl sulphide. This reaction, however, would require only 1 h at a pH value of 7.2.

themselves act as intermediates during the formation of their ethyl analogues.

(i) Miscellaneous compounds

A very wide range of other compounds are involved in determination of aroma and flavour of wine. These include acetals, such as 1,1-diethoxyethane, and secondary acetamides, such as *N*-ethylacetamide, as well as C_{12} to C_{18} alkanes, styrene and linalooloxides. The contibution made by these compounds is difficult to assess, but it is likely that the main role lies in contributing to the overall character of the wine.

8.3.3 Colour

The colour of red wine is derived initially from anthocyanin pigments (Figure 3.9, page 111). A 'full bodied', young red wine contains *ca.* 500 mg/l total anthocyanins, of which malvidin-3-glycoside (oenin) and its acylated derivatives comprise 60–80%. Additional phenolic metabolites are also present in young wine and include hydroxycinnamic acid esters and yellow flavonoid pigments. The most important flavonoids are the flavan-3-ols, (−)-epicatechin and (+)-catechin.

Anthocyanins are highly reactive as a consequence of the electron deficient flavylium nucleus. The pigment composition becomes more complex during maturation as the concentration of the original anthocyanins falls and new, more stable oligomeric pigments are formed which maintain the colour of the wine (Figure 8.3). This process commences immediately after extraction of the anthocyanins and is largely completed within one to two years. The colour hue and density of a young red wine is thus determined by a complex series of delicately balanced equilibria involving different anthocyanins and their different structural forms. Temperature, anthocyanin concentration, self-association and co-pigmentation are all important in the development and change of colour density.

---

* Co-pigmentation plays a crucial role in the colouration of red wines as well as fruit and flowers. Co-pigments are compounds which themselves have little or no colour. When added to an anthocyanin solution, however, the colour is greatly enhanced.

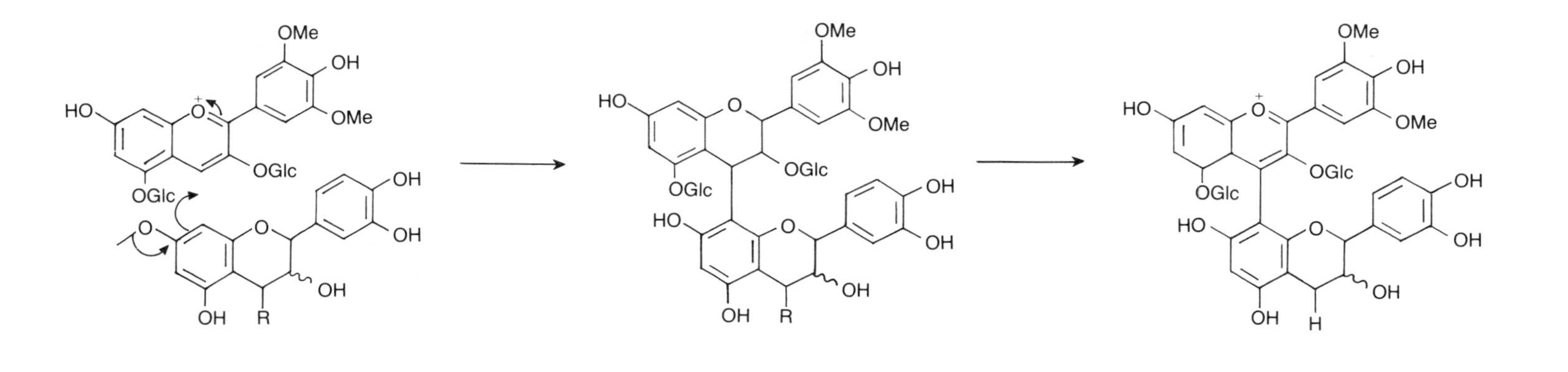

Figure 8.3 Changes in anthocyanin pigments during maturation.

The colour of mature red wine is a consequence of the changes amongst the anthocyanins and is also strongly influenced by the presence of yellow-orange pigments. These are formed by reaction between flavan-3-ols such as (−)-epicatechin and malvidin.

8.4 MICROBIOLOGY

8.4.1 Micro-organisms involved in the spoilage of wine and related beverages

(a) Acetic acid bacteria

Acetic acid bacteria, primarily *Acetobacter pasteurianus*, have been implicated in spoilage of wines, producing a vinegary or a 'mousy', sweet-sour taste and turbidity. The organism is common in the environment of wineries, but despite this is not usually a major spoilage organism.

Species of *Acetobacter* have also been implicated in spoilage of cider, although the scale of the problem has been much reduced in recent years, and sake, where spoilage is associated with vinegary taste, ropiness and turbidity.

(b) Lactic acid bacteria

Lactic acid bacteria are the most important bacteria involved in spoilage of wine and have also been associated with cider and sake. In wine, *Lactobacillus* and *Leuconostoc* are most commonly involved, growth resulting in souring, off-tastes, turbidity and, in some cases, slime production. Sweet wines are most commonly affected. *Pediococcus* is less common in wine but occasionally causes spoilage involving souring, turbidity and, in red wines, a loss of colour intensity due to fall in pH value. Some species of *Pediococcus* are responsible for ropiness.

(c) Moulds

Moulds are only rarely involved in spoilage of wine and similar products. Moulds can, however, develop on poorly sealed corks and cause mustiness in the wine. Very occasionally the mould hyphae penetrate the cork and a limited amount of growth occurs on the upper surface of the wine. Limited mould growth can also occur on the surface of wine where corking is grossly defective or in opened bottles which have been kept for excessive periods - an all too common fate of wine in public houses!

Mould pellicles can also be present as the result of growth on inadequately cleaned, and sporadically used bottling machinery or in the dregs remaining in the re-usable bottles used for some wine and cider.

(d) Yeasts

In most circumstances yeasts are the most common spoilage organisms of wine. A large number of genera have been implicated and there is some geographical variation.

Dominant species of spoilage yeast tend to differ between pre- and post-bottling wine. Wine stored in bulk tanks is most usually spoiled by species of *Dekkera* or *Saccharomyc. ludwigii*. In contrast, film-forming yeasts such as *Candida vini*, *C. zeylanoides*, *C. rugosa*, *Issatchenkia orientalis* and *Pichia membranaefaciens* are most commonly involved in spoilage of wine stored in barrels. *Zygosaccharomyces bailii* is the major cause of problems in bottled wine. This yeast is resistant to SO_2 and is able to grow in ethanol contents in excess of 12% and problems can be caused by an initial contamination with a very small number of cells. *Saccharomyces cerevisiae*, *C. rugosa*, *C. vini* and *P. membranaefaciens* may also cause spoilage. Yeast spoilage of wine is readily recognized due to cloud or sediment formation, together with gas production and taste defects.

Yeasts are also common spoilage organisms in cider. Spoilage by film yeasts is common in farmhouse cider, but is now rare in large scale industrial practice where fermentative yeasts are the major spoilage organisms. In past years, *Kloeckera apiculata* was the major concern, but most problems are now caused by *Saccharomyc. ludwigii*.

(e) Zymomonas

With the exception of an anecdotal account of spoilage of a sweet de-alcoholized wine, *Zymomonas* is not considered to cause problems in wine. The organism was, however, of considerable importance in the English cider industry as the cause of 'cider sickness' and '. . . literally shaped the way English cider was made'.

Changes in the manufacturing technology of sweet cider (see page 383) means that 'cider sickness' is currently rare, although not totally unknown. The spoilage pattern is distinctive, characterized by production of large quantities of gas, unpleasant aroma and flavour, a heavy turbidity and loss of sweetness. *Zymomonas* has also been implicated in a distinctive spoilage of French cider, *framboisement*, although the acetic acid and lactic acid bacteria have also been associated with this condition and the actual role of *Zymomonas* has been doubted.

8.4.2 Microbiological methods

Routine microbiological examinations of wine, cider, etc., are not required, with the exception of determination of the presence of yeast at bottling. A low level of contamination may lead to spoilage and membrane filtration is required to obtain the necessary sensitivity.

Strongly selective media are not required for enumeration of yeast since competitive organisms are not usually present in significant numbers (and would in themselves be cause for alarm!). Many media have been used including malt extract agar, oxytetracycline-glucose-yeast extract agar and WL nutrient agar. These media are generally effective in the recovery of yeast, but their value is limited by the inability to differentiate between strains of yeasts able to grow in (and spoil) the wine and those which die rapidly. To overcome this problem a 'significance' broth, containing ethanol and urea as selective agents, was developed. This medium is used, in conjunction with a resuscitation medium, in a membrane filtration technique which provides an effective means of predicting spoilage. The test does, however, require 3 days to obtain a result and a faster analysis is required at times of peak throughput. The detection time can be shortened by at least 24 h by combining the 'significance' test with an assay for ATP.

* Spoilage of wine and similar products is generally well understood and where unknown causes are cited this tends to reflect the inadequacy of investigations. Since 1988, however, some 20 cases have occurred in the French wine industry where the alcohol content of the wine has fallen from an initial 12.5-13% to *ca*. 10%, or less, on bottling. The cause is not known, but a micro-organism, probably a bacterium, is thought most likely and has been code-named *Edouard*. (*The Guardian*, September 7, 1992).

Impediometry has also been investigated as a means of detecting spoilage yeasts in wine. The major advantage is automation rather than rapidity of obtaining a test result and further work is required on media.

Examination for other micro-organisms is only required during investigations of spoilage or for other special purposes. In the case of spoilage the value of a microscopic examination should not be overlooked as the first stage of the investigation and often provides a presumptive identification of the cause. In some cases the example of brewery microbiologists is followed and a single selective medium, usually WL nutrient agar, used to recover all types of micro-organism likely to be present. Although convenient, better results are likely to be obtained by use of specialist media (Table 8.9). *Leuconostoc oenos* will not grow on standard media for lactic acid bacteria such as MRS medium and special media containing 15% ethanol is required.

Table 8.9 Media for use during the investigation of microbiological spoilage of wine and related products

Acetic acid bacteria
Acetobacter agar
Lactic acid bacteria
MRS agar (general purpose)
Garvie's medium (*Leuconostoc oenos*)
Yeasts
as for routine examinations
Zymomonas
WL nutrient medium

EXERCISE 8.1.

In the dairy industry it is now becoming common practice to use 'ripening cultures' during the manufacture of soft cheese from pasteurized milk. These cultures are derived from the raw milk microflora and contribute to the development of desirable flavour. Do you consider that the use of starter cultures containing yeasts additional to *Sacch. cerevisiae* could be beneficial in wine making? How could the fermentation be manipulated to ensure both adequate ethanol production by *Sacch. cerevisiae* and production of flavour compounds by other yeasts? Compare the relative advantages and disadvantages of this approach with the use of genetic modification to produce a 'perfect' yeast.

EXERCISE 8.2.

'Snakebite', a mixture of cider and lager, usually in equal proportions, is notorious for rapidly inducing drunkenness and aggression. For this reason, its sale is banned in some bars. Consider the physiological basis for the fact that a mixture of cider and lager should have a more rapid and profound effect than either drink alone. To what extent do you think this is psychological, stemming from the fearsome reputation of snakebite?

When cider and lager are mixed, a haze is formed. You are taking part in a phone-in at a local radio station, organized by your professional body as part of a 'science awareness week', and are asked by a listener how clear cider and lager produce a cloudy snakebite. You have 3 min to prepare an answer which is concise, easily understood by the radio audience and scientifically correct. Bear in mind the distractions of the music of the Velvet Underground and the inane chatter of the disk jockey.

EXERCISE 8.3.

You are employed as an assistant wine maker in a large Californian winery. The company produces bland, low-cost wines, which are largely retailed through catering outlets. The best known brand, *Midnight Express*, is a muscatel-type wine favoured by vagrants. The company has been slower than its competitors in introducing higher quality branded wines and is suffering a small, but continuing, loss of market share. It has been decided, as a matter of urgency, to introduce a range of distinctive, higher quality wines, retailed through supermarkets under a new brand name. The lack of expertise is recognized and the Chief Executive Officer, a computer enthusiast, has proposed development of an expert system to assist quality assurance. Establishing this system would involve incorporation of the accumulated experience of several European wine makers, together with published material, into a suitable computer program. Other executives have expressed unease at the proposal and you are asked to report on the feasibility and likely success of such a system. What specific problems do you anticipate in basing quality assurance on an expert system? To what extent do you see expert systems becoming used, in the near future, in the food industry as a whole?

9

ALCOHOLIC BEVERAGES: III. DISTILLED SPIRITS

OBJECTIVES

After reading this chapter you should understand

- The nature of congeneric and non-congeneric spirits
- The basic principles of distillation
- The technology of production of the major types of distilled spirits
- The nature of non-traditional spirit-based products
- Quality assurance and control
- The role of micro-organisms in determining the character of distilled spirits
- The chemistry of the flavour compounds of the major types of distilled spirits

---

9.1 INTRODUCTION

Distillation is one of the earliest examples of chemical technology. The process was known in China many hundreds of years before the birth of Christ and the first distilled beverage is believed to have been made from rice wine about 800 BC. The secret of distillation remained in China until the first few years AD, when the process was studied in Egypt. The art was learnt by Arabic chemists, who developed the first truly efficient still, the *alembic*. A direct descendent of the *alembic*, the pot still is in use today in the distillation of several spirits, including Scotch whisky.

Distillation was introduced into western Europe from North Africa by the Arabs. The art was of much interest to alchemists and to monks, who applied it to the manufacture of distilled beverages.

European alchemists thought that the distillate was a new element ('water of life') and spirits were considered to have a medicinal value. Spirits were widely used during the various epidemics which afflicted Europe and, in 1506, King James IV of Scotland granted the Guild of Barber Surgeons sole licence to distil *aqua vitae*. The belief in the medicinal powers of distilled spirits persists today, especially with respect to brandy and whisky, and during the later years of prohibition in the US laws were relaxed to permit manufacture of 'prescription whisky'.

BOX 9.1 **King cholera**

During the cholera years of the 19th century, the drinking of spirits, especially gin, was widely considered to be a predisposing factor in contracting the disease. At the same time brandy was amongst the wide range of suggested cures and, as a precautionary measure, a flask containing the spirit was carried by many of the wealthier citizens of London. An enterprising member of the Southwark *demi-monde* found herself able to obtain an unlimited supply of free brandy by the simple expedient of collapsing at appropriate times on Blackfriars Bridge and accepting the medicinal spirit offered by concerned by-passers.

The geographical spread of distilling continued as the European powers established colonies in the Americas. Rum was made in Barbados as early as 1630 and distilling began in North America later in the 17th century. Meanwhile improvements to the design of stills continued and major developments during the 18th century greatly improved the efficiency of distillation of the thick mashes used in Britain and Germany. The development of the continuous 'patent' still by Aeneas Coffey, in Dublin, in 1830 can be considered to be the last major technological innovation in distilling, although the design of continuous stills continues to be refined.

9.2 TECHNOLOGY

9.2.1 Types of distilled beverages

Distilled beverages can be prepared from any ethanol containing material. Prohibition tends to stimulate ingenuity, although many of the resulting drinks would not be considered palatable under

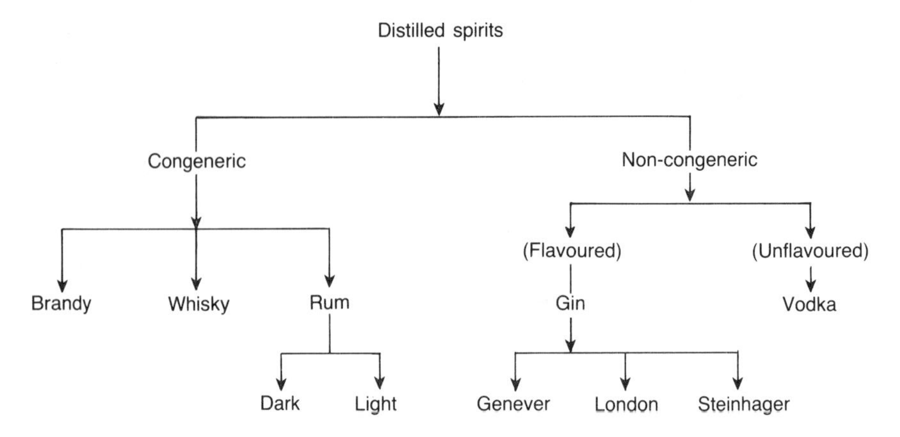

Figure 9.1 The main types of distilled spirit.

normal circumstances. Distilled beverages may be classified according to the starting material or by the type of still. In many cases, however, the most satisfactory approach is to consider two types of spirit. The first, represented by vodka and gin, are rectified and have no requirement for congeners. The raw material is spirit which may be obtained from any source. The second, is the congeneric or self-flavoured type, represented by whisky and rum, where a fermentable starting material is traditional and where conditions of fermentation and distillation are prescribed. Further subdivisions may be made according to the nature of the final consumer product (Figure 9.1).

9.2.2 Principles of distillation

Distillation involves the separation of the components of a solution on the basis of volatility at the boiling point (distillation point). The material to be distilled is a mixture of water, ethanol and other compounds of varying volatility. At one atmosphere total pressure the boiling point is the temperature such that the sum of the partial pressures exerted by each component equals one. Thus in a model ethanol/water system:

$$p_{eth} + p_w = 1$$

where p_{eth} is the partial pressure of ethanol and p_w is the partial pressure of water. (In practice there will be a contribution from other components.)

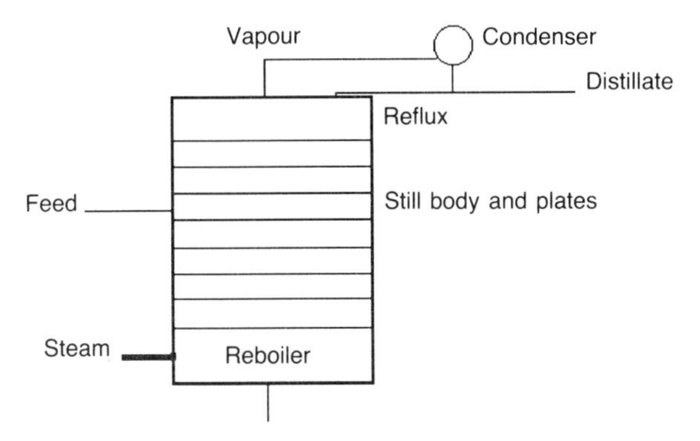

Figure 9.2 Principles of distillation.

In the ethanol/water system:

$$p_{eth} = \gamma_1 x_1 P_{eth}$$
$$p_w = \gamma_2 x_2 P_w = \gamma_2(1-x_1)P_w$$

where γ_1 is the activity coefficient of the more volatile component (ethanol), γ_2 is the activity coefficient of the less volatile component (water), x_1 is the mole fraction in the liquid phase of the more volatile component, x_2 is the mole fraction in the liquid phase of the less volatile component, P_{eth} is the vapour pressure of ethanol and P_w is the vapour pressure of water. The number of molecules of each component is related to its partial pressure by:

$$p_1/P_T = N_1/N_T = y_1$$

where N_T is the total number of moles of vapour, P_T is the total pressure, N_1 is the number of moles of component 1 in the vapour and p_1 is the partial pressure of component 1.

At the boiling point a vapour of specific combination exists in equilibrium with the distilling liquid. As distillation proceeds the vapour becomes richer in the more volatile components and the liquid poorer. It is possible to continuously enrich the vapour with the more volatile component by repeatedly condensing the vapour and partially vapourizing the liquid (Figure 9.2). In practice, this is achieved by use of a fractionating (rectifying) column.

A commonly used type of rectifying column consists of a tower with a series of plates, which may be integrated within the still or positioned separately downstream of the still. The plates are constructed so as to allow vapour to rise into the plates and the liquid

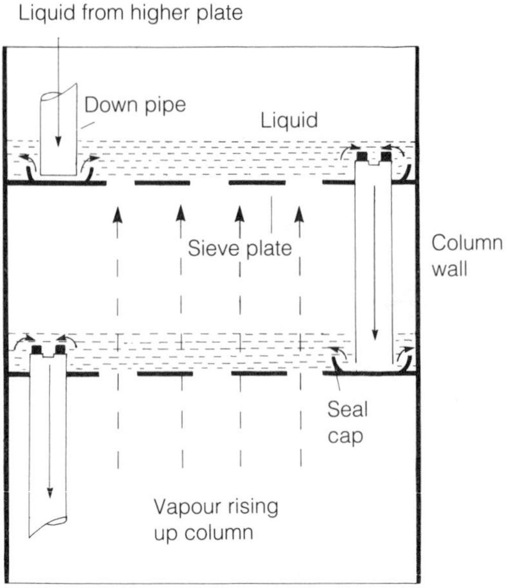

Figure 9.3 Design and operation of rectifying plates.

to fall down from the plates (Figure 9.3). As the liquid falls from above it comes into contact with rising vapour, from which the less volatile components tend to condense. At the same time the more volatile components of the liquid phase tend to be vapourized. Operation of a rectifying column thus makes it possible to obtain a distillate richer in the more volatile components than could be achieved by simple distillation and equilibration alone. Operation may be expressed mathematically as

$$F = D + W$$
$$Fx_F = Dx_D + Wx_W$$

where F is the feed introduced to the column, D is the distillate and W is the waste containing a high proportion of less volatile components. Rectifying columns may be packed with various shaped particles rather than plates, but the operating principle is identical.

9.2.3 Whisky

Whisky may be defined as the potable spirit obtained by distillation of an aqueous infusion of a malted barley and other cereals, that has been fermented with *Saccharomyces cerevisiae*. Whisky is

Table 9.1 The main types of whisky

Scotch malt	
materials	peated, malted barley
distillation	double: pot stills
maturation	charred oak casks, minimum three years
Scotch grain	
materials	maize or wheat and a small quantity of unpeated, malted barley
distillation	continuous still
maturation	used sherry casks, minimum three years
Irish malt	
materials	unmalted barley and unpeated, malted barley
distillation	triple: pot stills
maturation	sherry, or uncharred oak casks, minimum three years
American bourbon	
materials	maize, rye, unpeated, malted barley
distillation	continuous
maturation	at least 1 year
Canadian rye	
materials	rye, maize, unpeated, malted barley or malted rye
distillation	continuous
maturation	variable

now made in many countries, the product differing primarily in the nature and proportion of raw materials used in addition to malted barley. Differences in the final products also arise from the type of still used and the means of maturation (Table 9.1).

The economic importance of whiskies to the major exporting countries, such as Scotland, has resulted in legislation to define and protect the character of the whisky and to prevent imitation. In some cases legislation is extended to define the geographic origin of the whisky and 'Scotch' and 'Irish' whiskies are thus geographic descriptions rather than generic. Legal definitions applied in major producing countries are summarized in Table 9.2.

(a) Scotch whisky

Whisky distilling in Scotland has a long and often turbulent history. Whisky is a major export commodity and of very considerable economic importance to the whole of the UK. Consumption has fallen in recent years as a result of consumer preference for lighter

Table 9.2 Legal definitions of whisky in the main producing countries

Scotland (and Northern Ireland) Obtained by distillation in Scotland (Northern Ireland) from a mash of cereal grain, saccharified by diastase of malt and matured in cask for a period of at least 3 years.
Eire (pot still) Spirits distilled in pot stills in Eire, from a mash of cereal grains normally grown in that country, and saccharified by a diastase of malted barley. Minimum maturation 3 years.
United States 'Rye whiskey', 'bourbon whisky', 'wheat whiskey', 'malt whiskey', potable spirits, distilled at a strength not exceeding 160° proof USA, from a fermented mash, which contained not less than 51%, respectively, of maize, rye, wheat or malted barley grain and stored in charred, new oak containers.

spirits. Scotch whisky is of two types: malt whisky, in which the only cereal is malted barley and which is distilled in a pot still, and grain whisky, which is made from a mash of maize and a small proportion of malted barley and distilled in a continuous still. Malt whisky may be retailed unblended (single malt), in a blend with other malts or blended with grain whisky. Only very small quantities of grain whisky are retailed unblended. Most of the whisky in commerce is a blend of malt and grain whisky, although the best known single malts are considered to be definitive in terms of quality. Malts themselves are classified according to geographic origin as highland, lowland, Campbeltown, Islay or island whiskies.

The preparation of material for distilling malt whisky is essentially similar to brewing an unhopped beer and the material is, indeed, referred to as 'beer' (see Chapter 7). The malt used is of limited enzyme activity to ensure maximum fermentable extract is available in the mash. Malt is now made from single cultivars to avoid processing malts kilned at different stages of modification.

* Intensive plant breeding programmes have been established to develop improved cultivars for production of distillers' malt. A typical high yielding cultivar is Triumph, which produces a distillery yield of *ca.* 405 l/tonne of malt. This is a very significant improvement over the yield of the previous dominant cultivar, Golden Promise, which produced *ca.* 390 l/tonne of malt. (Dolan, T.C.S. 1987. *Food Technology International Europe*, 150–2).

The use of peat smoke in drying the malt is important in determining the Scotch whisky character. In traditional practice malting was conducted at the distillery, the malt being floor dried over coke and peat fires. Peat was used as fuel in the early part of drying so that the moist, green malt absorbed the peat flavour (the 'reek'). It is now more common for malting to be conducted on a larger-scale, centralized basis. The process is highly mechanized, the peat flavour being obtained by use of a secondary peat-fired drying kiln. The amount of peat used varies and can be an important factor in determining the character of the whisky. Many Islay whiskies, for example, are very heavily peated.

Wort preparation usually involves a decoction process, in which coarsely ground grist is mixed with hot water ('liquor') at 60–65°C. The resulting wort is then cooled to *ca.* 25°C and filtered before passing to the fermentation vessel. The mashing process is then repeated three to four times, the resulting wort being used as liquor during subsequent mashings. The process differs from brewery practice in that the wort is not boiled and thus the fermentation is not carried out under 'aseptic' conditions. The continuing activity of limit dextrinases in the unboiled wort also increases the quantity of sugars available for fermentation.

Traditional fermentation vessels are made of larch or Oregon Pine. Stainless steel has now largely replaced wood, although in some cases stainless steel vessels have been fitted with wooden liners. Strains of *Sacch. cerevisiae* used for distillery fermentations are chosen for high ethanol tolerance (12–15%) and for ability to hydrolyse oligosaccharides, such as maltotriose and maltodextranose, to glucose and thus maximize conversion of starch to ethanol. Only the largest distilleries have facilities for propagating yeast and in most cases the inoculum consists of dried, pressed yeast, supplemented with surplus brewery yeast supplied as a suspension in beer. In the past an inoculum level of 5×10^6 to 2×10^7 cells/ml was used, but in modern practice a higher level of *ca.* 10^8 cells/ml is used. This gives maximum yeast growth under anaerobic conditions.

Lactic acid bacteria are also involved in whisky fermentations. In Scotch whisky, these are not deliberately introduced, but are derived from the grain and the distillery environment. Species of *Leuconostoc* are usually present in greatest numbers, accompanied by *Lactobacillus brevis*, *Lb. casei*, *Lb. collinoides*, sub-species of *Lb. delbrueckii*, *Lb. fermentum*, *Lb. plantarum*, *Pediococcus damnosus* and sub-species of *Streptococcus lactis*. Growth of lactic acid bacteria occurs late in fermentation and is enhanced by nitrogenous compounds excreted by the yeast. There is also some removal of citric and malic acids produced during yeast fermentation. This 'late fermentation' is generally considered to improve the quality of the final whisky. Poor standards of hygiene, however, lead to excessive numbers of lactic acid bacteria in the wort and growth early in the fermentation. This results in an unacceptable reduction of ethanol yield, the possibility of over-acidification and the production of undesirable flavour compounds. The most important of these is hydrogen sulphide, although problems may arise due to metabolism of glycerol to β-hydroxypropionaldehyde. This is degraded to acrolein during distillation and imparts a pungent, peppery aroma to the whisky. Problems due to excessive growth of lactic acid bacteria have become less common in recent years following improvements in hygiene standards and the use of higher yeast inoculum levels.

The course of distillery fermentations is generally the same as that in breweries (see Chapter 7, pages 319–20). Until recently, however, the control of fermentation, especially in small distilleries was primitive. In the absence of effective cooling facilities, temperature rise resulting from sugar hydrolysis could lead to 'stuck' fermentations and necessitated the use of low gravity worts and low initial fermentation temperatures. Fermentation control has been considerably improved in most distilleries with resultant improvements in consistency, yield of ethanol and reduction of the incidence of faults in the final whisky.

Malt whisky is distilled in traditional pot stills. The basic design has changed little for centuries (Figure 9.4), although internal heating by steam coils is now common rather than the external furnace previously used.

This results in a significant reduction in the extent of pyrolysis of the still contents. The traditional means of condensing, the 'worm',

* Wort for whisky production contains a significant quantity of suspended solids. The significance of this for the fermentation has not been studied in detail. It is known, however, that the growth rate of *Sacch. cerevisiae* is increased and that excretion of ethanol and glycerol is enhanced. There is also a marked increase in production of isobutanol and 2-methylbutanol and a general increase, with the possible exception of *n*-propanol, in production of other higher alcohols.

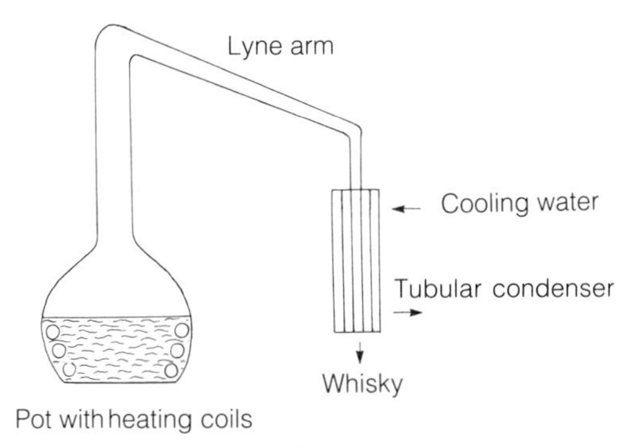

Figure 9.4 Basic design of a pot still.

has also been replaced by tubular heat exchangers, which are more effective and of higher thermal efficiency. Various modifications exist to the basic design, including the fitting of water jackets to the heating vessel and return loops from the first stage of condensation. Some distilleries fit a 'purifier', a circular vessel cooled by running water, which is interposed between the neck of the still and the condenser. This removes some of the less volatile components of the vapour. Some undesirable sulphur-compounds are removed by fixation to the copper walls of the still and, for this reason, the use of copper as a constructional material continues.

Scotch whisky is distilled twice (Figure 9.5). The first is a non-selective process, which takes 5–6 h. A three-fold concentration occurs, the distillate being referred to as 'low wine'. The residue

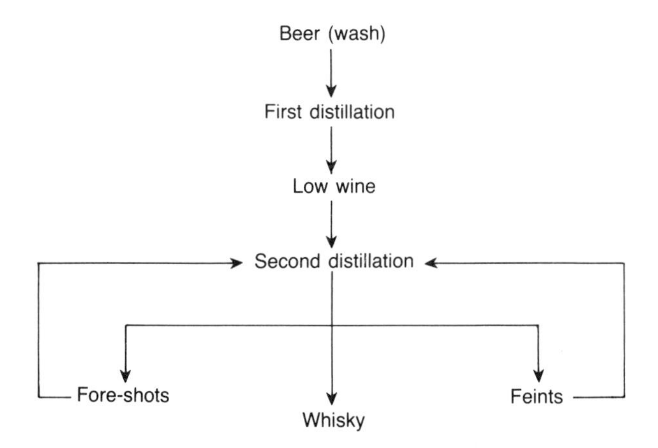

Figure 9.5 Distillation of Scotch whisky.

('pot ale') is now commonly converted to animal food. The second stage of distillation is selective and requires a high level of control and operator expertise. The first distillate fraction consists of low boiling point components (the 'fore-shots'), which are recycled. Whisky is collected as the second fraction, when the distillate is 25–30° overproof as determined by a hydrometer. Choosing the right time to commence and cease collection of whisky is the responsibility of the distiller, the appearance of a blue tinge when the distillate is added to water is used as an empirical aid. After termination of whisky collection, the distillate is known as the 'feints', which are recycled. Distillation of the feints continues until all alcohol is removed from the low wine. The residue in the still, the 'spent lees', are rejected or used as animal food.

Whisky leaving the still enters a spirit safe, which is sealed under the control of the Excise authorities. The raw spirit is then diluted with water and filled into charred oak casks for maturation. The period of maturation is usually much longer than the legal minimum of 3 years. During maturation, a number of chemical changes take place (see pages 436–8) and there is loss, by evaporation, of liquid. In Scotland, casks are traditionally stored in unheated warehouses with a high air humidity. The relative loss of ethanol under these conditions is higher than that of water and the ethanol content falls during maturation. The chemical changes occurring during maturation are now better understood and there is now a trend to control conditions during maturation to ensure that desirable compounds are formed.

Production of grain whisky involves the use of unmalted cereals as the major source of fermentable carbohydrate. In the past maize has been used but, during the 1980s, wide fluctuations in the price of maize led to some distillers using wheat flour. Use of wheat flour requires modifications to the manufacturing process to avoid unacceptable differences in taste.

Both wheat and maize starches have a high gelatinization temperature and are usually precooked before incorporation in the mash. Batch cooking, under pressure, is still used by smaller distillers and typically involves heating for 1–1.5 h to a temperature of 120°C and a pressure of *ca.* 1.5×10^5 Pa. The starch is then held at 120°C for a further 1.5 h before being blown direct into the mash tun containing the malted barley. Cold water is added to lower the temperature to 60–65°C.

Various types of continuous cooker are available. A commonly used process involves forming a starch/water slurry at *ca.* 50°C. The slurry is pumped through stainless steel tubes, where it is heated by direct steam injection to *ca.* 150°C, at a pressure of 4.5×10^5 Pa for 10 min. The slurry is then vacuum cooled. Efficient cooling is essential for both batch and continuous cooking to prevent the mash solidifying. It is also necessary to minimize the time at *ca.* 82°C, at which temperature lipid-carbohydrate complexes are formed. These complexes are not available for fermentation and formation leads to a loss of up to 2% in spirit yield.

Malt barley is present in the mash at a level of 10-15%. This level is higher than that required for provision of diastatic enzymes and is used to provide flavour. In recent years, green (unkilned) malt has become widely used and permits high spirit yields at relatively low levels.

The mashing and fermentation processes used for grain whisky are very similar to those for malt whisky, although the scale of production is usually larger. Lactic acid bacteria are considered to be of less importance for flavour and are suppressed as far as possible to maximize ethanol yield. Formation of α-acetolactate (ALA) during fermentation results in a serious defect in the final spirit due to conversion of ALA to diacetyl during distillation. Production of ALA can be minimized by the use of higher fermentation temperatures. A late reduction in the pH value is also beneficial since spontaneous decarboxylation of ALA proceeds rapidly at pH 3.5. If necessary, conversion of ALA to diacetyl can be minimized by the use of inert gases to protect the wash from oxygen.

Grain whisky is produced by continuous distillation. Many distilleries still use the relatively simple Coffey still, although four and five column stills similar to those used for neutral spirits (see pages 426-8) have been introduced to Scotland. The Coffey still has changed little since its introduction, with the exception of a much higher level of instrumentation and automation. The still consists of two main components, the beer still ('analyser'), in which the ethanol is stripped from the beer, and the rectifier (Figure 9.6), in which the undesirable volatiles are removed from the ethanol. Beer is prewarmed by rising vapour in the coiled pipe which passes through the rectifying column and is then sprayed into the top of the beer still. Precise control is required to optimize the ethanol content of the spirit drawn off the spirit plate in the rectifier. The

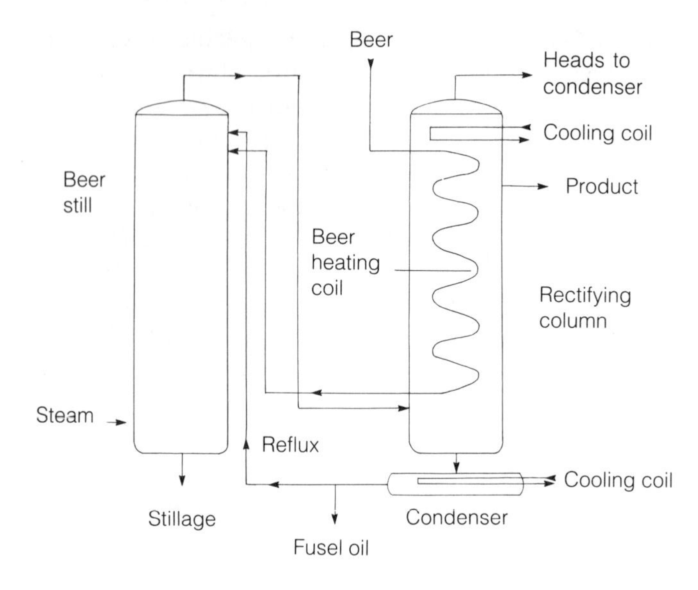

Figure 9.6 Basic design of a Coffey-type continuous still.

ethanol content should not exceed 94.17% at 20°C, higher or lower concentrations affect the proportion of congeners and have an adverse effect on quality of the final product.

Grain spirit is matured in used sherry casks and is used for blending with malt whisky. In the case of single malts, it is only necessary to adjust the ethanol content with water and add caramel to produce the desired depth of colour before filtration and bottling. Water with a low content of dissolved minerals, or which has been demineralized, must be used to avoid precipitates within the whisky.

Blending involves a very large degree of expert judgement. A typical blended Scotch whisky usually contains 30–40% malt whisky and comprises 20–30 individual types of malt and grain whisky. After blending, the whisky is re-filled into used casks, either with or without adjustment of the ethanol content, and stored for a further 6 months to 'marry' the flavours. The colour is adjusted with caramel and the whisky clarified by filtration. Pretreatment with charcoal may also be used. Asbestos-containing filter media have largely been replaced by cellulosic materials and chill-filtration is now widely used to prevent instability due to tannins. Glass bottles remain in general use for consumer products, but polyethyleneterephthalate (PET) bottles are being introduced for use in catering establishments and for special purposes, such as 'miniature' bottles for use on aircraft.

(b) Irish whiskey

Irish whiskey is also of two types, malt and grain. Manufacture of grain whiskey in Ireland uses methods which are effectively identical to those used in Scotland.

> BOX 9.2 **History is past politics**
>
> In Northern Ireland, politics and religion are closely intertwined and, sadly, never far from every day life. Even choice of whiskey has political and religious connotations. 'Bushmills' being favoured by Protestants and 'Jamiesons' by Catholics. Both brands are, in fact, produced by the same group of companies.

Manufacture of Irish malt whiskey is also very similar to Scotch malt whisky, although there are some significant differences. Irish whiskey is made from a mash containing a high proportion (up to 60%) of unmalted barley. The remainder of the cereal, malted barley, is not smoked over a peat fire and must have a high level of enzymes. Methods of wort production and fermentation are generally similar to those used in Scotland. The use of unmalted barley, however, requires the use of stone or hammer mills to obtain the correct degree of grinding. Unmalted barley is usually 'sweated' to raise the moisture content to *ca.* 14% and then dried to 4.5% moisture before grinding, but these procedures can be eliminated by use of wet milling.

Irish malt whiskey is distilled in a pot still, which differs from its Scottish equivalent in having a distillation vessel which is a flattened sphere in shape. The Irish pot still also has a longer lyne arm, which is cooled by water in a surrounding tank, the condensate being returned to the distillation vessel. Irish whiskey is distilled three times (Figure 9.7). The first distillation in the 'wash' still produces two fractions. The first of these, 'strong low wines' is collected for transfer direct to the final 'spirit' still, while the second, 'weak low wines', is passed to an intermediate 'low wines still'. This still also produces two fractions, 'strong feints', which

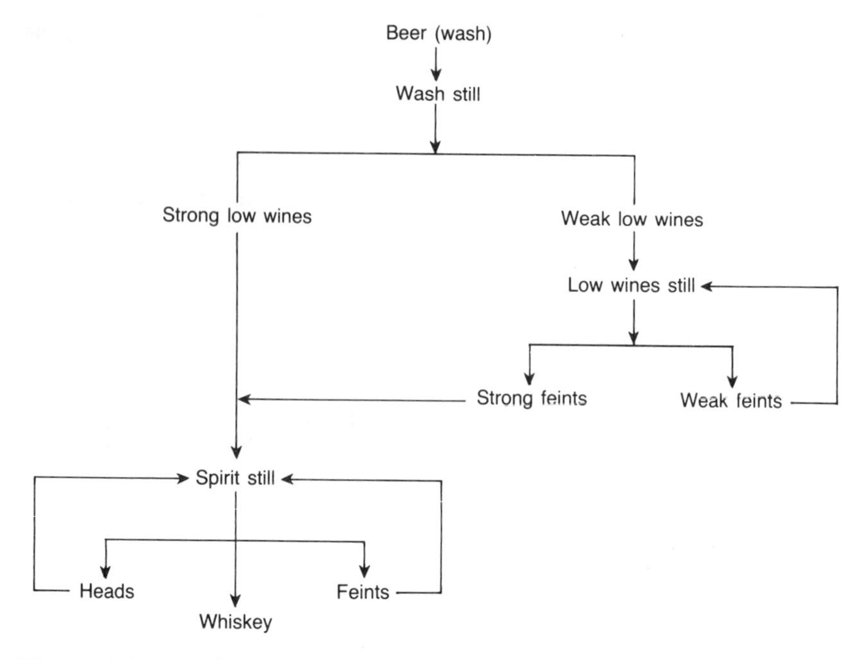

Figure 9.7 Distillation of Irish whiskey.

are mixed with the strong low wines and passed to the spirit still and 'weak feints', which are recycled. Whiskey is drawn as the second of three fractions from the spirit still, the first ('heads') and third ('feints') being recycled.

Irish whiskey lacks the peat characteristic of Scotch whisky, but is more strongly flavoured and has a heavier body. Irish whiskey is also of higher ethanol content. In recent years a lighter whiskey, more closely resembling Scotch whisky, has been produced in response to consumer preference for lighter spirits.

(c) North American whiskey

Although a number of types of whiskey are produced in the US two types, Bourbon and rye, are dominant. A similar situation exists in Canada. Bourbon whiskey is of greatest importance and a large number of brands are available. Each has its own characteristic flavour and considerable care must be taken during production to ensure consistency of the end product. Some brands are defined as 'sour mash' and must meet a number of legal requirements during production (see page 416).

BOX 9.3 **Only in America**

American distillers protect the identity of their brands very strongly and legal action is taken against attempts to market whiskies with similar names. This is quite understandable since passing-off substitutes can have severe commercial effects. The attempt by the distillers of Jim Beam™ whiskey to prevent the marketing of the Irish Beamish™ stout does, however, appear extreme, especially since the Beamish family have been plying the brewer's trade in Cork since the 17th century. It is almost inconceivable that the products could be confused, although the lawyers will, no doubt, grow rich.

Bourbon whiskey is made from maize, unpeated malted barley and another cereal, usually rye. As little as 51% maize may be used, but 60–70% is more common. Traditionally made malt is used at a level of *ca*. 15%, but malt made using gibberellic acid, which has twice the α-amylase content, is permitted and used at lower levels. In this case glucoamylase from *Aspergillus niger* may also be added. The choice of third cereal, the 'small grains' depends on the desired flavour of the matured whiskey. Although rye is most common, barley and wheat are also used and each impart a unique flavour to the distillate.

Grain is prepared for cooking and mashing by grinding in a hammer mill. The degree of grinding represents a compromise between a fine grind, which has a high fermentation efficiency, and a coarse grind, which aids recovery of economically important, non-fermentable by-products. Further preparation of the grains and malt varies according to the desired product characteristics. The usual practice is to prepare a slurry of the milled grains using dealcoholized beer ('backset stillage') from a previous production cycle. A portion of the malt, usually 1–5%, is added at this stage ('premalting'). This promotes liquefaction of the starch through enzyme activity during the earlier stages of cooking and improves the efficiency of cooking. There is also degradation of proteins to peptides and amino acids. Both batch and continuous cooking are in use and some distilleries continue to use cooking at atmospheric pressure at *ca*. 100°C, which is considered to produce the highest quality product. After cooking the grains are cooled to 63°C and the remaining malt ('conversion malt') added. After transfer to the fermenting vessel, yeast is added along with water and backset stillage. Filtration of the mash is not practised in the manufacture of Bourbon whiskey.

Yeast strains used in Bourbon production are selected for production of congeners involved in producing flavour characteristics rather than a high ethanol yield. Strains must, however, be capable of producing at least 6% ethanol in a medium containing 11% starch. Strains are initially selected to meet the requirements of the distillery and maintained on agar slopes. The pitching inoculum is prepared by sequential inoculation of increasing quantities of mash. Inoculation is at 2% to inhibit undesirable bacteria.

Historically, fermentation vessels for Bourbon whiskey were made of woods, such as cypress, redwood and larch, but stainless steel is now preferred. It is recognized that the use of stainless steel results in a slightly different taste, but this has no effect on consumer preference. Where a 72 h fermentation is used, it is usual to control the temperature at a maximum of 32–35°C. This ensures a high quality product by maintaining yeast activity throughout the fermentation period.

Lactic acid bacteria, especially *Lactobacillus* and *Pediococcus*, are of importance in bourbon whiskey fermentations, although the role is secondary to that of yeast. In many cases lactic acid bacteria are derived from malt and other raw materials. A typical development pattern suggests that a population of 10^2 cfu/ml in the mash entering the fermentation vessel, develops to 10^3 cfu/ml in the young beer and 10^8 cfu/ml at the end of fermentation. In some cases indigenous lactic acid bacteria are not considered sufficiently reliable and the mash is inoculated with single, or mixed, strain cultures of lactic acid bacteria with desirable properties.

---

* Bourbon whiskey described as 'sour mash' must have a minimum of 20% backset stillage in the fermenting vessel. In addition, the yeast must be propagated in soured yeast mash and the fermentation continued for 72 h. Some distilleries continue to rely on souring by indigenous lactic acid bacteria, but in most cases a mash of similar composition to the main mash is soured to a pH value of 3.8 to 4.0 by a pure culture of *Lactobacillus delbrueckii* ssp. *delbrueckii*. This organism may be cultured at 50°C, is a fast acid producer and produces no undesirable flavours. After sufficient acid is produced the yeast mash is pasteurized at 80°C and cooled before inoculation. This method of propagation is used even where the whiskey is not the sour mash type as a means of controlling undesirable indigenous bacteria.

At the termination of fermentation, the beer is drawn off and passed to the stills. Aeration is minimized to eliminate, as far as possible, aldehyde formation. A number of types of still are used, the vast majority of which are continuous in operation. Distillation columns are widely used, usually in combination with a 'doubler'. The distillation column consists of a vertical shell containing a series of sieve plates arranged in three sections: stripping, entrainment removal and rectifying. Alcohol and congeners are removed from the beer in the lowest stripping section, the vapour passing to the entrainment removal section where particulate matter is removed. Refinement takes place in the rectifying section. A copper mesh 'demister' is sometimes fitted above the rectifying section to remove volatile sulphur compounds from the vapour. Some of the vapour is condensed in the first 'dephlegmator' condenser, the remainder passing to the second 'vent' condenser. Most of the distillate from the dephlegmator condenser is collected for transfer to the doubler, while the remainder, together with distillate from the vent condenser is recycled to the rectifying section of the still.

The doubler is a simple pot still fitted with a copper condenser. The distillate from the distillation column is redistilled and some higher boiling point congeners removed. This imparts a cleaner flavour to the final matured product. The aldehyde content is also reduced by selectively removing the first, 'head' fraction of the distillate.

More complex, five-column stills have come into wider use. Many types are available, but the basic design is similar to that used in production of neutral spirits (pages 426–7). There are, however, important differences in operation, in that selectivity is set to retain desirable congeners.

Bourbon whiskey is matured in charred, white oak barrels, the optimum temperature for maturation being 21–30°C. The rate of maturation can vary considerably and frequent checks must be made. Bourbon whiskey may be retailed unblended or blended with other bourbons or other types of whiskey. The blending process is basically the same as that used for Scotch whisky (see page 412). Bourbon whiskey may have a high solids content and requires a multi-stage filtration before bottling. This includes chill filtration at −9°C.

Most other whiskies produced in North America are similar to Bourbon. American rye whiskey, for example, is made from a mash produced by atmospheric cooking at *ca.* 100°C or by infusion mashing at *ca.* 70°C. Corn whiskey differs from Bourbon in ageing procedures, being matured either in aged Bourbon barrels or new uncharred oak barrels. Each of these whiskies is widely used in blends. Canadian whiskey is generally similar to that made in the US, a common blend consisting of 10–20% bourbon- or rye-type whiskey and grain whiskey. In both countries a 'light whiskey' is produced, which is made using a distillation process similar to that used for neutral spirits.

9.2.4 Rum

Rum is the distilled alcoholic beverage made from sugar cane. Rum has been known for centuries and is made wherever the climate is suitable for sugar cane cultivation. Production is centred, however, on the West Indies.

> BOX 9.4 **A gunpowder plot**
>
> Rum is, according to tradition, the favoured drink of the lower ranks of the Royal Navy. The daily rum allowance was continued until recent years and the naval connection is reflected in the brand names of many British rums. The buccaneers of the West Indies also had a taste for rum and those who 'sailed on the plundering account' commonly drank a mixture of rum and gunpowder before embarking on their adventures. A mixture of rum, gunpowder and graveyard soil is also allegedly consumed as part of voodoo initiation rites in Haiti.

The most commonly used raw materials are cane juice, syrup and molasses. Cane juice is most suitable for manufacture of light rum and is prepared by pressing finely ground sugar cane. Cane juice can be fermented without further processing or, for higher quality rums, after heating and clarifying. Alternatively corn syrup may be produced by vacuum concentration of cane juice. Molasses are used for production of rum with heavy aromas. Molasses is the mother liquor remaining after the separation of sugar from cane by repeated crystallization. The by-product status of molasses means that prices are relatively low and the product is of good keeping quality. Depending on the variety of cane, the growing climate and the means of production, molasses contains 50–60% sugar. Molasses contains a number of compounds which contribute to the flavour of rum (see page 441) and also may contain inhibitors of yeast. Inhibitory substances include fatty acids and, in over-heated molasses, hydroxymethylfurfural.

Molasses requires pretreatment before fermentation. This varies somewhat according to the type of molasses, but involves an initial clarification, which is necessary to prevent blocking of distillation equipment. The first stage of clarification involves precipitation with alumina and calcium phosphate, or addition of sulphuric acid. Sulphuric acid pretreatment may, however, significantly lower the sugar content. Final impurities are then removed by centrifugation. Molasses often carries a high microbial load and it is usual to apply a heat treatment equivalent to pasteurization. The molasses are then diluted with water to lower the viscosity and standardize the sugar content at 10–12 g/100 ml, the pH value adjusted to *ca.* 5.5 and ammonium sulphate or urea added as yeast nutrient.

In the manufacture of the heavy, fruity, dark rums typical of Jamaica, it is common practice to fortify the diluted molasses with 'dunder'; the lees from a previous distillation which have been aged and ripened by indigenous micro-organisms.

Yeasts for rum fermentations must be able to produce a high ethanol content and the correct aroma content. In small-scale production, especially in Jamaica, it is still common practice to rely on 'spontaneous fermentation', the resulting rums having a highly characteristic and individual flavour. In other cases pure cultures of yeasts are used. Fast-fermenting strains of *Sacch. cerevisiae* are preferred for light rums, but *Schizosaccharomyces pombe* is usually chosen for heavy rums. Molasses is the best source of yeast and can contain a wide range of genera including *Candida*, *Hansenula*, *Kloeckera*, *Pichia*, *Saccharomyces*, *Saccharomycodes*, *Schizosaccharomyces* and *Torulopsis*. Pure cultures of yeast are propagated in molasses and inoculated at a high level of 4–10%.

Fermentation time and temperature varies considerably. Light rum fermentations using *Sacch. cerevisiae* are usually at a temperature of 28–33°C for 28–36 h. Heavy rum fermentations using *Schizosacch. pombe* are usually at 30–33°C, although temperatures as high as 37°C may be used. Where bacteria play an important role in the fermentation, temperature should not exceed 30°C. Fermentations with *Schizosacch. pombe* tend to be relatively slow and times may be 72 h or longer. Fermentations lasting as long as 12 days have been described when the wash contains a high proportion of dunder and is of high glucose content.

Bacteria are also important in rum fermentations and butyric acid bacteria, such as *Clostridium butyricum*, *Cl. pasteurianum* and *Cl. saccharobutyricum* are important in production of volatile acids in heavy rum. There may also be synergistic relationships with yeast.

Butyric acid producing clostridia are most effective when added as pure cultures. The pitching rate is usually *ca.* 20% of that used for yeast, the clostridia being added to the fermentation when the ethanol content is *ca.* 4.0% and the sugar concentration below 6 g/100 ml wash. An alternative procedure is to preferment a portion (15–30%) of the wash with pure cultures of clostridia.

The role of other bacteria in rum fermentations remains obscure. Lactic acid bacteria are present and can cause problems by lowering the pH value of the wash below the optimal 5.5–5.8. Species of *Leuconostoc* are often the dominant lactic acid bacteria and can cause problems due to slime formation. Acetic acid bacteria, predominantly *Acetobacter* are also present. Such bacteria are usually the causes of spoilage of alcoholic beverages, but may be beneficial in light rum where acetic acid plays a role in production of the final flavour. Acetic acid production does, however, lead to a reduction in ethanol yield.

At the end of fermentation, the yeast cells are settled out and the wash clarified by centrifugation. Pot stills are generally used for distillation of heavy rum in English- and French-speaking areas, while continuous stills are normal for light rums.

---

* Many species of *Clostridium* ferment carbohydrates to acetic and butyric acids, CO_2 and H_2. The initial stage of the butyric fermentation involves production of pyruvate from sugars *via* the Embden–Meyerhof pathway. Pyruvate is subsequently oxidized to acetyl-CoA and CO_2, with ferredoxin as electron acceptor. Hydrogen is then produced during the reoxidation of ferredoxin by hydrogenase. This system is usually known as the 'clastic' system. In some cases neutral products, including butanol and acetone, are produced through a modification of the normal pathway of butyric acid formation. In the past, *Cl. acetobutylicum* was used for the industrial manufacture of acetone.

In small-scale production for local consumption a single pot still may be used. This produces a very crude rum which is usually drunk after dilution with lime juice. Commercial-scale distilling involves distillation in three successive stills; the wash still, the low wine still and the high wine still. The final distillation is selective, rum being taken from the second fraction, the first and third fractions being recycled. A higher level of purification may be obtained by fitting a rectifying column after the final high wine still. Rectifying columns are not, however, in general use in Jamaica.

Continuous stills of traditional design usually have three columns. In the first, the 'beer' or 'exhaustive' column, alcohol is stripped from the wash. The vapour is condensed and part of the distillate recycled. The remaining distillate is then diluted with water to 20–50% alcohol and passed to the second 'purifying' column. This column removes low boiling point volatiles, the water/alcohol mixture being taken off the bottom of the column. The third stage involves transfer to the rectifying column, where the mixture is reconcentrated to the desired ethanol content and further purification attained by removal of highly volatile components and fusel oils.

Three column continuous stills remain in use for distilling light rum, but an important modification is being widely adopted. This involves conversion of the purifying column to an 'extractive' column by introducing water at the head and permits production of a very high purity rum.

Heavy rum is matured in oak casks, which are usually charred. Maturation periods of 10–12 years are common and may be as long as 15 years. Before bottling, the ethanol content is standardized and the colour adjusted with caramel. In a few cases, spices and, less commonly, fruit juice is added. Light rums are matured in uncharred oak casks. There is usually little change during maturation and periods are short. Depending on the desired colour, caramel may be added. In the case of colourless (white) rum, filtration through charcoal is often used to remove any colour and also to 'polish' the flavour. Products of this nature form a bridge between the congeneric spirits and the non-congeneric spirits, such as vodka.

9.2.5 Brandy

Brandy is the spirit distilled from wine, while the very similar calvados is distilled from cider. Brandy is produced in all viticultural regions and spirits can also be distilled from fruit wine. In terms of quality, the best known brandies are cognac and, to a lesser extent, armagnac. Both of these brandies are produced by distillation of white wine from geographically defined regions of France.

The characteristics of different brandies are defined by soil, climate, variety of grape, management practice during cultivation, vinification and storage of wine, distillation and maturation. Historically, only white wine was used for brandy manufacture, but red wine has been used in some areas for many years. Base wine-making procedures are restricted to the minimum necessary for production of a suitable wine for distilling. The must is not clarified and, for a number of reasons (Table 9.3), SO_2 is never added during base wine manufacture. Musts should have a low content of sugar and tannins and a high content of malic acid to yield wines of high acidity and relatively low ethanol content. In traditional wine-producing areas, base wine fermentation involves naturally occurring yeasts. Species of *Saccharomyces* are of greatest importance, but *Saccharomyc. ludwigii* can also be present in large numbers. In non-traditional wine making areas, increasing use is being made of pure cultures of *Sacch. cerevisiae*. In many cases a malo-lactic fermentation (MLF) follows the alcoholic fermentation. Although a variety of lactic acid bacteria can be isolated from the must and new wine, *Leuconostoc oenos* is responsible for the MLF and comprises virtually 100% of the lactic acid bacterial microflora at the end of this fermentation. A population of *ca.* 10^8 cfu/ml of *Leuc. oenos* is attained by the end of the MLF (usually 8–10 days after completion of the alcoholic fermentation), but this falls to *ca.* 10^6 cfu/ml before distillation. For many years the MLF was thought to have an adverse effect on the quality of the brandy. Opinions

Table 9.3 Problems caused by the presence of SO_2 in base wine for brandy manufacture

1. pH value may be lowered to an unacceptable level by sulphuric acid production
2. The presence of SO_2 favours formation of acetal by reaction between ethanol and acetaldehyde
3. The presence of SO_2 favours formation of acetaldehyde during fermentation and impairs the aromatic quality of the brandy
4. Damage to the still can result from the formation of a corrosive acid sulphonate by reaction between SO_2 and acetaldehyde

have changed somewhat and a number of distillers consider that the production, during distillation, of ethyl lactate from lactic acid formed during the MLF is beneficial in conferring 'subtlety'. It is essential, however, that the MLF is completed before distillation, otherwise a brandy of poor aroma is produced.

Base wine is stored on the lees prior to distillation and the presence of yeast is beneficial in providing a reducing atmosphere and thus protecting against oxidative deterioration. The acidity of the wine protects against bacterial spoilage, but a storage temperature of 0–5°C is necessary to prevent metabolism of glycerol to acetic acid and subsequent acrolein formation. Acetic acid bacteria can be isolated from base wine during storage, but acetification is not a problem in practice. A surface film containing yeasts, such as *Candida*, *Hansenula* and *Pichia*, may develop in the later stages of fermentation or storage and can detract from the bouquet of the brandy.

Grape residue and other heavy particulate material is removed from the base wine directly before distillation, but the yeast cells remain. The design of stills vary somewhat according to the producing region. Pot stills are used for distilling cognac brandy. A two-stage distillation is used. The first is non-selective and produces a spirit of 28% ethanol content, the *brouillis*. This is redistilled in a selective distillation to yield three fractions. The first fraction, the 'heads', contains much of the acetaldehyde present and is discarded, the second fraction drawn is the brandy and the third fraction is recycled. Distillation of cognac is gentle to avoid entrainment of particulate matter. The presence of yeast is considered beneficial and fatty acids and endocellular esters released during distillation contribute to the character of the brandy. Excessive

* Until recently there has been no alternative to distillation as a means of producing spirits. The feasibility of the recently developed pervaporation process has, however, been demonstrated for brandy production. Pervaporation is a separative technique using a synthetic membrane. The process may be defined as a physical method for separation of the volatile components of the volatile constituents of a mixture of liquids, by evaporation through a dense membrane, the downstream face of which is under low pressure obtained by application of a vacuum. The selection of molecules is governed by chemical affinity between the membrane and the constituents. Brandy produced using pervaporation is of low furfural and long carbon chain ester content. The development of improved membranes must, however, take place before commercial application may be contemplated. (Moutonet, M. *et al.* 1992. *Food Science and Technology*, **25**, 71–3).

quantities of fatty acids are deleterious, but substantial removal results from formation of salts with the copper walls of the stills. For this reason attempts to replace copper with stainless steel and other materials have met with limited success, although stainless steel stills are used for production of some low quality brandy.

Armagnac brandy is either distilled by a double distillation in pot stills, or by continuous distillation in a two- or three-column still.

Brandy is matured in new oak barrels, there being some variation according to the producing area. The wood used and its treatment is considered to be of considerable importance in determining the character of the brandy. In the case of Cognac brandy, oak wood comes from Limousin or the forest of Troncaus. Trees have a minimum age of 40 years and staves are split, not sawn. The wood is aged by exposure to air for long periods, ageing being required to eliminate the more astringent tannins and for the partial oxidation of polyphenols. Moulds develop which partially degrade lignin and darken the wood. After the initial maturation the brandy is transferred to used barrels to avoid excessive enrichment with tannins. Barrels used for maturing very old brandies are made from hard wood with very dense fibres. Armagnac brandy is also aged in new oak barrels, but peduncular oak, which is richer in tannins, is used. Very considerable losses due to evaporation occur, especially in the first 12 months of maturation.

9.2.6 Gin, vodka and related spirits

Gin and vodka, although technically and organoleptically distinct, typify non-congeneric spirits. Such products are based on neutral spirits, choice of which involves a balance between quality and economics. In contrast to congeneric spirits, such as rum and whisky, maturation usually plays no part in development of the final product characteristics. In the case of gin, the spirit is flavoured with juniper and other 'botanicals', while with vodka, the flavour is modified by filtration through charcoal. Dutch gin (genever), however, owes its character to strongly congeneric base spirits and, like the German spirits, korn and doppelkorn, undergoes a short maturation period and is to some extent self-flavoured. Spirits such as genever and korn, together with some types of white rum (see page 421), represent a bridge between congeneric and non-congeneric spirits.

(a) Production of neutral spirits

Virtually any carbohydrate containing material can be used in the production of neutral spirits. Grain is probably most widely used, but there is also extensive production of spirits from potatoes and molasses. The choice of fermentation substrate usually reflects the availability of raw materials and the overall local economic situation. Waste sulphite liquors from the wood pulp industry, for example, are used as a substrate for neutral spirit production in Finland. Other unusual substrates include pineapple waste, artichokes and whey. Alcohol is also produced from whey by a relatively new process involving *Kluyveromyces lactis*. Economic factors dictate, however, that the driving force for development comes from industrial uses for alcohol, including fuel (gasohol), rather from use in potable spirits.

In Europe and the US, wheat and maize are the main substrates for gin and vodka production. Rye is also used for vodka production in central Europe and for production of korn and similar beverages. Potatoes are widely used as a source of neutral spirits for beverages such as aquavit and, in Poland and parts of the Russian Republic, for vodka. Many distilleries in central Europe are equipped to handle both grain and potatoes, depending on which is economically most favourable.

Means of preparing the fermentation mash obviously vary according to the nature of the raw materials. Infusion mashing is possible with small grains, such as rye, but cooking is required for maize and some other cereals. Either batch or continuous cookers can be used and premalting is common practice, especially in the US (see page 415).

Malt is traditionally used for conversion of the starch to sugars, but has no role in flavour. The minimum possible is used and green, unkilned malt of high enzyme activity is chosen. Continuous cooking processes can be extended to include conversion. This involves cooling the cooked grain, adding a malt slurry and

* The need for premalting can be avoided by use of a jet heater which raises the temperature of the starch above the gelation point very rapidly. Holding time is controlled by the plug flow of the mash through the heater. The cooked mash is cooled by vacuum cooling where heating is at atmospheric pressure or by flash cooing where heating is at high pressure.

blending, before passage to a conversion tube. A residence time of *ca.* 10 min is sufficient for amylolysis to approach equilibrium. The mash is then cooled to *ca.* 27°C and transferred to the fermentation vessel. Industrial enzymes are now widely used in conversion and can directly replace malt. The most widely used enzymes are a heat-stable α-amylase from *Bacillus subtilis* and an amyloglucosidase from *Aspergillus niger*. The most effective usage is addition of the α-amylase at 83°C followed by the amyloglucosidase at 55–60°C. Thermostable α-amylase is sensitive to low pH values and, if backset stillage is added to the mash, addition should be made when α-amylase action is largely completed. Alternatively a pH-stable, fungal α-amylase should be used at *ca.* 60°C.

Potatoes also require cooking, batch pressure cookers which can also be used for cereals being most common. As with cereals, conversion involves either green malt or enzymes. Molasses undergo a pretreatment similar to that used in rum manufacture.

Yeasts are selected to have good performance in the particular type of mash used. The main criteria are fast fermentation rate, high ethanol yield, high ethanol tolerance and ability to ferment carbohydrates at relatively high temperatures. A general rule is that the yeast should be capable of producing *ca.* 14% ethanol within 60 h. Pure cultures are used at a pitching rate of 1.5–2.0%, the yeasts being maintained and propagated in the distillery. In the past a maximum fermentation temperature of 33°C has been used, but it is now common practice to initiate the fermentation at 32°C and to control the temperature at 38°C. Over-heating can be a serious problem and temperatures in the fermentation vessels must be carefully controlled.

In large-scale manufacture of gin and vodka, the base neutral spirit is distilled continuously using multiple column stills. Such stills effectively remove all congeners from the spirit. Batch distillation remains in use for small-scale manufacture of neutral spirits for beverages such as korn.

Modern continuous stills for neutral spirit manufacture employ four or five columns (Figure 9.8). Such stills produce a spirit of very high purity, while also being of high overall efficiency. The first column, the beer still, strips alcohol from the mash and must be designed and constructed to handle liquid of high solids content. Distillate from the beer still passes to the second column, the

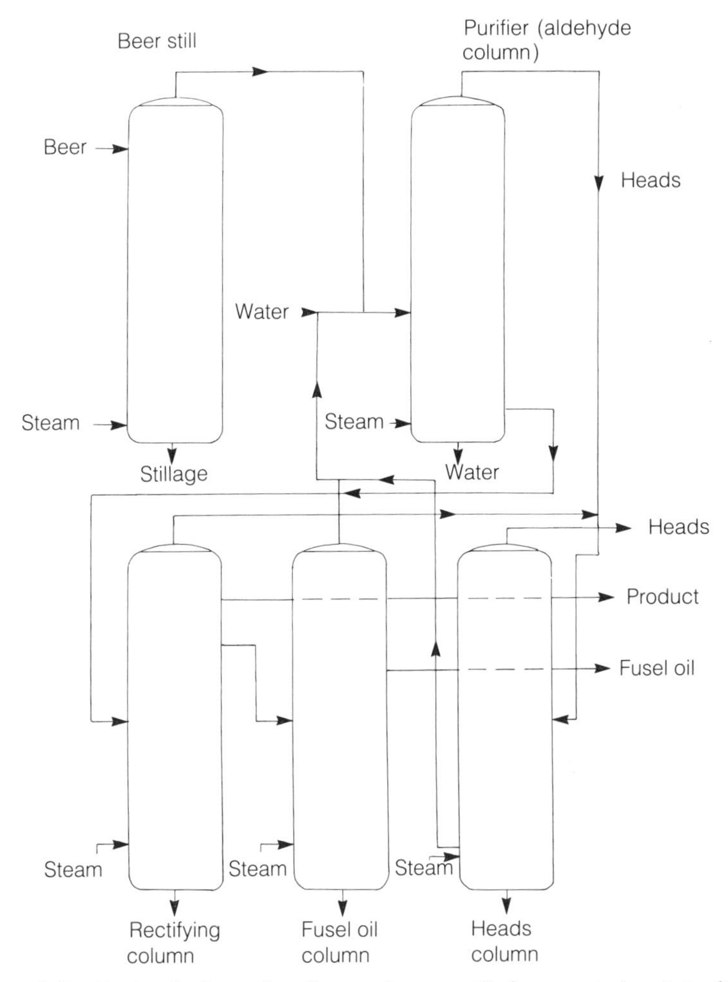

Figure 9.8 Basic design of a five column still for neutral spirit distillation.

'purifier' or 'aldehyde' column, where conditions are such that differences in volatility between ethanol and aldehydes, volatile esters, etc., are maximized. This permits a more effective separation of these components than can be achieved in the rectifying column. Suitable conditions are achieved by injecting water into the purifier column, the ethanol/water mixture being removed at the base and pumped to the rectifying column. The heads taken off with the vapour consist of the aldehydes, volatile esters, acetals, diacetyl and methanol. In a four-column still the heads pass direct to the fusel oil column, but five column stills employ a separate 'heads concentrating' column, where ethanol is reclaimed from the low boiling point volatiles. The rectifying column concentrates the ethanol in the feed from the base of the purifying column. High purity ethanol is drawn off this column. Fusel oil, intermediate and high boiling point esters, acids and phenolics as well as some low boiling point compounds are removed and pass to the fusel oil column. Water is drawn off from the base of the rectifying column and used for dilution in the aldehyde and, where fitted, the heads concentrating column. Ethanol is recovered from fusel oil and compounds of similar volatility in the fusel oil column. In a four-column still the quantity of ethanol recovered is relatively large and is removed from the still for sale as second grade spirit. The smaller quantity recovered in a five-column still is recycled to the purifying column.

(b) Gin

Gin may be defined as any spirit made from a relatively pure base alcohol of fermentation origin, flavoured with materials of plant origin of which juniper (*Juniperus communis*) is most important. Gin is produced in many parts of the world and there are distinct regional variations. The most important gin in international commerce is London dry gin, which is no longer produced exclusively in London, but which has a highly distinctive character.

BOX 9.5 **Dreaming, dreaming all the time of Plymouth gin**

During the high days of sin and gin, many English towns had a local gin distillery. In most cases the product was an imitation of London dry gin, but some characteristic local variations developed. Of these only Plymouth gin has survived and is historically recognized as a favourite drink of Naval officers. Until the 20th century, other types of gin were much less socially acceptable, many of the problems of working class poverty and ill health being attributed to excessive consumption of low priced and low quality gin.

London dry gin is traditionally made from a base of grain spirit, although some imitative types are made from molasses spirit. Each brand of London gin uses a unique combination of botanicals in addition to juniper. Coriander seeds and angelica root are also common, but the total number of types of botanical is very large (Table 9.4).

Table 9.4 Botanicals used as flavouring of London dry gin

Juniper berries	Coriander seeds
Angelica root	Sweet orange peel
Bitter orange peel	Lemon peel
Cinnamon bark	Cassia bark
Cardamom seeds	Nutmeg
Orris root	Liquorice root
Caraway seed	Aniseed
Fennel seed	

Distillation of London dry gin takes place in a pot still. The design of the still varies, but in all cases the degree of rectification is very low and the strength of the distillate rarely exceeds that of the base spirit. Neutral spirit and botanicals are placed in the still and allowed to infuse overnight. Alternatively the botanicals may be placed in trays above the liquid surface, but in this case extraction is less efficient. Gin is drawn from the still as the second of three fractions, the head and tail fractions passing to a separate still for recovery of ethanol. The still may be operated to produce a distillate which only requires dilution before bottling as a consumer product. Alternatively a concentrated gin can be produced for transport to a remote bottling site, possibly overseas. In this case a higher level of botanicals is required. Gin is not blended in the sense of whisky, but spirit from different distillation cycles is often combined to improve consistency.

The Dutch gin, genever, is of two types; *oude* (old), which has a distinctive grain spirit characteristic combined with juniper and other botanicals, and *jonge* (young), which is much less strongly flavoured. *Oude* genever is the most important in international trade and is often retailed in characteristic stone bottles.

Genever is made from a strongly flavoured base spirit, the *moutwijn*, distilled from a fermented mash containing equal proportions of barley, maize and rye. The fermentation, which lasts 48 h, is initiated under aerobic conditions. The fermented mash then undergoes two consecutive batch distillations, which raise the ethanol content to *ca*. 46%. The distillations are non-selective and there is no rectification. *Moutwijn* is produced by a small number of specialist distilleries and sold in bulk to genever manufacturers.

Oude genever is currently produced by a further distillation of a blend of *moutwijn* with botanicals or a blend of *moutwijn* with a distillate of botanicals in neutral spirit. In addition to juniper, caraway seed is of importance in determining flavour, while coriander and aniseed are also common flavourings. *Oude* genever undergoes maturation before bottling. The proportion of neutral spirit is much higher in *jonge* genever and there is normally only a short maturation.

The German steinhager differs from most other types of gin in being flavoured with fermented juniper berries, the *wacholderlutter*. Manufacture of the *wacholderlutter* is a specialist trade, which retains a considerable craft element. Dried, crushed juniper berries are the sole source of fermentable carbohydrates and a mash is made, in warm water, of berries, yeast and a nitrogen source, usually ammonium phosphate. The fermentation is slow and difficult and the temperature must be carefully controlled at 25°C to complete the operation in 8–14 days. Under fermentation conditions, acetification can occur. This must be controlled by tightly covering the fermenters to raise the CO_2 concentration and inhibit *Acetobacter*. After fermentation is complete, the mash is distilled in a simple pot still. Distillation is non-selective and essential oil of juniper is recovered with the distillate. This is removed by decanting. Steinhager is produced by mixing 2–10% juniper distillate with neutral spirit and redistilling in a pot still. The steinhager is removed as the middle of three fractions.

(c) Vodka

Although there are many types of vodka, the most common in international commerce is that produced from essentially flavourless spirit and which obtains its character almost entirely from ethanol. Vodkas flavoured with fruits, such as cherry, and other materials including bison grass, pepper and honey are also manufactured. Some of these vodkas are produced and consumed only in restricted localities, but others have entered international trade.

* The water used to dilute all types of spirits is of major importance in obtaining a high quality end product. Naturally soft or artificially softened water is used of high purity. In the case of vodka the sodium ion content of the water is of particular importance. It is generally considered that the presence of Na^+ ions imparts a desirable mellow character to the vodka.

The fermentation substrate for production of the base spirit is less important than the degree of purification during distilling. Cereals are most widely used in western Europe, while in central and eastern Europe, potatoes are a traditional source of base spirit. Manufacture of unflavoured vodka involves dilution of selected base spirits and treatment with activated charcoal. Treatment involves agitation of the spirit with powdered charcoal in a tank, or passage of the spirit through a series of columns containing granular charcoal. Treatment with charcoal reduces the concentration of undesirable flavour compounds, such as diacetyl, by up to 90%.

Some types of flavoured vodka are produced by redistilling an infusion of fruit, etc., and neutral spirits. A simple batch still is used and there is usually no rectification. Alternatively an infusion of vodka and fruit may be made and simply filtered before bottling or the flavour may be imparted by addition of an essence. In the case of bison grass and pepper vodkas, the source of flavour is placed in the bottle and is present at time of consumption.

9.2.7 Liqueurs

Liqueurs may be defined as distilled spirits, which have been sweetened and flavoured with substances of compatible taste. Colouring may also be added. A few types are unsweetened. Manufacture of traditional liqueurs involves blending the distilled spirit with a sugar syrup containing small quantities of essences or herbs.

BOX 9.6 **Over the sea to Skye**

The formulation of many traditional liqueurs is shrouded with secrecy. Some are undoubtably of ancient origin and those made by monastic orders, such as Benedictine, are probably derived from medicinal spirits. The recipe of the Scotch liqueur, Drambuie, was allegedly given to the Mackinnons of Skye by an aristocratic bodyguard of Prince Charles Edward as a reward for assisting the Young Pretender to escape after the 1745 uprising. The mystique and mystery surrounding the origins of many liqueurs is, of course, of great help to the modern marketing man!

Fruit flavoured spirits, such as apricot, peach and cherry brandy and sloe gin may also be classed as liqueurs and are produced simply by steeping the fruits in the appropriate spirit, or by adding an essence (*cf.* fruit vodkas). A sugar syrup may also be added.

In recent years a number of new types of liqueur have been marketed. In a number of cases these are fruit-flavoured and minimally sweetened and exploit consumer taste for lighter spirits and natural fruit flavours. As such these products can be regarded as evolutionary developments of traditional types of liqueur. In contrast, the increasingly popular cream liqueurs do represent a novel technology.

Cream liqueurs are compound products, formulated from milk fat, distilled spirit, sodium caseinate and sugar. Emulsifiers, flavouring and colouring may also be present. A strongly congeneric spirit is required to balance the flavour of the dairy components and impose a recognizable character. Whisky is the most common spirit base for cream liqueurs, but products made from brandy and rum are also available. Most are of similar overall composition, typically containing *ca*. 15% butterfat, 20% sugar, 5% sodium caseinate and 14% alcohol. Low alcohol liqueurs containing less than 10% alcohol are also available. These are prone to spoilage by species of *Lactobacillus* and pasteurization of the finished product is necessary.

Cream liqueurs are oil in water emulsions. Manufacture involves production of an emulsion which will remain stable to point of consumption and which is resistant to creaming and formation of a 'fat-plug' in the bottle neck. The simplest method of manufacture involves adding cream, sugar and spirit to a solution of sodium caseinate and homogenizing twice at a temperature of 55°C and a pressure of 30 MPa. The liqueur is then cooled, stabilized, flavoured and coloured before bottling. Emulsion instability, caused by calcium-induced aggregation, is a serious potential problem at

---

* Advocaat is a liqueur consisting of egg yolk, sucrose and spirit, the ethanol content of traditional types being 17.5%. The use of unpasteurized egg yolk is preferred to maintain the characteristic appearance and mouth-feel. Concern has been expressed over the possible presence of pathogens, primarily *Salmonella*, in egg yolk and the consequent safety of advocaat. Inoculation experiments showed rapid death of both *Salmonella* and *Staphylococcus aureus* in the presence of 17.5% ethanol, and it was concluded that, in traditional advocaat, risk to consumers is insignificant. Advocaat of reduced ethanol content should, however, be made from pasteurized egg yolk. *Bacillus cereus* survived for at least 6 months and has been reported to spoil advocaat.

higher storage temperatures. The extent of the problem may be minimized by adding citrate salts as calcium sequestrants. A more satisfactory solution, however, is the use of butter oil, which is effectively calcium-free, as the source of butterfat.

9.2.8 Novel distilled beverages

Sales of many distilled beverages are static or declining and there is thus considerable pressure to offset this trend by new product development. This can be difficult, since the nature of many spirits is strictly defined by law. Further, the nature and patterns of consumption of spirits limit the scope for development of added-value product. An area which has been exploited, however, is the development of premixes. These may either be a combination of a distilled beverage with a non-alcoholic 'mixer', such as gin and tonic, or a more exotic combination of one, or more, distilled beverages with bitters or other flavouring to form a 'cocktail'. In either case stabilizers, such as gum arabic, may be used to prevent layering of the components. Preprepared mixer drinks are packaged in wide necked containers with ring-pull caps. A variety of packaging is used for premixed cocktails often reflecting the novelty nature of the product. One brand, for example, is packed in test tube shaped containers, allegedly to facilitate smuggling into dry venues.

BOX 9.7 **All in the name**

It was originally thought that convenience would be the major driving force for purchasing pre-prepared mixer drinks, such as gin and tonic, whisky and ginger ale, etc., and little attention was originally paid to the brand of spirit. Consumer response was initially modest, however, due to doubts over the quality of the spirit. It is now recognized that, while convenience remains a strong sales point, the use of a respected brand of spirit is essential to marketing success.

A further, rather bizarre and essentially pointless, development is jellified distilled spirits. A gel structure is imparted by appropriate use of stabilizers, colouring added and the product packed in small pots. Concern has been expressed over the possibility of children confusing such products with sweets.

9.2.9 Quality assurance and control

Despite the fact that manufacture of most congeneric spirits involves a considerable craft element, formal systems of quality assurance are necessary in manufacture of all distilled spirits. Distillation is, of course a crucial stage and the need for control has been recognized for many years. Where batch distillation is used, considerable emphasis is placed on operator skill, supported by simple hydrometry. This is considered perfectly acceptable where personnel of sufficient skill and experience are available. Continuous stills, especially of the four- and five-column types do require a higher level of control, although the importance of operator skills should not be underestimated.

The importance of good quality raw materials is now generally recognized and, in some situations, is controlled by legislation. Much greater attention is now paid to the fermentation, which should be as closely controlled as fermentations for beer and wine production.

In the past, there has been a tendency to consider maturation largely as a function of time, but there is now greater emphasis on controlling conditions to obtain a product of desired characteristics. To be fully effective, this requires a greater knowledge of the course of maturation and the use of easily analysed 'markers' of progress.

Many spirits, especially whiskies, are blended for consumption. Until recently, blending has depended entirely on expert skills, but there is now increasing interest in use of objective systems based on gas–liquid chromatography and other instrumental methods of analysis. It appears likely, however, that expert blending will remain of importance for many years.

9.3 CHEMISTRY

9.3.1 Flavour and aroma compounds

A very large number of flavour- and aroma-active compounds have been identified in congeneric distilled spirits. Consumer perception of flavour and aroma of spirits, however, comes not from recognition of specific compounds, but from an overall interpretation of the compounds present, their balance and their interactions. This is a similar situation to non-distilled alcoholic beverages and many foods. Despite the organoleptically distinct nature of the different distilled beverages, the same flavour components tend to be of importance in each case. The basis of the flavour of distilled beverages is primarily derived from volatile compounds, such as higher (fusel) alcohols, fatty acid esters and some carbonyls. The characteristics of each type arise from differences in the balance between the various individual components, while the nuances which distinguish between closely related spirits tend to arise from minor components.

Flavour and aroma compounds in distilled beverages have two main sources, the fermentation process and the wood barrels in which the spirit is matured. Excretion of organoleptically significant substances by yeast is well recognized in beer and wine fermentations. Higher alcohols, quantitatively the largest group in congeneric spirits, are produced during fermentation by either an anabolic biosynthetic pathway from sugars, or by a catabolic process from exogenous amino acids (Figure 9.9). In each case oxo acids act as intermediates. Fusel alcohols tend to produced by the catabolic route in the early stages of fermentation and by both catabolic and anabolic routes during the later stages.

In addition to higher alcohols, a number of other congeners are produced during yeast fermentation. The most important of these are carbonyl compounds, especially aldehydes, esters and furfural.

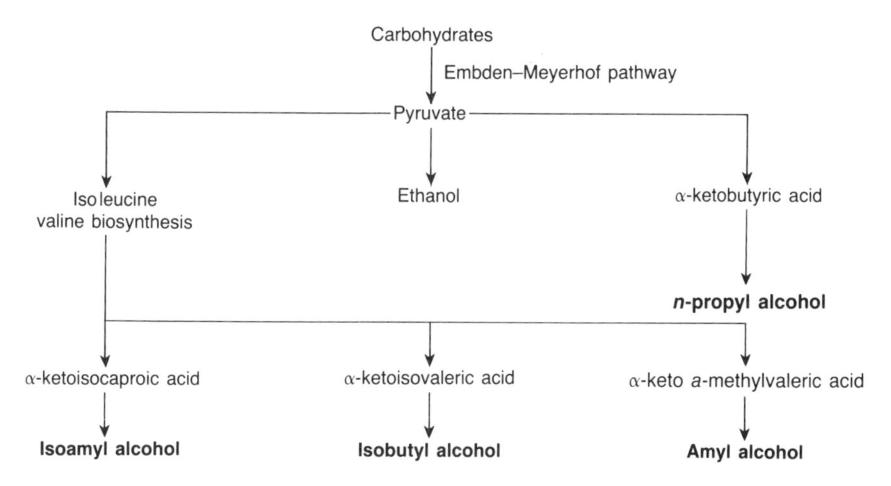

Figure 9.9 Formation of higher (fusel) alcohols during yeast fermentation

Different strains of yeast vary in the quantities of congeners produced during fermentation and cultural conditions are also important. Production of congeners is enhanced by increased inoculum size, agitation and higher fermentation temperatures. As a general rule, conditions which increase the fermentation rate favour congener production.

The wood of barrels used for maturation is the second most important source of flavour and aroma compounds in congeneric spirits. This contribution lies both in substances extracted from the wood during the maturation period and in promoting the chemical modification of precursors. Substances extracted from the wood vary according to the type of wood used, charring and the ethanol content of the maturing spirit. More than 120 compounds have been identified in white wood extract. Some result from the oxidative degradation of ethanol–lignin, formed by reaction between ethanol and lignin. Others, such as 2-hydroxy-3-methyl-2-cyclopentenone and maltol are produced during charring. Oak wood is also a significant source of acetic acid. Barrel woods promote oxidation and esterification and enhance formation of acetaldehyde, acetal and ethyl acetate.

Other micro-organisms, including lactic acid bacteria and, in rum, species of *Clostridium*, can also act as a source of flavour and aroma compounds. The overall role is considerably less important than that of yeast, but can be significant in imparting specific characteristics. Some flavour compounds are also derived direct from fermentation substrates.

(a) Whisky

As with other distilled spirits, higher alcohols are of major importance in flavour and aroma of whisky, there being considerable quantitative variation according to the type. Overall levels are higher in Scotch whisky than in other types (Table 9.5). Levels generally increase during maturation. Whisky typically contains

---

* In commerce, it is sometimes necessary to trace the geographical origin of Scotch whisky. A number of means have been suggested including the use of gas chromatography. The simplest means of obtaining a broad grouping, is determination of 3-methyl-1-butanol. Levels are highest in highland whisky, followed by Islay, Campbeltown and lowland whiskies.

Table 9.5 The fusel alcohols of three types of whisky (g/100 l at 80° proof)

Fusel alcohols	Scotch	Irish	Canadian
Total	102	49	31
n-propyl	20	25	2
isobutyl	34	15	7
(+)-amyl	48	9	22

relatively high levels of 3-methyl-1-butanol (isoamyl alcohol) and 2-methyl-1-butanol (optically active amyl alcohol). These alcohols are accompanied by lower levels of 2-methyl-1-propanol (isobutanol) and 1-propanol. Characteristically, 1-butanol and 2-butanol (sec. butanol) are present only at low concentrations.

The major volatile fatty acid in whisky, as in other spirits, is acetic acid which comprises 50–95%. Acetic acid is mainly formed from ethanol in an oxidative reaction in which acetaldehyde is an intermediate. There is also a contribution from barrel wood. Most of the acetic acid is formed during the first 6 months of maturation and levels then fall somewhat due to reaction with ethanol to form ethyl acetate. Capric, caprylic and lauric acids are the most important of the minority acids. Part of the characteristic flavour of Bourbon and Irish whiskeys may, however, be due to butyric acid.

Scotch whisky contains higher levels of palmitoleic acid (and its ethyl ester) than other whiskies and it has been suggested that the 'stearin-like' odour of Scotch is derived from ethyl esters of long-chain fatty acids. Quantitatively the most important ester is ethyl acetate, formation of which continues at an even rate throughout maturation. Highest levels are found in Bourbon, which can contain as much as 380 mg/l, and Canadian whiskey. Other esters, such as ethyl caprate, are present but only at levels of 2–10 mg/l.

Acetaldehyde is the most important carbonyl compound and is present with other short chain aldehydes. These include furfural, which can be present at levels as high as 20–30 mg/l, and which is an important aroma compound, imparting a cereal or grain character. Acrolein may also be present, formed during distillation by degradation of β-hydroxypropionaldehyde (see page 408). Acrolein imparts a peppery taste and is desirable in small quantities, but is a fault in large. There is some doubt, however, about the stability of acrolein in whisky and, at least some, appears to react with ethanol and be removed as 1,1-diethoxy-2-propene and 1,1,3-triethoxy-propene.

Ethanol-lignin, formed by reaction between ethanol and some of the lignin from barrel wood undergoes degradation to ethanol, coniferyl alcohol and sinapic alcohol. These alcohols are oxidized to the corresponding aldehydes, coniferaldehyde and sinapaldehyde and then to vanillin and syringaldehyde. Vanillin is of particular importance in whiskies other than Scotch. Wood also contributes gallic acid, protocatechuic acid and syringic acid, levels of gallic acid being relatively high in Scotch whisky. A characteristic lactone, β-methyl-τ-octalactone, appears in whisky during storage in oak barrels and is sometimes referred to as 'whisky lactone'.

Pentose and hexose sugars, as well as glycerol, have a role as congeners. Arabinose, galactose, rhamnose and xylose are assumed to be formed by degradation of wood-derived hemicellulose. Glycerol is produced by yeast during fermentation, but a more significant source is the degradation of wood triacylglycerols, probably during charring. Fructose is also present, but its origin is not clear.

Phenols are a quantitatively small group of flavours and aroma compounds. Some are produced during malt kilning by degradation of precursors, such as cinnamic, *p*-coumaric and ferulic acids. Ferulic and *p*-coumaric acids are also degraded by yeast decarboxylases to 4-vinyl- and 4-ethyl phenols. Cresols are responsible for the distinct character of Scotch whisky made from peated malt. Cresols are not found in whiskies made from unpeated malt.

(b) Rum

Rum has a rather more complex aroma than whisky, containing 497 identified aroma compounds compared with 269 in whisky. The levels of higher alcohols in rum vary widely. Levels are lower than in other congeneric spirits in many brands of rum in interna-

---

* A number of polysulphides and thiophenes can be detected in whisky, but probably have little role in determining the flavour of the matured products. The composition does, however, tend to reflect production conditions and thus give an indication of origin. The ratio of dimethyl disulphide to dimethyl trisulphide is a relatively accurate measure of the maturity of a whisky. Determination can be useful in quality control. (Leppaner, O. *et al.* 1983. In *Flavour of Distilled Beverages: Origin and Development*, (ed. Piggott, J.R.). Ellis Horwood, Chichester, pp. 206–14).

tional commerce, but high levels are found in rum produced on a small scale, as a result of primitive distillation techniques. The yeast used for fermentation also has an effect on levels of fusel alcohols and *Schizosacch. pombe*, used for heavy rum fermentation, produces relatively small quantities. It has also been suggested that fusel alcohol production during fermentation is suppressed by the high levels of nitrogen added to the molasses.

The composition of higher alcohols in rum is basically the same as in other distilled beverages, although some may be derived from molasses rather than fermentation. Jamaican rum contains 2-methyl-1-butanol, 2-butanol, 3-methyl-1-butanol and 2-methyl-1-propanal in significant quantities. This profile is similar to that of Scotch whisky, but rum is characterized by very high levels of 2-butanol. The high level of this alcohol has been attributed to bacterial metabolism, possibly involving lactic acid bacteria. Rum also contains higher levels of 1-propanol and 1-octen-3-ol may also be of significance.

The total aldehyde content of rum varies from 50–90 mg/ml. In common with other spirits, acetaldehyde is the most abundant carbonyl, but the higher aliphatic aldehydes propionaldehyde, isobutyraldehyde, 2-methylbutyraldehyde and isovaleraldehyde are of considerable importance in aroma. Ketones, which are highly aroma-active, are generally present at significantly higher levels than in whisky and brandy. The most common are acetone, 2-butanone, 3-penten-2-one, 2-pentanone, 4-ethoxy-2-butanone and 2,3-butanedione, which has been reported at levels of 0.4–4.4 mg/l.

Acetals are formed during distillation by reaction between an aldehyde and an alcohol. In the first stage, an aldehyde molecule adds a molecule of alcohol to form a labile hemiacetal. This then combines with a second alcohol molecule to yield a stable acetal. Levels of acetal in white rum are similar to those in other distilled spirits, but are very high in dark Jamaican rum. There is also a greater diversity of types of acetal, possibly due to the higher ethanol content during maturation. The diethyl acetal of acetaldehyde is present at the highest concentration, but contribution to aroma may be limited.

Levels of volatile fatty acids vary widely from as little as 100 mg/l in light Puerto Rican rum to more than 600 mg/l in heavy Martinique rum. Heavy rum also contains a greater variety of volatile acids. As with other spirits, acetic acid is quantitatively most important. Acetic acid is of particular significance in light rum and some may be derived from acetification of ethanol by acetic acid bacteria. Rum, in general, contains higher levels of butyric, propionic and *n*-valeric acids than other spirits, but slightly lower levels of hexanoic and higher acids. Minor differences between rums may result from variation in levels of 2-ethyl-3-butyric, octanoic and decanoic acids. Although most of the fatty acids in rum are saturated, unsaturated acids have also been detected. Linoleic, oleic and palmitic acids have been found in both Jamaica and Martinique rums, while heptenoic acid has been found only in Jamaican rums.

Ethyl esters of fatty acids are of considerable importance in rum, although the concentration is very variable. Ethyl acetate is the only ester present in high concentrations in light rum and is generally the most abundant volatile ester in all types. The relative proportion of esters in different rums also varies widely. The concentration of ethyl esters of acids with six or more carbon atoms ranged from 0.1 to 0.5 mg/l in Caribbean rums to 1.5 to 20 mg/l in American (New England) rums. Caribbean rums contain relatively high levels of ethyl octanoate, while ethyl decanoate appears to be of greater importance in New England rums.

The aroma complex of rums can be classified into three fractions depending on the boiling point. The light aroma fraction includes ethyl, isoamyl and isobutyl esters, including ethyl-3-phenylpropanoate, isoamyl lactate and isobutyl octanoate. These esters have fruity aromas and are very important in dark, heavy rums. Esters in the middle aroma fraction range from ethyl hexanoate up to ethyl laurate. These esters are also important in the flavour of heavy rum, but of little significance in light rums (Figure 9.10). The main components of the heavy aroma fraction are esters of long chain carboxylic acids, such as ethyl palmitate. The quantities are small and the importance limited.

The minority flavour and aroma compounds of rum are generally similar to those of other spirits. Phenolics are present and eugenol

* The use of increasingly sophisticated gas chromatography has enabled several 'new' volatile fatty acids to be identified in rums. The anti-bacterial sorbic acid has been identified in white, but not in dark rum. Available evidence suggests that sorbic acid is derived from the sugar cane juice.

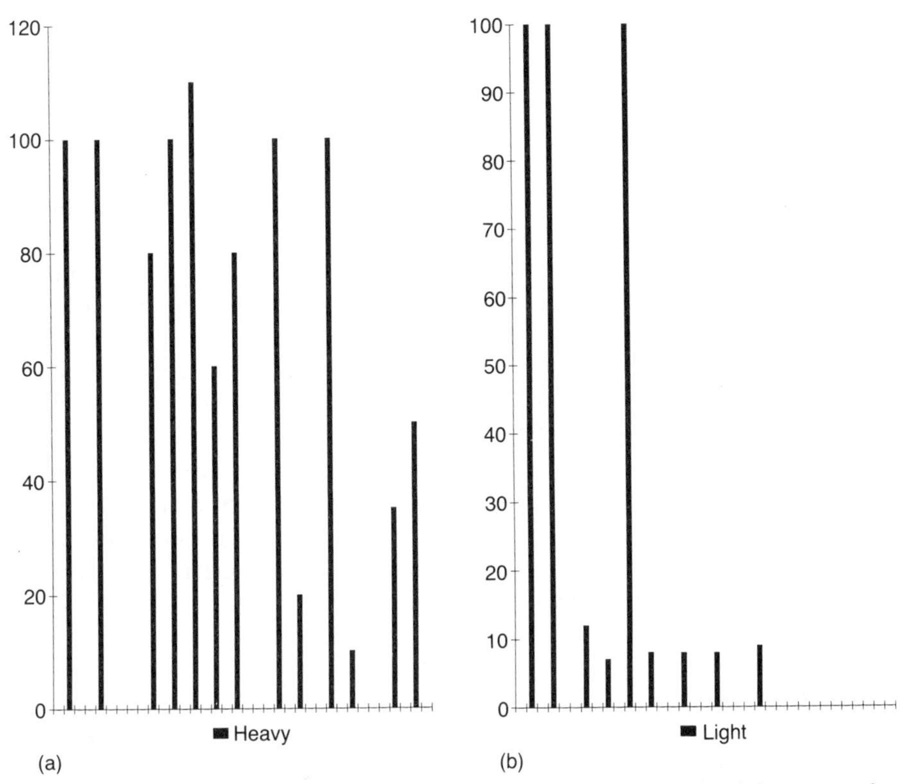

Figure 9.10 Gas chromatogram: middle aroma fraction of heavy and light rum.

and 4-methylguaiacol may be of greater significance in rum than in whisky. Propylguaiacol, 4-vinylguaiacol, 2,6-dimethoxyphenol and 4-ethyl-2,6-dimethoxyphenol may also be important. Rum also contains cresols. These are probably derived from molasses, which may also be the the source of the relatively high levels of terpenoids.

Both τ- and δ-lactones are involved in determining rum aroma. Formation by pyrolysis of hexoses is important in heavy rum. Sulphur compounds are present in rum and, while most have little role in aroma, formation of diethyl disulphide by oxidation of ethanethiol can lead to distinct off-flavours.

* The carboxylic acid 2-ethyl-3-methyl butanoic acid is found in rum, but has not been detected in whisky or brandy. The origin is uncertain, but may be bacterial fermentation. This supposition is supported by the finding of a possible precursor; the methyl ester of the hydroxy acid.

(c) Brandy

Brandy has the most complex aroma and flavour, with at least 546 compounds identified. Although yeast fermentation is the most important source of aroma-active compounds, the raw material used in fermentation, grapes, is of greater significance than in other spirits. Brandy is matured for longer periods than whisky or rum, and the wood of barrels is also an important source of tannins, polyphenols, lignins, proteins, pectins and minerals. Maturation may be considered to take place in two stages: during the first 5 years acidity increases, acetal is formed, extracted tannins progressively oxidized, with associated deepening of colour and hemicelluloses hydrolysed. Between 10 and 30 years the aroma is concentrated by loss of water and ethanol by evaporation, a reduction in ethanol content of 6–8% occurring over 15 years. Cognac acquires a sweetness due to the relatively low ethanol content and the high level of sugars. In Armagnac, this is overlaid by astringency due to the high level of tannins.

The principal higher alcohols of brandy are 3-methyl-1-butanol, 2-methyl-1-propanol, 2-methy-1-butanol and *n*-propanol. These alcohols are always present, but in high concentrations have an adverse effect on the bouquet. Methanol is present at only low levels and 2-butanol and *n*-butanol are absent from high quality brandy, although 3-methyl-2-butanol is present in relatively high concentrations.

Acetaldehyde is the most important carbonyl compound in brandy. Most of that formed during fermentation is removed during distillation, but there is further formation during maturation. Furfural is formed during maturation and all brandies contain diethyl acetal.

The concentration of fatty acids correlates with the yeast biomass in the wine during distillation. Both propionic and butyric acids are desirable in all types of brandy. In cognac, caproic, caprylic, nonanoic, lauric, myristic and stearic acids play an important role in aroma. In Armagnac fatty acids are largely esterified and those higher than capric acid are present only in low concentrations.

* The presence of 2-butanol in brandy is considered indicative of the growth of lactic acid bacteria in the grapes, or must. This implies the use of poor quality raw material. There is no evidence, however, that 2-butanol has any significant adverse effect on flavour or aroma.

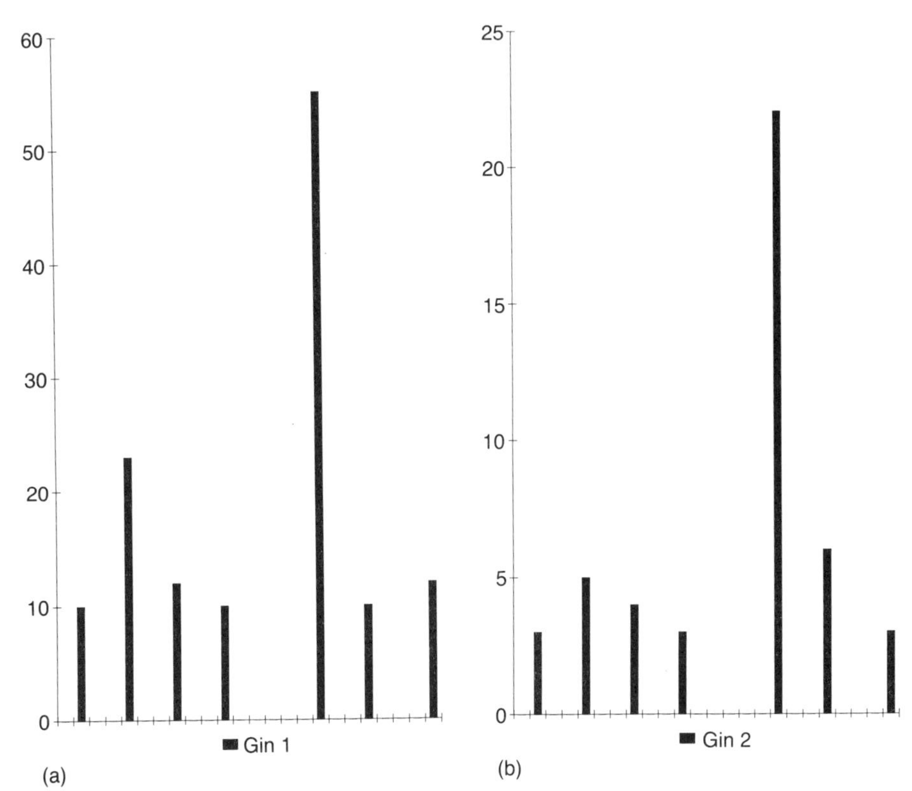

Figure 9.11 Gas chromatogram: two types of London dry gin.

Ethyl acetate is present in all brandies, but is usually at higher concentrations in armagnac. The most important esters with respect to flavour and aroma are ethyl, hexyl and isopentyl myristate, monoethyl succinate, capryl enanthate, ethyl oleanate and ethyl phenylcaproate. The latter ester is an indicator of quality.

Although it is often considered that the strain of yeast used is the most important factor in determining the composition of the aroma complex of brandy, volatile compounds from the grape are important in determining character. Monoterpene polyols and glycosides derived from the grape are of considerable importance as masked flavour compounds. These components are of low volatility and do not distil over with the spirit, but hydrolysis to free monoterpenes, which are highly flavour-active, readily occurs. Under mild conditions of distillation the main products are the alcohols linalool, geraniol, nerol and α-terpineol. In the presence of ethanol the corresponding ethyl esters are also formed. A greater range of pro-

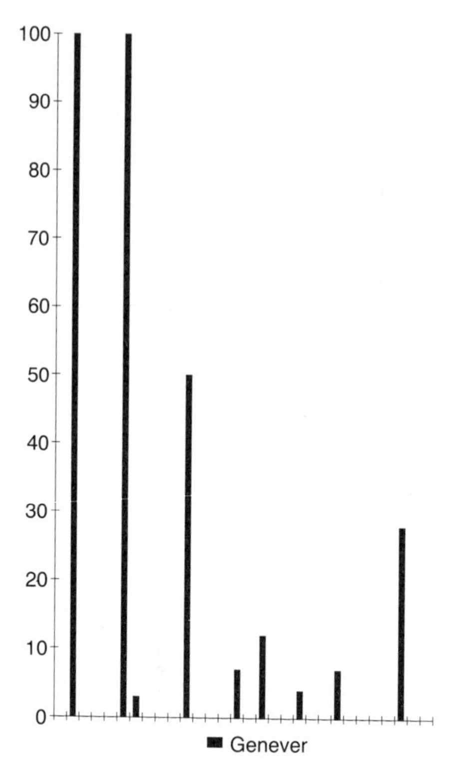

Figure 9.12 Gas chromatogram: genever.

ducts is formed when distillation is more vigorous. These include 1,4- and 1,8-cineoles, 1-terpineol, β-terpineol and τ-terpineol.

(d) Gin and vodka

In gins such as London dry gin, flavour is almost entirely derived from botanicals. The peaks of gas chromatograms of unconcentrated gin samples correspond to botanical constituents and reflect the flavourings added (Figure 9.11). Other compounds can be detected when samples are concentrated, but their nature remains obscure. In genever, which is made from a highly congeneric base and which undergoes some maturation, fusel alcohols such as 3-methyl-1-butanol and 2-methyl-1-propanal are detectable together with other congeners such as furfural. In most cases these are present at higher levels than botanical constituents (Figure 9.12) and while the contribution to flavour may be less, all are present at levels detectable organoleptically. Steinhager contains no detect-

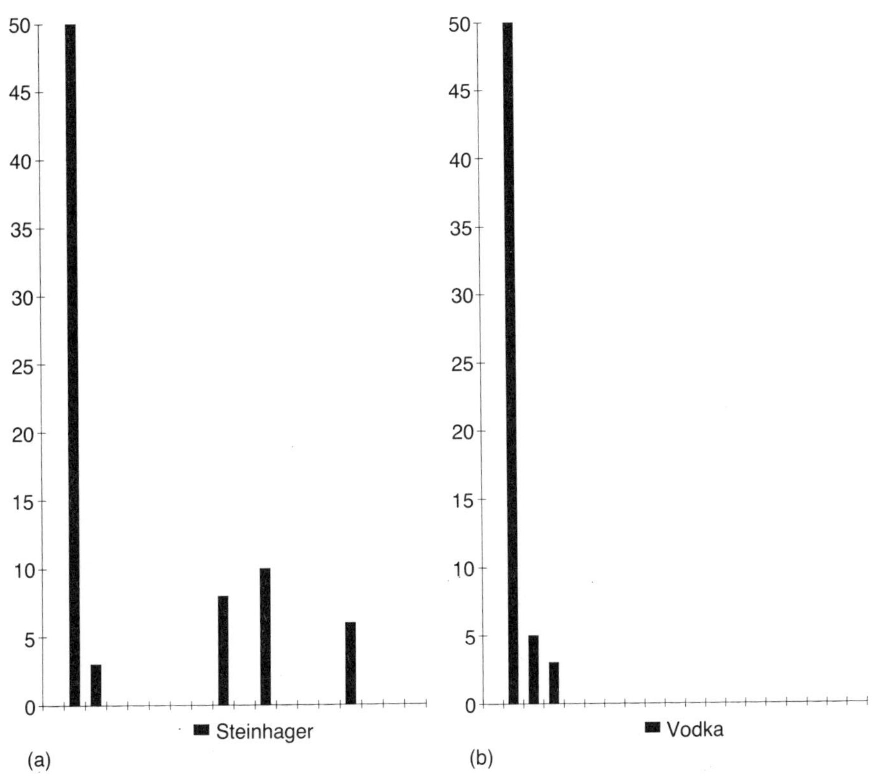

Figure 9.13 Gas chromatogram: steinhager and vodka.

able congeners and only traces of botanicals, while in unflavoured vodka, only ethanol is detectable (Figure 9.13).

The flavour-active compounds of juniper are concentrated in the oil and consist primarily of mono- and sesquiterpenes. The most important of these is 4-terpineol, but a large number of other compounds have been detected. These include α-pinene, β-pinene, myrcene, limonene, α-terpinene, τ-terpinene, *p*-terpineol, α-terpineol and sabine. Ethyl caprylate and a saturated hydrocarbon of unknown identity have also been detected in juniper oil.

The flavour-active compounds of coriander are primarily monoterpenes, of which τ-terpene-*p*-cymene, d-limonene and α-pinene are quantitatively most important. Significant quantities of linalool, geraniol, camphor, geranyl acetate, borneol, nerol, decyl aldehyde and linalyl acetate are also present.

Table 9.6 Main flavour-active components of some botanicals used in gin manufacture

Angelica	
pentadecanolide	
Caraway	
D-carvone	
Cassia	
cinnamic aldehyde	cinnamic acid
cinnamyl acetate	benzaldehyde
methyl salicylate	salicylaldehyde
methyl-*o*-coumaraldehyde	coumarin
Cinnamon	
cinnamic aldehyde	eugenol
α-pinene	caryophyllene
linalool	furfural
β-phellandrene	*p*-cymene
methylamyl ketone	nonyl aldehyde
hydroxycinnamaldehyde	benzaldehyde
coumaraldehyde	
Orange peel	
d-limonene	linalool
octanal[1]	decanal
dodecanal	α-sinensal

[1] In orange peel the aldehydes octanal, decanal, dodecanal and α-sinensal are present in relatively low concentrations, but are the most important flavour-active components.

In addition to the two main botanicals, juniper and coriander, a range of other flavour sources may be used in gin manufacture. Flavour-active components of the most important of these are summarized in Table 9.6.

EXERCISE 9.1.

As a graduate food scientist, you are employed by a small distillery producing a premium-quality, single-malt whisky. Your responsibilities are not well defined, but your duties include management of the microbiology laboratory and propagation of the dried yeast used as starter culture. Shortly after you joined the company, some 5 years previously, the older, wood-lined fermenting vats were replaced by stainless steel equipment and a new high-performance (high alcohol) yeast strain was introduced. Since that time it has been found that with some batches a 'musty' aroma becomes detectable by *ca.* 50% of tasters after 3 years maturation and there is concern that the character of the mature whisky is adversely affected. The problem has been attributed, on the basis of circumstantial evidence, to the changes made to the fermenters and the yeast. Draw up a plan to determine the cause of the problem, bearing in mind the particular difficulties resulting from the long time interval between the manufacture of the whisky and development of the problem during maturation. In this connection, how you would attempt to define a marker, which would permit early detection of the problem. Your colleagues variously have suggested that the problem could be prevented by raising the temperature of the fermentation (currently controlled at 30°C), lowering the temperature of the fermentation or by reducing the quantity of the whisky collected during the second distillation. Would you feel justified in recommending that any of these suggestions should be adopted as an interim measure before the problem is fully understood?

EXERCISE 9.2

Consider the requirements for stabilizers used to jellify distilled spirits. How do these differ from the requirements for use in ice cream?

Imagine that, as a food technologist, you are asked to develop a stabilizer system for a jellified spirit which, after shaking in its container, forms a stable aerated gel. Outline the required properties of the stabilizers and discuss the technical feasibility of such a product.

EXERCISE 9.3.

Old recipe books can be a valuable source of inspiration for the technologist faced with developing variants of existing products. As technologist for a small gin distillery, you have successfully developed a 19th century recipe for elderflower gin as a modern beverage. The gin is flavoured by a simple infusion process, frozen elderflowers imported from central Europe are normally used, supplemented by fresh elderflowers when available from local sources. Problems arise due to variations in the strength of elderflower flavour in the gin. This appears to occur on a random basis and is unconnected with the source of the elderflowers. Organoleptic differences between different batches of flowers are minimal. Discuss the different methods of instrumental analysis which may be used to detect differences between the batches of elderflower and describe how you would apply these on a routine basis.

Bibliography

GENERAL

Chemistry

Fennema, O.R. 1985. *Food Chemistry*. Marcel Dekker, New York.

Microbiology

Doyle, M.P. (ed.) 1989. *Foodborne Bacterial Pathogens*. Marcel Dekker, New York.

Varnam, A.H. and Evans, M.G. 1991. *Foodborne Pathogens: An Illustrated Text*. Wolfe Publishing, London.

Fundamentals of food and beverage processing

Charm, S.E. 1978. *Fundamentals of Food Engineering*, 3rd edn. AVI, Westport, CT.

Earle, R.G. 1983. *Unit Operations in Food Processing*, 2nd edn. Pergamon Press, Oxford.

Kessler, H.G. 1981. *Food Engineering and Dairy Technology*. Verlag A. Kessler, Freising.

Rehm, H.-J. and Reed, G. (eds) 1983. *Biotechnology, vol. 5. Food and Feed Production with Micro-organisms*. Weinheim, Basel.

Quality assurance

ICMSF 1986. *Micro-organisms in Foods, 2. Sampling for Microbiological Analysis: Principles and Specific Applications*. Blackwell Scientific Publications, Oxford.

ICMSF 1988. *Micro-organisms in Foods, 4. Application of the Hazard Analysis Critical Control Point (HACCP) System to Ensure Microbiological Safety and Quality*. Blackwell Scientific Publications, Oxford.

NRC 1985. *An Evaluation of the Role of Microbiological Criteria for Foods and Food Ingredients*. National Academy Press, Washington, DC.

Sutherland, J.P., Varnam, A.H. and Evans, M.G. 1986. *Colour Atlas of Food Quality Control*. Wolfe Publishing, London.

CHAPTER 1. MINERAL WATER AND OTHER BOTTLED WATERS

Brace, J. 1988. *The Quality of Bottled Waters with Particular Reference to Microbiological Standards, Shelf-life and Suitability for Infant Feeding*. MSc Thesis, University of Reading, UK.

Hunter, P.R. 1993. The microbiology of bottled natural mineral waters. *Journal of Applied Bacteriology*, **74**, 345-52.

Hunter, P.R. and Burge, S.H. 1987. The bacteriological qualities of bottled natural mineral waters. *Epidemiology and Infection*, **99**, 439-43.

Robins, N.S. and Ferry, J.M. 1992. Natural mineral water - a Scottish perspective. *Quarterly Journal of Engineering Geology*, **25**, 65-71.

Ward, R.C. and Robinson, M. 1989. *Principles of Hydrology*, 3rd edn. McGraw-Hill, London.

CHAPTER 2. FRUIT JUICE AND NECTARS

Downing, D.L. (ed.) 1989. *Processed Apple Products*. Van Nostrand Reinhold, New York.

Hicks, D. (ed.) 1989. *Production and Packaging of Non-carbonated Fruit Juices and Fruit Beverages*. Blackie, Glasgow.

Kimball, D.A. 1991. *Citrus Processing. Quality Control and Technology*. AVI, Westport, CT.

Nagy, S. *et al.* (eds) 1988. *Adulteration of Fruit Juice Beverages*. Marcel Dekker, New York.

Nelson, P.E. and Tressler, D.K. (eds) 1980. *Fruit and Vegetable Juice Processing Technology*. 3rd edn, AVI, Westport, CT.

CHAPTER 3. SOFT DRINKS

Green, L.F. (ed.) 1978. *Developments in Soft Drinks Technology*, vol. 1. Elsevier Applied Sciences, London.

Hicks, D. (ed.) 1989. *Production and Packaging of Non-carbonated Fruit Juices and Fruit Beverages*. Blackie, Glasgow.

Houghton, H.W. (ed.) 1981. *Developments in Soft Drinks Technology*, vol. 2. Elsevier Applied Sciences, London.

Houghton, H.W. (ed.) 1984. *Developments in Soft Drinks Technology*, vol. 3. Elsevier Applied Sciences, London.

Woodroof, J.G. and Phillips, G.F. 1981. *Beverages: Carbonated and Non-carbonated*. AVI, Westport, CT.

CHAPTER 4. TEA

Willson, K.C. and Clifford, M.N. (eds) 1991. *Tea, Cultivation to Consumption*. Chapman & Hall, London.

CHAPTER 5. COFFEE

Clarke, R.J. and Macrae, R. (eds) 1985. *Coffee*, vol. 1. *Chemistry*. Academic Press, London.

Clarke, R.J. and Macrae, R. (eds) 1987. *Coffee*, vol. 2. *Technology*. Academic Press, London.

Clarke, R.J. and Macrae, R. (eds) 1986. *Coffee*, vol. 3. *Physiology*. Academic Press, London.

Clarke, R.J. and Macrae, R. (eds) 1986. *Coffee*, vol. 4. *Agronomy*. Academic Press, London.

Clarke, R.J. and Macrae, R. (eds) 1987. *Coffee*, vol. 5. *Related Beverages*. Academic Press, London.

Clifford, M.N. and Willson, K.C. (eds) 1985. *Coffee. Botany, Biochemistry and Production of Beans and Beverage*. Chapman & Hall, London.

CHAPTER 6. COCOA

Minifie, B.W. 1989. *Chocolate, Cocoa and Confectionery, Science and Technology*, 2nd edn. AVI, Westport, CT.

Wood, G.A.R. and Lass, R.A. 1985. *Cocoa*, 4th edn. Longman, London.

CHAPTER 7. ALCOHOLIC BEVERAGES: I. BEER

Briggs, D.E. Hough, J.S., Stevens, R. and Young, T.W. (eds) 1981. *Malting and Brewing Science*, vol. 1. *Malt and Sweet Wort*. Chapman & Hall, London.

Hough, J.S., Briggs, D.E., Stevens, R. and Young, T.W. (eds) 1982. *Malting and Brewing Science*, vol. 2. *Hopped Wort and Beer*. Chapman & Hall, London.

Lewis, M.J. and Young, T.W. 1993. *Brewing*. Chapman & Hall, London.

Neve, R.A. 1990. *Hops*. Chapman & Hall, London.

CHAPTER 8. ALCOHOLIC BEVERAGES: II. WINE

Amerine, M.A., Berg, H.W. and Kunkee, R.E. 1979. *Technology of Wine Making*, 4th edn. AVI, Westport, CT.

Farkas, J. 1988. *Technology and Biochemistry of Wine Making*. Gordon and Breach, Montreux.

Rose, A.H. (ed.) 1977. *Economic Microbiology*, vol. 1. *Alcoholic Beverages*. Academic Press, London.

CHAPTER 9. ALCOHOLIC BEVERAGES: III. DISTILLED SPIRITS

Piggott, J.R. (ed.) 1983. *Flavour of Distilled Beverages*. Ellis Horwood, Chichester.

Rose, A.H. (ed.) 1977. *Economic Microbiology*, vol. 1. *Alcoholic Beverages*. Academic Press, London.

Index